Springer Climate

Series Editor

John Dodson, Institute of Earth Environment, Chinese Academy of Sciences, Xian, China

Springer Climate is an interdisciplinary book series dedicated to climate research. This includes climatology, climate change impacts, climate change management, climate change policy, regional climate studies, climate monitoring and modeling, palaeoclimatology etc. The series publishes high quality research for scientists, researchers, students and policy makers. An author/editor questionnaire, instructions for authors and a book proposal form can be obtained from the Publishing Editor. **Now indexed in Scopus® !**

Laely Nurhidayah • Amy Young
Wiwandari Handayani
Editors

Climate Change, Labour and Migration in Indonesia

Impacts on Women and Children

Springer

Editors
Laely Nurhidayah
Research Center for Law
National Research and Innovation Agency (BRIN)
Jakarta, Indonesia

Amy Young
Childlight East Asia and Pacific Hub
School of Social Science, University of New South Wales
Sydney, Australia

Wiwandari Handayani
Department of Urban and Regional Planning
Faculty of Engineering
Diponegoro University
Semarang, Indonesia

ISSN 2352-0698 ISSN 2352-0701 (electronic)
Springer Climate
ISBN 978-3-032-15333-3 ISBN 978-3-032-15334-0 (eBook)
https://doi.org/10.1007/978-3-032-15334-0

This Springer imprint is published by the registered company Springer Nature Switzerland AG
The registered company address is: Gewerbestrasse 11, 6330 Cham, Switzerland

Foreword

Addressing Sea Level Rise Need Holistic, Integrated, and Participatory Approach from Local Level

Sea level rise and land subsidence present significant challenges for Indonesia and many other parts of the world. Coastal communities—particularly fishermen, women, and children—are among the most vulnerable. The impacts include frequent flooding, loss and damage to land and property, disruption of daily life, and, in many cases, forced displacement and migration to safer areas. These challenges threaten not only livelihoods but also the social fabric and resilience of affected communities.

The Intergovernmental Panel on Climate Change (IPCC) has emphasized the urgent need for governments to implement adaptation measures to ensure that coastal areas remain habitable in the future. In response, the Indonesian government has made considerable efforts to develop adaptation plans and policies. These efforts involve cross-sectoral collaboration, ideally integrating the work of various ministries to avoid siloed approaches. For example, coordination between the Ministry of Environment—as the National Focal Point for the UNFCCC—the National Development Plan Agency, the Ministry of Agriculture, the Ministry of Health, the Ministry of Maritime Affairs and Fisheries, the Ministry of Public Works and Housing, and the Ministry of Women's Empowerment and Child Protection is essential in aligning legal, policy, and infrastructure measures that address both environmental protection and social resilience, climate resilience development, and ensuring that these will be implemented fairly, in a gender-inclusive manner, and will strengthen the position of women and vulnerable groups.

This book presents the latest developments in research on sea level rise and its multifaceted impacts on labor, health, education, displacement, disaster risk reduction, housing and infrastructure, and ecosystems such as mangroves. It explores both the consequences and adaptation strategies related to sea level rise from a range of disciplinary perspectives, including law, social work, urban planning, and gender studies. By addressing this complex issue through an interdisciplinary lens, the book offers readers valuable insights into the far-reaching implications of sea level rise—highlighting the need for comprehensive and holistic policy responses. Notably, it is the first edited volume to compile diverse perspectives on how women and children experience the impacts of sea level rise and how their communities build resilience in the face of this slow-onset disaster

This book is divided into three key sections: climate change, labor, and migration. It highlights the far-reaching impacts of sea level rise, emphasizing that the issue extends beyond environmental concerns to deeply affect labor markets and trigger population displacement. By exploring the intersection of these themes, the book underscores the urgent need for integrated responses that address both the environmental and socio-economic dimensions of climate change. We hope this book can be useful for policymakers and readers to get more deeper understanding for the complex problem and trigger more collaborations and efforts to bridge sciences, practices, and policies, and local initiatives in addressing sea level rise and land subsidence.

Director Climate Change Adaptation
MoE of Republic Indonesia
Jakarta, Indonesia

Franky Zamzani

Preface

It has been an honour and a pleasure to not only write the preface for this book, 'Climate change labour and migration in Indonesia: Impacts on women and children' but also to collaborate with a dedicated team of researchers who are passionate about advocating for those impacted by climate change. It is a humbling experience to work with so many experts from diverse disciplines such as law, engineering, urban planning, social work, anthropology, public health and gender studies, and to include the experiences of our local non-government workers.

The impacts of climate change, especially on emerging generations, can seem overwhelming. Climate change is shifting livelihoods in Southeast Asia, and with it, social and cultural protections that have underpinned family and kampong life. These effects receive a great deal of attention when they relate to a rapid onset disaster, like the Boxing Day Tsunami of 2004; however, when the impacts are tied to slow-onset disasters such as sea-level rise, they can become obscured and thus normalized.

This book plays an important role in highlighting the social impacts of slow-onset disasters in coastal Java and translating the legal and policy implications into actionable recommendations for policymakers throughout the Asia-Pacific, and beyond. The social impacts of climate change intersect and thus can become cumulative, especially for those already vulnerable in society.

It is a strength of the book that it documents how compelled labour, migration, and climate change interact and intersect. Policy interventions and responses have become increasingly fragmented and siloed. This collection strongly encourages developing a holistic response that encourages policymakers from diverse ministries to work together, to minimize the unintended consequences that fragmentation can bring.

This book provides an important platform for the voices of women and children living in coastal Java who are experiencing daily flooding. We were deeply touched by our visits to the communities and immensely grateful to the community members for sharing their experiences with the research team.

The editors commend all the hard work that went into developing this useful and insightful volume and look forward to the impact it will undoubtedly have for researchers, practitioners, and readers.

Jakarta, Indonesia Laely Nurhidayah
Sydney, Australia Amy Young
Semarang, Indonesia Wiwandari Handayani

Acknowledgements

The edited volume would not have been possible without research funding support from the Koneksi funding scheme [2023–2024 grant number: 1147/CRG/2023/10-GU]. Koneksi generously provided funding to do research in the four cities of Jakarta, Semarang, Pekalongan, and Demak. The project allowed researchers from Griffith University, Universitas Diponegoro (UNDIP) Semarang, and Badan Riset dan Inovasi Nasional (BRIN) Jakarta to do collaborative and interdisciplinary work to address the problem of sea level rise in coastal Java with a focus on women and children.

Special thanks to professors and colleagues from Griffith University, early career researchers from BRIN, UNDIP scholars, and NGOs Pattiro and IKUPI, whose contributions to this project have been invaluable at all stages of this research. This included guidance in methodologies, roundtable discussions, surveys, in-depth interviews, kick-off meetings, stakeholder forums, and writing all the reports, articles, and chapters.

We also convey our gratitude to local communities in our research sites for their invaluable discussion and help during the fieldwork, and government officials who have supported the project throughout the process. Thank you to the Springer team for their support and patience. To all those who we cannot mention here individually, we are forever indebted.

Laely Nurhidayah
Amy Young
Wiwandari Handayani

Contents

About the Contributors

National Research and Innovation Agency (BRIN) Team

Tuti Alawiyah is Assistant Professor and Department Head, Social Welfare Department, Faculty of Social and Political Science, University of Muhammadiyah Jakarta, Indonesia. She graduated with master and doctoral degree in Social Work from the University of Texas at Austin in 2013. She has conducted a number of research projects since her early research career and has worked with a number of women's organizations in Indonesia including KAPAL Perempuan, Koalisi Perempuan Indonesia, and Central Board of 'Aisyiyah. Additionally, Dr. Alawiyah has strong program management and stakeholder engagement skills as she has held several senior positions in international development programs. She managed a multi-year, large-scale, national program on tuberculosis elimination program funded by The Global Fund, worked for The Raoul Wallenberg Institute as Deputy Director for regional Asia-Pacific office, and was a Deputy Team Leader for NGOs' Strengthening Services Program under DFAT-Australia. Contact her at tuti.alawiyah@umj.ac.id

Vera Bararah Barid is a Junior Researcher at the Research Center for Law, Institute of Social Sciences and Humanities (ISSH), National Research and Innovation Agency (BRIN), Indonesia. She holds a Master of Law from the University of Indonesia, majoring in Criminal Justice System. Her research interests focus on labor law, including social and legal protection for workers and wages. She is particularly concerned with the rights of persons with disabilities in accessing justice and explores how climate change affects labor rights and vulnerable groups. She can be reached at vera003@brin.go.id

Ahmad Fathony is a researcher at the Research Center for Law, National Research and Innovation Agency (BRIN). He is involved in law and human rights research. He earned a Master of Laws degree from Gadjah Mada University in 2025 and a Bachelor of Laws degree from Bung Karno University. He has been involved in several research projects and received funding from both domestic and international sources. He previously worked as a researcher at the Law and Human Rights Research and Development Agency, Ministry of Law and Human Rights. Contact him at ahma114@brin.go.id

Adis Nur Hayati is a researcher at the Research Center for Law, National Research and Innovation Agency (BRIN), Indonesia. She earned her Bachelor's degree with a concentration in Economic Law from Universitas Indonesia in 2018, and later obtained her Master of Laws in the same field from the same university in 2022. With a background in economic law, she has authored several articles in the field and is now exploring legal theory, as well as environmental law through her study on planned relocation. Contact her at adis011@brin.go.id

Inayah Hidayati focuses on conducting research and policy recommendations in population studies at the Directorate of Human Development Policy, Population, and Culture, National Research and Innovation Agency (BRIN). She holds degrees in human geography from Universitas Gadjah Mada and Universitas Indonesia. Her research interests include demography, population mobility, and human geography. Since beginning her academic career, she has authored more than 20 peer-reviewed journal articles, books, conference proceedings, and technical reports, and continues to contribute to national and international research collaborations. Contact her at inay001@brin.go.id

Dewi Analis Indriyani is a researcher at the Research Center for Law, Indonesian Research and Innovation Agency (BRIN). Dewi obtained a bachelor's degree in law from the Universitas Muhammadiyah Malang and a master's degree in law from Universitas Airlangga (business law) and the Universitas Indonesia (intellectual property law). Dewi is excited and involved in several research related to intellectual property, vulnerable groups, and human rights. Currently, Dewi is a member of the Legal Theory and Doctrine research group since 2024, and the research pursued in the future will focus on Legal Theory and Doctrine. Contact her at dewi043@brin.go.id

Ade Angelia Yusniar Marbun is a junior researcher at the Research Center for Law, National Research and Innovation Agency (BRIN) in Jakarta, Indonesia. A graduate of Lampung University with a Bachelor of Law degree, she is passionate about exploring how legal frameworks can support environmental governance, climate change adaptation, and sustainable development. Her work bridges law, policy, and environmental protection, with a focus on creating solutions for a more resilient future. She can be reached at adea004@brin.go.id

Josefhin Mareta is a researcher at the Research Center for Law, Research and Innovation Agency (BRIN). She earned a bachelor of law from Atma Jaya Yogyakarta University, specializing in economic law, and a master of national resilience from University of Indonesia. Her research interests include economic law, intellectual property law, labour law with a particular focus on issues related to vulnerable groups. She has published journal articles, books, and conference proceedings and is involved in national and international research collaborations. Contact her at jose003@brin.go.id.

Nawawi is currently working as the Head of Research Center for Population, Institute for Social Sciences and Humanities, National Research and Innovation Agency, BRIN. His research focuses on the issues related to human resources and employment development in Indonesia. He has published books and articles both as national and international publication. He obtained his PhD from the school of Politics and International Studies University of Leeds, United Kingdom in November 2019. He can be contacted at: nawa003@brin.go.id

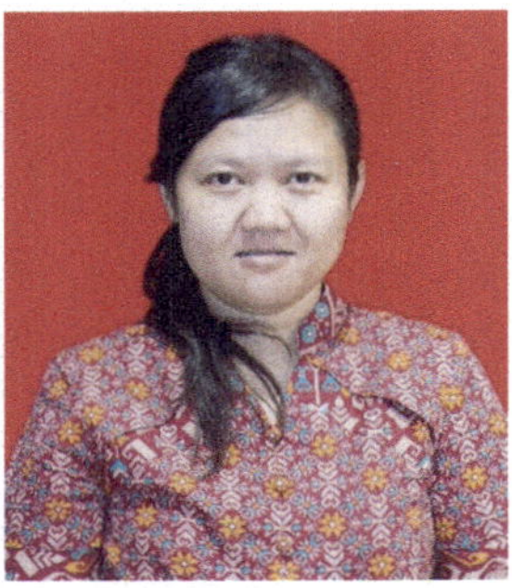

Laely Nurhidayah is a researcher at the Research Center for Law, National Research and Innovation Agency (BRIN), Jakarta, Indonesia. She leads the law, environment, climate change, society, and natural resources research portfolios in her research center. She was awarded her PhD in Law from Macquarie University, Australia in 2015. Her PhD thesis was entitled "Transboundary Haze Pollution in the ASEAN Region: An Assessment on the Adequacy of the Legal and Policy Framework in Indonesia". In addition, she was obtained her Master of Laws (Advanced) from TC Beirne School of Law, the University of Queensland Australia and received Bachelor of Law from Jenderal Soedirman University in Indonesia. She can be reached at lael003@brin.go.id

Oktriyanto is a dedicated researcher at the Research Center for Population—National Research and Innovation Agency (BRIN), focusing on the complex dynamics of family issues in Indonesia. He possesses a unique multidisciplinary academic background, beginning with a Bachelor's degree in Geography from Universitas Gadjah Mada (2007), followed by a Master's in Family and Child Development from IPB University (2015). Currently, he is advancing his expertise as a Doctoral candidate in Social Welfare at Universitas Padjadjaran. His research bridges the gap between spatial perspectives and social interventions, aiming to strengthen family resilience and social well-being across the archipelago. Through his work at BRIN, he contributes evidence-based insights to shape effective population policies. Driven by a commitment to social impact, he welcomes academic discourse and professional collaborations. For inquiries or potential research partnerships regarding family studies and social welfare, he can be reached via his official email at oktr001@brin.go.id

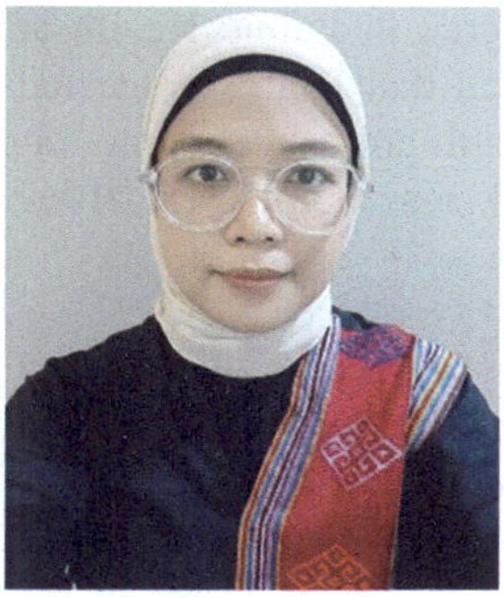

Resti Pujihasvuty, S.Si., M.APS, is a researcher at the Population Research Center, National Research and Innovation Agency (BRIN), Indonesia, specializing in social demography and family dynamics. She earned her Bachelor's degree in Statistics from Padjadjaran University (2008) and her Master of Applied Population Studies from Flinders University, South Australia (2013). She is currently enrolled in a doctoral program in Public Health at the Faculty of Public Health, University of Indonesia. With over 15 years of experience in population and family research, she has contributed extensively to national studies on family development, stunting reduction, reproductive health, aging, and family empowerment. She was involved in the formulation of Family Development Indicators with the National Population and Family Planning Board (BKKBN) and has led and collaborated on research projects focusing on parenting practices, family economic resilience, the well-being of older adults, and multiple vulnerabilities of women with disabilities (gender and disability). Her work applies a life-course perspective to advance evidence-based policies and strengthen family resilience. She can be contacted at rest017@brin.go.id or reztea03@gmail.com

Aditya Rahmadhony is a researcher at the Research Center for Law, National Research and Innovation Agency (BRIN). He earned a Bachelor of Laws from Padjadjaran University, specializing in Constitutional Law, and a Master of Laws from Pakuan University. He is currently pursuing a Doctor of Laws program at Diponegoro University. His research interests include constitutional law, legal theory, and legal politics, with a particular focus on issues related to legislation, child marriage, vulnerable groups, and indigenous communities. He is actively involved in legal research, academic discussions, and the publication of scientific articles. He can be contacted via email: adit030@brin.go.id

Diah Puspita Sari is a Sexuality Research Group and Reproductive Health (Sexuality and Reproductive Health) researcher at the Research Center for Population, National Research and Innovation Agency (BRIN). She earned her medical degree (MD) from YARSI University and a Master of Science (MSc) from Universitas Indonesia in Population and Labor Studies. She is currently pursuing a doctoral degree in Public Health at Universitas Indonesia. Research that Diah has carried out related to reproductive health and social demography. Diah can be contacted via email: diah017@brin.go.id

Agus Suntoro is an Associate Researcher at the Legal Research Center, National Research and Innovation Agency (BRIN), Indonesia. He holds a Master of Laws degree, specializing in environmental and human rights issues. He previously worked for ten years at the National Commission on Human Rights and for two years at the Supreme Court Research Center. He earned his law degree from Jenderal Soedirman University in Purwokerto, a small, cool city in the south of Central Java Province. His contact is agus154@brin.go.id

Nurrahman Aji Utomo serves as a researcher at the Legal Research Center within the National Research and Innovation Agency. He attained his undergraduate degree from the Faculty of Law at Sebelas Maret University and pursued his postgraduate studies in the Master of State Law program at Gadjah Mada University. His concentration lies in the realms of Law, Human Rights, and the advocacy for Vulnerable Groups. Before his tenure at the Legal Research Center, BRIN, he served as a researcher at the National Commission on Human Rights (Komnas HAM) and the Center for Research and Development of Law and Justice, Supreme Court of the Republic of Indonesia. A selection of publications is accessible at https://scholar.google.co.id/citations?hl=en&user=S7MDFrAAAAAJ. The author is also reachable through the email address nurrahman.aji.utomo@brin.go.id.

Universitas Diponegoro Team

Diana Nur Afifah is a Professor at the Department of Nutrition Science, Faculty of Medicine, Universitas Diponegoro, Indonesia. She earned her PhD in Food Science from Bogor Agricultural University with a sandwich program in South Korea. Her research focuses on functional foods, fermented foods, resistant starch, and food-based interventions for metabolic diseases. She has authored numerous peer-reviewed publications and led various research projects on nutrition and health, including emergency food development and protein hydrolysates for stunting prevention.

Mega Anggraeni has over 13 years of work experience on the topics of climate change, environmental management, resilience city, and waste management. As the manager of Initiative for Urban Climate Change and Environment (IKUPI- Inisiatif Kota untuk perubahan Iklim), she joined to several collaboration programs and research in the field of climate change adaptation, resilience cities, as well as water and waste management in collaboration with national and international partners including World Resources Institute, JICA, The Netherlands Government (RVO), CRISIL, Mercy Corps Indonesia, IGES, GIZ, 100 Resilient Cities, Yayasan Kota Kita, Yayasan BINTARI, some universities, and governments. To date, Mega has moved with research, training, and community empowerment on climate resilience, water, and solid waste management issues. Please contact at: mega.anggraeni12@gmail.com

Landung Esariti graduated from Bachelor of Urban and Regional Planning of Universitas Diponegoro Indonesia in 1999. She continued her master study at Master of Property Studies in University of Queensland, Brisbane, Australia in 2004. She works at a lecturer at Urban and Regional Planning Department of Universitas Diponegoro Indonesia since 2000. She is a member of Urban Development Laboratory. Her research interest focuses on urban housing management, poverty reduction programs, gender and empowerment, gender and disaster planning. Please contact at: landungesariti@lecturer.undip.ac.id

Wiwandari Handayani is a Professor at the Department of Urban and Regional Planning, Universitas Diponegoro. Since 2011, she has been engaged in work and research concerning urban resilience and governance. Her involvement includes participation in the Asian Cities Climate Change Resilience Network (ACCCRN) and the Resilient Cities Network (R-Cities), supported by the Rockefeller Foundation, the Global Center of Spatial Methods for Urban Sustainability (SMUS), and the Networking Unit Paradigm Shift (NUPS) that centres on migration and climate change. She has also received several international research grants, including from KONEKSI (Australia), KOICA (Republic of Korea), DAAD (Germany), and the LUCE Foundation (United States). Please contact at: wiwandari.handayani@live.undip.ac.id

Ira Irawati PhD (cand.) is a lecturer in the Department of Urban and Regional Planning at Insitut Teknologi Nasional, Indonesia, where she explores how urban areas can be transformed into climate-resilient environments for their stakeholders. Her research examines the interactions between urban form, environmental conditions, and residents' behavior, which collectively influence the urban housing market, foster healthy urban development, and guide decision-making for climate adaptation and mitigation. By integrating spatial analysis with behavioral surveys, she aims to achieve a comprehensive understanding of these dynamics. Furthermore, Ira Irawati is dedicated to aligning the economic value of environmental services with the principles of sustainable urban planning.

Dina Auliya Kharisma graduated from the Bachelor of Urban and Regional Planning Program at Universitas Diponegoro, Indonesia in 2024. She currently works as a research assistant at the Urban Development Laboratory, Department of Urban and Regional Planning, Universitas Diponegoro. Her research focus includes housing and settlement, disaster risk management, and gender mainstreaming in urban development. Please contact at: dinaauliyak@gmail.com

Bulan Prabawani is a professor at the Department of Business Administration, Universitas Diponegoro, Indonesia, and currently serves as an expert of the SDGs Center at the same university. Her research and community engagement focus on sustainable business practices, community-based adaptation, environmental governance, and inclusive development. She has led various interdisciplinary research projects related to business sustainability, social innovation, and sustainable development. Prof. Prabawani actively collaborates with local governments, NGOs, and international partners to promote climate adaptation strategies that are inclusive, participatory, and grounded in local knowledge. Please contact at: bulan@live.undip.ac.id

Datuk Ary A. Samsura is an assistant professor and urban planner with a background in both Spatial Planning and Engineering Policy Analysis. He is interested in how stakeholders in urban and regional development processes make decisions to improve the effectiveness and efficiency of the development process and also the quality of life in their area. Particularly, he focuses on collaboration or cooperation and competition among different stakeholders with conflicting interests, objectives, and positions from public and private sectors by implementing different scientific methodological tools including game-theoretical modelling and experiment and also Multi-Criteria Decision Analysis. He is also interested in issues related to local climate mitigation and adaptation, sustainable development management, and collaborative planning.

Sariffuddin is an Assistant Professor at Universitas Diponegoro, Indonesia. His research focuses on spatial data science, econometrics, and big data analytics to understand urban dynamics, property markets, and climate risks. He applies geospatial modeling, remote sensing, and machine learning—such as GWR and hedonic models—to explore housing market segmentation, coastal vulnerability, and urban inequality. Please contact at: sariffuddin@live.undip.ac.id

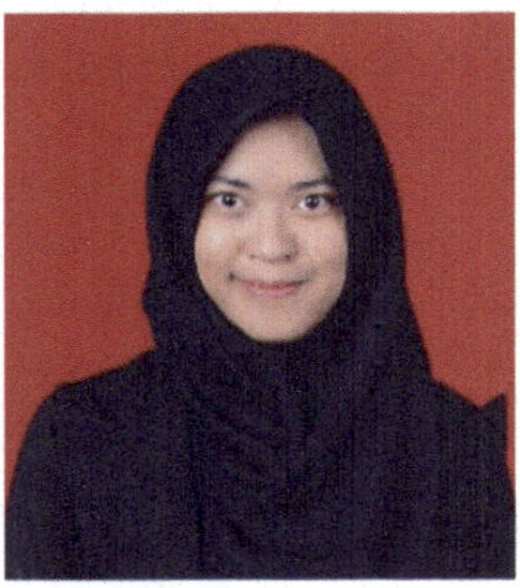

Intan Hapsari Surya Putri completed her undergraduate and doctoral studies in Urban and Regional Planning at Diponegoro University, Indonesia. She has postdoctoral experience at the Technical University of Berlin, Germany, and at Indonesia's National Research and Innovation Agency (BRIN). She has worked as a research assistant at the Center for Urban and Resilience Research (CURE) at Diponegoro University and has been actively involved in various national and international research collaborations, including the KONEKSI research partnership with Griffith University and National Research and Innovation Agency (BRIN). Her research focuses on urban resilience, disaster risk management, and urban and regional planning. Her scholarly work has been published in both national and international journals, covering key topics such as coastal community adaptive capacity, disaster mitigation strategies, and the resilience of urban environments.

Citra Tatius has graduated with a Bachelor's degree from the Department of Urban and Regional Planning at Universitas Diponegoro. She is currently working as a research assistant at the Center for Urban and Regional Resilience Research (CURE), Universitas Diponegoro. With over two years of experience, her research focuses on urban and community resilience towards climate change. She has contributed to international research projects, including the research funded by KONEKSI (Australia). She has also been involved with capacity building programs on enhancing urban resilience with Resilient Cities Network (R-Cities) and Semarang City Government. Please contact at: citratatius@gmail.com

Team from Griffith University and University of New South Wales

Jennifer Boddy is Dean (Sustainability and Rankings) at Griffith University, where she leads strategic initiatives aligned with the Sustainable Development Goals. An internationally recognised scholar in social work, she specialises in domestic violence, disaster response, and environmental justice. Her research highlights the interdependence of human well-being and ecological health. With over 100 publications and extensive industry engagement, she has secured significant research funding and supervised numerous doctoral students. Professor Boddy's work reflects a deep commitment to social justice and sustainability. Please contact at:j.boddy@griffith.edu.au

Amanda Rose Collins is a researcher with the Disrupting Violence Beacon, Griffith University, Australia, where she is also a doctoral candidate under an Australian Government Research Training Scholarship in Social Work. Amanda's current research is focussed towards gaining better understanding of the relationship between personal trauma and professional practice among social workers. She has experience working as a researcher and academic, as well as working across diverse areas of practice, including therapeutic work with people from a refugee background and with young people in out-of-home care. She holds a Master of Social Work (with distinction) from Griffith University and a Bachelor of Social Science, majoring in Sociology, from Southern Cross University, Australia. Please contact at: amanda.collins@griffithuni.edu.au

Sara E. Davies is a Professor of international relations, Griffith Asia Institute, Griffith University and Deputy Director (Indo-Pacific Research) of the Australian Research Council Centre for the Elimination of Violence Against Women. Sara leads the Context Pillar, the Griffith Node, and co-leads the Impact of Shocks on Violence Against Women workstream for CEVAW. Sara's research examines how to scale up prevention and protection capacity for at-risk populations during shocks including conflict, mass atrocities, and infectious disease outbreak events. Please contact at: sara.davies@griffith.edu.au

Febi Dwirahmadi is a Senior Lecturer in Global Health at Griffith University's School of Medicine and Dentistry and Program Coordinator for Indonesia at the Centre for Environment and Population Health. He completed his PhD at Griffith in 2016 on disaster risk reduction and climate change adaptation partnerships to build urban community resilience to floods in Jakarta. With nearly a decade of experience in disaster and emergency management, he has worked with the International Federation of Red Cross and Red Crescent Societies and Islamic Relief. His expertise includes disaster risk reduction, climate change adaptation, global health security, and inclusive resilience-building. Please contact at: f.dwirahmadi@griffith.edu.au

Celeste Harris is a Senior Research Assistant at the Disrupting Violence Beacon and a doctoral candidate at the Centre of Excellence for the Elimination of Violence Against Women (CEVAW), Griffith University in Australia. Her current research focuses on the emerging field of climate violence, particularly the intersection of gender-based violence and climate-induced disasters. Celeste has an extensive multidisciplinary background as a social worker and environmental scientist. She has worked across multiple sectors, including academic research, not-for-profit organisations, government, and intergovernmental agencies. She holds a Master of Social Work from Griffith University, Australia and a Bachelor of Science (Honours) from the University of Sydney, Australia. Please contact at: celeste.harris@griffith.edu.au

Ana Borges Jelinic is an Associate Research Fellow with the Disrupting Violence Beacon, Griffith University, Australia. Her work focuses on migration, trauma, and violence against women. She graduated in Psychology with Honours at PUC-SP, Brazil, in 2005. She holds post-graduate qualifications from Griffith University and the University of Queensland (UQ) in Australia, including her PhD, awarded in 2020. In 2022 she was the recipient of the Andrew Little Award for teaching excellence at the Australian College of Applied Psychology. ORCID iD - 0000-0002-2421-4367 Please contact at: a.borges@griffith.edu.au

Jianqiang (Joe) Liang is a China-born social work academic with expertise in the psychosocial well-being of vulnerable individuals and groups across the lifespan development. He is the Program Director for the Bachelor of Social Work programs, and a key member of the international faculty at Griffith University. Joe's research interests are program evaluation in adolescents and young people's mental health and psychosocial well-being, family and domestic violence, education and employment, particularly those who are culturally and linguistically diverse, migrants and refugees, and international university students. Joe also has a strong track record in scholarship of learning and teaching in international social work. Please contact at: j.liang@griffith.edu.au

Elena Marchetti is a co-Director of the Disrupting Violence Beacon. She is also a member of the Australian Research Council, College of Experts and a Fellow of the Queensland Academy of Arts and Sciences. She was the Deputy Chair of the Queensland Sentencing Advisory Council from November 2016 to August 2025. Elena completed her PhD in 2005 on how the Royal Commission into Aboriginal Deaths in Custody considered the experiences of Aboriginal and Torres Strait Islander women whose deaths in custody were investigated by the Royal Commission. Please contact at: e.marchetti@griffith.edu.au

Donna McAuliffe is Professor of Social Work in the School of Allied Health, Sport and Social Work at Griffith University. Donna has a social work background in working with communities impacted by various forms of disadvantage and specialises in the ethical considerations involved in research and practice. She has been part of the Koneksi project in Indonesia assisting with community forums and roundtables focused on impacts of climate change on women and children. Please contact at d.mcauliffe@griffith.edu.au

Patrick O'Leary is an internationally recognised researcher on social work, gender-based violence, and child protection. Professor O'Leary is a Chief Investigator for the ARC Centre of Excellence for the Elimination of Violence Against Women (CEVAW). He is also Co-Director of the Disrupting Violence Beacon at Griffith University, and Co-Chief Editor of International Social Work Q1 journal. Professor O'Leary has worked with UNICEF, Terre des hommes, ECPAT International, and Islamic Relief Worldwide. He serves on numerous boards and advisory positions in the government and non-government sectors including being on the Advisory Group for National Strategy to Prevent and Respond to Child Sexual Abuse for the Australian Government, non-executive Director of DVConnect, and member of the Practice Advisory Committee for Survivors and Mates Support Network. Please contact at: p.oleary@griffith.edu.au

Susan Harris Rimmer focuses on international human rights law, climate justice, national security, foreign policy, and gender equality in the Griffith Law School. She was appointed as a General Member (sessional) in the Administrative Review Tribunal in March 2025. With Professor Sara Davies, Susan is co-convenor of the Griffith Gender Equality Research Network. Sue also leads the Climate Justice theme of the Griffith Climate Action Beacon. She is the founder of the EveryGen coalition (www.everygen.online), which seeks to amplify the voices of current and future generations and highlight the long-term impacts of today's policy decisions. Please contact at: s.harrisrimmer@griffith.edu.au

Lauren Terzis is a Lecturer in the School of Allied Health, Sport and Social Work at Griffith University, Australia. She holds a PhD (2019) and a Master of Social Work from Saint Louis University, as well as a Bachelor of Science from Illinois State University, both located in the United States. Her research focuses on the immigrant paradox, also known as the healthy immigrant effect, with a particular emphasis on financial behaviours among immigrant populations. Please contact at: l.terzis@griffith.edu.au

Yudi Wahyu is a Research Assistant at the Disrupting Violence Beacon, working on the KONEKSI project "Forced Labour and Climate Change: Keeping a Focus on Women and Children." He has over twelve years' experience in evidence-based policy research at the SMERU Research Institute in Indonesia, covering infrastructure, child welfare, health, education, climate change, and democracy. He holds an MA in Political Science (HARD-SEA, Gadjah Mada University and University of Oslo) in 2010 and a BA in Political Science from the University of Indonesia in 2004. Please contact at: y.fajarmargonowahyu@griffith.edu.au

Amy Young is the project lead for the Koneksi scheme-funded project 'Forced labour and climate change: Keeping a focus on women and children', which informed this volume. She is a Research Fellow with the Childlight East Asia and Pacific Hub in the School of Social Sciences at the University of New South Wales and an adjunct Research Fellow with Griffith University's Disrupting Violence Beacon. Dr Young is passionate about promoting women and children's rights through research. Please contact at: amy.young@unsw.edu.au

Chapter 1
Collaboration as a Key to Combatting the Social Impacts of Climate Change

Amy Young, Laely Nurhidayah, and Wiwandari Handayani

Indonesia dan Australia,
Saling berjumpa dalam kerja sama.
Mari berdiskusi membangun asa,
Agar anak bangsa dapat terus sejahtera.
Indonesia and Australia,
Meet each other in cooperation.
Let's discuss building hope,
So that the people can continue to prosper

This poem, written by Yudi Wahyu, illustrates the hope and collaborative spirit that anchors the volume. In 2023, researchers from Indonesia and Australia came together to commence world-first research into the intersections between forced labour, migration and climate change. The impacts of climate change are felt globally, and international collaboration is key to addressing transnational social impacts such as migration and exacerbated impacts of violence. Research plays a pivotal role in ensuring that mitigation and adaptation measures are evidence-based, as well as suited to the needs of local communities. The recognition that international research partnerships are required to ensure global change has seen the growth of funding schemes that aim to unite researchers across borders with a shared purpose. The Koneksi knowledge partnership platform connects Australian and Indonesian organisations and supports innovative responses to climate change. Taken from the Bahasa Indonesia word for connection, the scheme aims to support equitable and sustainable knowledge transfer between Australia and Indonesia. The Koneksi

A. Young (✉)
School of Social Science, University of New South Wales, Sydney, NSW, Australia
e-mail: amy.young@unsw.edu.au

L. Nurhidayah
Research Center for Law, National Research and Innovation Agency (BRIN), Jakarta, Indonesia

W. Handayani
Department of Urban and Regional Planning, Diponegoro University, Semarang, Indonesia

L. Nurhidayah et al. (eds.), *Climate Change, Labour and Migration in Indonesia*, Springer Climate, https://doi.org/10.1007/978-3-032-15334-0_1

scheme supported the research presented in this volume and was the catalyst for the partnership that has grown amongst the authors.

Indonesia is Australia's closest neighbour by sea, and it was the connection to the sea that sparked the idea for this volume. Sea-level rise and other slow-onset disasters impact both nations, yet the daily impacts are more visible in coastal Java, where not only are sea levels rising due to anthropomorphic climate change, but the land is also subsiding, exaggerating the impacts. Coastal Java provides a case study to unpack the intersections of climate change, migration, immobility and labour. Four case sites in Java were selected, as illustrated in Fig. 1.1—Northern Jakarta, Semarang, Demak and Pekalongan—allowing the authors in the volume to explore what the impacts of climate change look like for women and children engaged in different industries and differing experiences of migration.

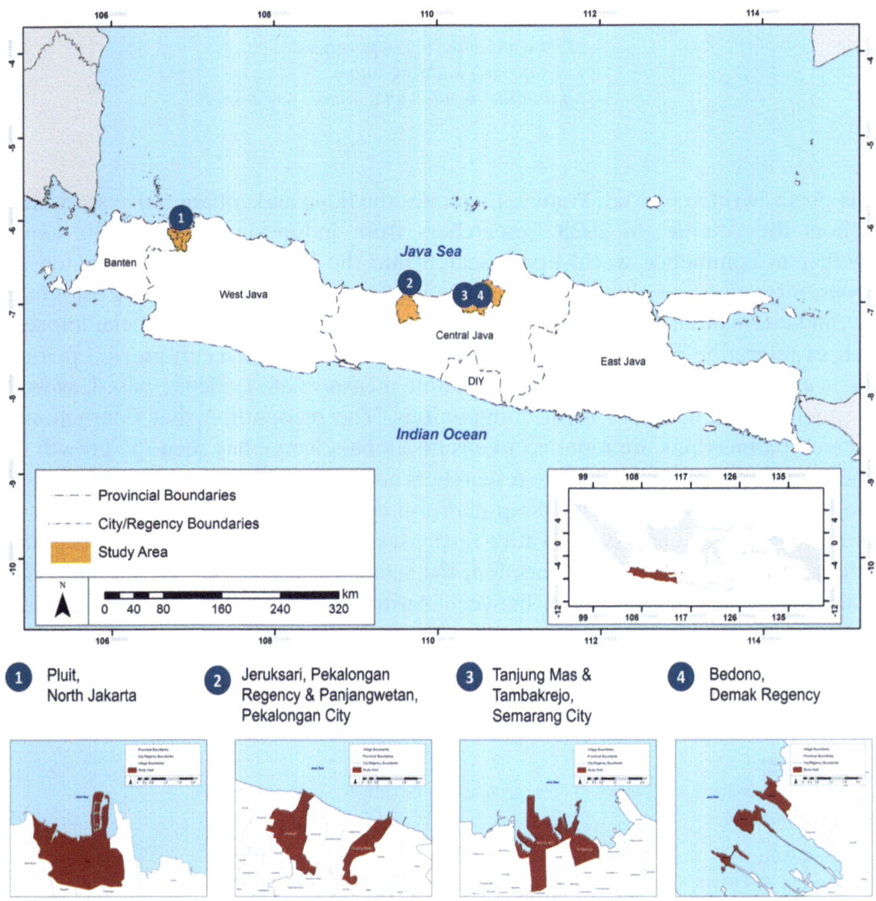

Fig. 1.1 Maps of the study area (BIG, 2024 [edited])

The first site is in Muara Angke Kampong, Pluit, North Jakarta. Located in the capital city of Indonesia, this area covers approximately 72 hectares and is categorized as a densely populated fishing settlement. Residents are heavily dependent on the sea, with 93% of the population engaged in the fishery sector. Within this sector, men predominantly work as fishers, while women work as mussel peelers. With the occurrence of climate change, this neighbourhood is experiencing significant impacts from sea level rise and land subsidence.

Pekalongan, Central Java, was chosen as the second location for the study. The study encompasses both the urban village within the city, specifically Panjang Wetan, and the rural area within the regency, Jeruksari. Pekalongan is known for its batik industry. The Pekalongan sites are significantly affected by tidal inundation, resulting in various disruptions to their livelihoods. These include a substantial decline in turnover due to damage to fish markets and fishponds, as well as a high rate of poverty and stunting amongst children.

Semarang City, the capital of Central Java Province, has been selected as the subsequent area of study. This study includes two urban villages within Semarang, namely Tanjung Mas, with particular attention to the Tambak Lorok Settlement Area and Tambakrejo. Tambak Lorok Kampong is a fishing settlement in the Semarang coastal area, where the majority of the population is employed in the fishery and construction sectors. This community features a fishing port and auction house, underscoring the predominance of the fishery sector in the local economy. Tambakrejo is an urban village situated along the East Flood Canal River. Despite its urban location, the village is predominantly engaged in the fishery sector, with a significant number of women employed in milkfish enterprises and as prawn peelers. While both locations are susceptible to tidal flooding, Tambakrejo is also vulnerable to pluvial flooding due to its proximity to the river.

Demak was chosen as the final study area, with Bedono Village as the focus. The community of Bedono was primarily engaged in the fisheries sector, with individuals working as fishers and fishpond farmers. Women predominantly participate in the fishery processing industry and as vendors. Children are also involved in fishing activities, often accompanying their parents to the sea. Since the 1980s, the village has been affected by tidal flooding, which has caused significant damage to houses, roads, and fishponds, thereby disrupting the community's livelihood. The population of this village has decreased by 6% from 2017 to 2022.

Throughout the volume, the authors passionately describe the sites, the people they met and interviewed there, and the visible impacts of daily inundation. Prior to the commencement of the research, the authors of this volume took a trip to each area and informally met local women's groups. This experience helped shape the project. For authors from Australia, daily coastal inundation was something they had intellectually conceptualized and seen television footage of but had not fully experienced. How women have had to adapt to everyday activities, including raising houses and roads, has become a visible reminder of how normalized the impacts of climate change can be. All of these adaptations cost money, often for those who have the least. This initial trip also highlighted the rapid changes impacting the physical environment. For instance, when chatting to women running a bakery in

Pekalongan regency, the research team noted that Google Maps showed the area around us as fields, when all that could be seen was water at least a metre deep. These rapid physical changes to environments shift community life through adaptation and migration. This volume unpacks community and social changes, especially for women and children, where adaptation strategies may tip into rights violations. While interconnected, the book is loosely divided into chapters focusing on climate change, forced labour and migration.

The Nexus of Climate Change, Migration and Forced Labour

Climate change exacerbates violence and human rights violations for women and children, especially in the global south, and could internally displace more than 216 million people by 2050 (Clement et al., 2021). Women and children are differentially impacted during the process of climate change-induced migration (Chindarkar, 2012; Bhatta et al., 2015). Globally, women are more likely to receive lower incomes, less formal education and autonomy to make individual decisions, reducing options for adaptation but also increasing their chances of being targets for exploitation (Hetherington & Boddy, 2013). A United Nations (UN) report on gender, climate and security (2020) has identified the need for greater attention to the intersections between gender and migration. This volume contributes to addressing this gap in the academic literature and policy responses.

Migration due to climate change can be spontaneous or planned as a local climate change response strategy, and it can be temporary, seasonal or even permanent (nationally or internationally) (Bohra-Mishra et al., 2014). The unsettling nature of the migration process is often associated with vulnerability to worsened well-being and exploitation (Burrows et al., 2011), particularly for women and children who routinely report sexual and family violence in the process of migration or once they arrive at temporary shelters (Scott, 2020). Population immobility is a subset of the literature focused on migration and is particularly relevant in climate change research, as it adds to the understanding of why communities may continue to reside in disaster-prone areas (Amin et al., 2021). Amin et al. (2021) critique the migration literature for its lack of attention to population immobility, especially in relation to climate change.

The availability of work acts as a motivator to migrate from disaster-impacted locations and a motivator to stay. Like migration (and sometimes in association with migration), work-related strategies are often part of mitigation and resilience strategies to address the impacts of climate change. However, the need to work and the impact of climate change on people's livelihoods (particularly in farming and fishing communities) render people vulnerable to work that is unsafe, unregulated, involuntary and dangerous, or even to forced labour. The risk of forced labour, particularly for women and children, is a relevant topic to be studied to better understand and plan for the mitigation of climate change impact.

Why Focus on Women and Children?

When it comes to the connection between gender and climate crises, some concerns in this context are as broad as the impact of heat waves on pregnant women; women as the largest group of small farmers likely to lose land to erosion, excessive heat or displacement; women and reproductive issues; and women as key front line workers in health, including during climate-related emergencies (Turquet et al., 2023). There are risks of forced marriages guaranteeing better land/family survival during relocation. Around 12 million more young girls are thought to have been married off as the frequency of natural disasters has increased (Atkinson and Bruce, 2015) and sexual abuse in displacement centres (Scott, 2019). Many of these concerns extend to women, girls and even children of all genders, and we can observe that many of these issues are intimately connected to displacement and migration and work and livelihoods, making the discussion of forced labour and forced migration even more relevant. Further, disasters related to climate impact children's physical and mental health and disrupt their social protections (Kousky, 2016). Disasters can interrupt children's schooling through displacement, pushing children into work earlier (Kousky, 2016).

The Focus on Climate Change in this Collection

Climate change has significantly affected coastal areas around the globe, particularly through sea level rise, leading to increased flood risk and human migration. Chapters in this edited collection focus on coastal adaptation strategies developed to protect populations affected by sea level rise. Women's agencies provide important leadership in embedding climate change adaptations into local communities. Handayani et al. provide a thoughtful assessment of Indonesian government policies on climate change adaptation, especially in response to sea level rise. This chapter unpacks the social, economic and environmental impacts of infrastructure projects aiming to mitigate climate change and provides a comprehensive overview of the Indonesian policy landscape in response to climate change. Nurhidayah and Borges highlight that Indonesia's current legal and policy framework is inadequate for addressing the impacts of climate change on coastal communities. In particular, it fails to effectively respond to sea-level rise and the resulting human mobility. They argue that legal and regulatory reforms are needed, as there are currently no sufficient protections for internally displaced people.

Landung et al. investigate gender-responsive adaptation strategies to climate change. They examine gender adaptation practices in three countries which have similar challenges to Indonesia, including economic disparities, limited access to health and education and patriarchal norms. They suggest that the success of gender-responsive adaptation strategies contributes to increasing women's access to training, education and resources, strengthening their participation in decision-making.

Analis et al. examine the impact of climate change on the education sector in Pekalongan, Central Java. They suggest that daily flooding, as an impact of sea level rise, impacts not only school infrastructure but also learning processes. Analis et al. apply an epistemic injustice lens to their research, concluding that students need to be informed about the impacts of sea level rise on their education. Alawiyah et al. discuss the role of women in climate change adaptation in Jakarta. They found that women play a crucial role in civic engagement by actively including their communities and cooperative organizations. Further, women are also active in climate change mitigation in communities through mangrove rehabilitation and improving the economic status and livelihood of their families.

Suntoro et al. assess the impact of hard infrastructure development, such as sea walls, in coastal Java to address sea level rise. Despite the benefits of preventing flooding, sea walls also have social, economic and environmental impacts. Major roads and seawalls, for example, have caused damage to the mangrove ecosystem and affected fishermen, as access to fishing has been blocked by hard infrastructure. This chapter recommends that there is a need for greater collaboration between policymakers, practitioners, and local communities to integrate nature into "mainstream" infrastructure planning and design, provide win-win solutions to communities, and reduce the negative implications on their livelihoods.

Hidayati et al. describe the role of local leaders in climate change adaptation in Demak, Central Java. Effective climate change adaptation at the local level is influenced by local leadership, who can become a central figure in mobilising resources, integrating local knowledge into policies and practice, and assisting in adapting and building resiliency in the local communities. They discuss the transformative power of grassroots leadership in mangrove conservation and the significant role of local leaders in helping the community relocate while maintaining social cohesion. Sarrifudin et al. explore the complementary role of property buyers in choosing homes based on their risk of flooding, known as autonomous adaptation. Understanding buyer behaviours in the property market will assist spatial planners in navigating urban growth. Dwirahmadi's chapter examines climate change by applying disaster management strategies using both top-down and bottom-up approaches. The health impacts on children from climate change and social inequality are illustrated by Rahmadhony et al. in their chapter, which focused on stunting cases in Java. This chapter focuses on the paradox of the abundance of protein-rich seafood in areas of coastal Java, which also reports high rates of stunting amongst children. This apparent contradiction highlights the disruption that climate change, migration for work and shifts in traditional livelihoods can have on the health of vulnerable children. Rahmadhony et al.'s work takes a deep dive into policy successes and failures in combatting climate change exacerbated stunting in Demak and Semarang, with key findings to guide researchers and policymakers.

The Focus on Forced Labour in this Volume

In the age of artificial intelligence, accurate translation between languages is often seen as a given skill, rather than a technically nuanced skill. It is when we come to translate complex and sensitive topics across languages and cultures that translators become truly valued. Over the course of the research, how forced labour was conceptualised by the project team shifted as it was adapted to reflect local understandings. In Bahasa Indonesia, there are multiple translations for the English term 'forced labour'. The interest in forced labour, especially amongst women and children, emerged from the Australian research contingent, who are specialists in violence prevention and human rights violations. This report understands the difference between forced and compelled labour. While the International Labour Organisation (ILO) defines forced labour as *any work or service which is extracted from a person under menace or penalty for which the person has not offered voluntarily* (adapted from ILO, 1930 (no. 29)), this volume focuses on compelled labour, which we are defining as labour practices that are socially mandated and which negatively impact wellbeing. In a collectivist society such as Indonesia, social mandates can be powerful drivers of social cohesion but also have unintended consequences for those who hold less social power in communities. The consequences for women's and children's health and education can be significant and inhibit well-being in the long term.

The chapters in this volume focus on hazardous labour and its health impacts on women and children, as well as working children. While working children can play an important role in family cohesion, especially in family owned businesses, this needs to be balanced by a child's right to education, play and safety. Suntoro et al. build on Rahmadhony et al.'s chapter, extending the discussion on changes in women's livelihoods in fishing communities in coastal Java. Suntoro et al. examine the economic strategies women are taking in coastal Java to mitigate and adapt to climate change impacts. This important chapter provides examples of how local governments can partner with women's organisations to create appropriate mitigation and adaptation programmes that not only address the impacts of climate change but also focus on economic empowerment.

Informal sector workers are especially vulnerable to compelled and hazardous labour, as they work outside of formalised legal agreements. Mareta et al. note that informal sector work is often gendered, and in their chapter, they examine the impacts on women's physical health and well-being when working extended hours in the informal sector in villages that experience persistent tidal flooding. This chapter encapsulates the intersection between slow-onset disasters and forced labour, providing readers with an insight into how the financial burden from daily flooding has pushed women into economic instability, leaving them with little choice but to engage in work that impacts their health. They are compelled into these precarious situations through family obligations, with choices further limited by a lack of formal education and little social support. Mareta et al.'s chapter gives voice to women's experiences in Pekalongan, translating their stories into calls for action. Bararah Barid et al. have produced a chapter to complement Mareta et al.'s work, highlighting the experiences of fishermen who are reliant on their wives' incomes in the informal sector. Changes in monsoonal patterns due to climate change have impacted the livelihoods of fishing villages, with Bararah Barid et al. calling for urgent collaboration between government stakeholders and the private sector to provide avenues for further education, training and capacity building. The chapters in this volume that have a primary focus on labour highlight the need to partner with local communities, build on their strengths, and provide them with skills to facilitate safe working environments.

The Focus on Migration in this Collection

The chapters in this volume contain diverse experiences of people who are migrating due to climate change and those who are immobile and living in hazardous conditions. Migration and immobility are themes woven throughout the volume, with specific chapters discussing the topic in depth. Hidayati and Liang focus on immobility due to sea-level rise, evaluating the policy response for those living in increasingly hazardous conditions. This chapter takes an intersectional focus on adaptation for immobile communities, focusing on the adaptation strategies developed by both communities and governments. Local adaptive strategies, such as raising houses, are a stark and visible reminder when visiting the coastal villages of Java regarding the risk of daily flooding. Hidayati and Liang call for the integration of local solutions, such as economic diversification, into government strategies. While Hidayati and Liang's chapter focuses on policies designed to facilitate those choosing to stay in their communities to remain safe, Hayati et al. focus on those who are forcibly relocated. This chapter unfolds the emerging challenges associated with the implementation of relocation programmes in response to tidal flooding, drawing insights from the case of Demak Regency, Central Java. Hayati et al. identify human rights violations that arise from relocation programmes. These challenges require comprehensive regulatory measures to address them effectively.

In addition to focusing on immobile populations and those forcibly relocated, this volume also gives visibility to those who migrate to find work. Nurhidayah et al. focus on compelled migration for populations from regional locations into Jakarta and the associated challenges for policymakers. This is a challenge for countries worldwide, with key learnings for policymakers highlighted in the chapter. How local governments can play a role in policy and mitigation strategies is the focus of Wahyu's chapter. Wahyu undertakes an in-depth scoping of policy responses to climate change and migration and provides commentary on the need for a greater focus on women and children in policy responses. Migration can be a contentious space for local communities, with conflicts arising if policies do not include adequate social protection.

What to Expect from this Book

This collection of chapters offers a transdisciplinary perspective on the social impacts of climate change, with the majority of contributing authors writing from the global south. Given the greater vulnerability to climate change impacts on populations in the global south, it is integral for authors from these regions to be actively included in research and discussions on climate change and its impacts. Unpacking where cross-cultural understandings of forced labour, migration, gender, and childhood converge and diverge enriches global literature and can begin to build consensus on suitable approaches to addressing global challenges. The inclusion of authors from different disciplines also provides complementary perspectives, allowing a holistic understanding to be developed. This volume provides perspectives from urban planning, social work, law, engineering, public health, anthropology and psychology. The majority of the authors are also women, highlighting the leadership of the three female editors provided not only to this collection but throughout the research process. This collection of papers provides policymakers and researchers with evidence-based recommendations on how to address the social impacts of climate change in coastal communities in a way that engages communities and keeps a focus on women and children. The research that underpinned this volume included interviews and surveys with women and children, as well as roundtables with community members and policymakers, and extensive policy scoping. This research was designed to highlight the wisdom of communities and actions that are already being undertaken. The stories contained in this volume offer not only insight into daily challenges but also capture the resilience and innovation of communities experiencing sea-level rise. At the end of each field visit, the researchers were encouraged by community members to take group photos. This always included making a gesture, either of strength, love or resilience. While the group photos haven't been included here, other images from the field sites have been used to depict the realities faced in coastal Javanese communities and to motivate the call to action contained within the final chapter of this collection.

References

Amin, M. N., Roy, K., & Hossain, M. S. (2021). Determinants of climate change adaptation knowledge and techniques in Bangladesh. *Environmental Science and Policy, 118*, 1–10. https://doi.org/10.1016/j.envsci.2021.01.008

Atkinson, H. G., & Bruce, J. (2015). Adolescent girls, human rights and the expanding climate emergency. *Annals of Global Health, 81*(3), 323–330. https://doi.org/10.1016/j.aogh.2015.08.003

Bhatta, G. D., Aggarwal, P. K., Poudel, S., & Belgrave, D. A. (2015). *Climate-induced migration in South Asia: Migration decisions and the gender dimensions of adverse climatic events. Journal of Rural and Community Development, 10*(4) https://journals.brandonu.ca/jrcd/article/view/1177

BIG. (2024). *Badan Informasi Geospasial.* Accessed on 1/10/2024 from: https://www.big.go.id

Bohra-Mishra, P., Oppenheimer, M., & Hsiang, S. M. (2014). Nonlinear permanent migration response to climatic variations but minimal response to disasters. *Proceedings of the National Academy of Sciences of the United States of America, 111*(27), 9780–9785. https://doi.org/10.1073/pnas.1317166111

Burrows, M. T., Schoeman, D. S., Buckley, L. B., Moore, P. J., Poloczanska, E. S., Brander, K. M., Brown, C. J., Bruno, J. F., Duarte, C. M., Halpern, B. S., Holding, J., Kappel, C. V., Kiessling, W., O'Connor, M. I., Pandolfi, J. M., Parmesan, C., Schwing, F. B., Sydeman, W. J., & Richardson, A. J. (2011). The pace of shifting climate in marine and terrestrial ecosystems. *Science, 334*(6056), 652–655. https://doi.org/10.1126/science.1210288

Chindarkar, N. (2012). Gender and climate change-induced migration: Proposing a framework for analysis. *Environmental Research Letters, 7*(2), 025601. https://doi.org/10.1088/1748-9326/7/2/025601

Clement, V., Rigaud, K. K., de Sherbinin, A., Jones, B., Adamo, S., Schewe, J., Sadiq, N., & Shabahat, E. (2021). *Groundswell Part 2: Acting on internal climate migration* (Report No. 36248). World Bank. http://hdl.handle.net/10986/36248

Hetherington, T., & Boddy, J. (2013). Ecosocial work with marginalized populations. In M. Gray, J. Coates, & T. Hetherington (Eds.), *Environmental social work*. Routledge.

Kousky, C. (2016). Climate change and risk management: Challenges for insurance, adaptation, and loss estimation. *Risk Analysis, 36*(4), 586–598. https://doi.org/10.1111/risa.12500

Scott, M. (2019). Accountability for state failures to prevent sexual assault in evacuation centres and temporary shelters. In. Kinnvall, C. & Rydstrom, H. *Climate Hazards, Disasters and Gender Ramifications*. Routledge. https://doi.org/10.432/9780429424861

Scott, M. (2020). *Climate change, disasters, and the refugee convention*. Cambridge University Press.

Turquet, L., Tabbush, C., Staab, S., Williams, L., & Howell, B. (2023). *Feminist climate justice: A framework for action*. UN Women. https://www.unwomen.org/en/digital-library/publications/2023/11/feminist-climate-justice-a-framework-for-action

United Nations. (2020). *Gender, climate & security: Sustaining inclusive peace on the frontlines of climate change*. United Nations.

Chapter 2
Climate Change and Development Policy in Indonesia: From Survival to Transformation

Wiwandari Handayani, Citra Tatius, and Mega Anggraeni

Introduction

Climate change is a pressing issue in Indonesia due to the country's unique characteristics as an archipelagic nation, comprising over 17,000 islands and extensive coastal areas. The impacts of climate change pose significant risks to biodiversity, agriculture, and the livelihoods of millions of people. In response, the Indonesian government has developed policies aimed at adapting to and mitigating climate change. Through international agreements, such as the 2015 Paris Agreement, Indonesia is obligated to contribute to reducing greenhouse gases (GHGs) to limit the global temperature increase to under 2 °C—and if possible under 1.5 °C—from the pre-industrial level. These targets are included within sectoral plans such as the National Action Plan for Reducing Greenhouse Gas Emissions (RAN-GRK), the National Action Plan for Climate Adaptation (RAN-API), and the Medium-Term Development Plan (RPJMN), all of which integrate climate resilience into national development goals.

There is also a growing emphasis on shifting policies towards sustainable practices across various sectors, such as agriculture, forestry, and fisheries. These efforts seek to reduce vulnerability to climate change while promoting economic growth. Engaging local communities in climate adaptation strategies through climate awareness campaigns and capacity-building programmes has been recognised as crucial

W. Handayani (✉)
Department of Urban and Regional Planning, Faculty of Engineering, Diponegoro University, Semarang, Indonesia
e-mail: wiwandari.handayani@live.undip.ac.id

C. Tatius
Center for Urban and Regional Resilience Research (CURE) – Faculty of Engineering, Diponegoro University, Semarang, Indonesia

M. Anggraeni
Initiative for Urban Climate Change and Environment (IUCCE), Semarang, Indonesia

L. Nurhidayah et al. (eds.), *Climate Change, Labour and Migration in Indonesia*, Springer Climate, https://doi.org/10.1007/978-3-032-15334-0_2

for promoting sustainability and environmental stewardship. Additionally, Indonesia has been actively participating in global climate negotiations, including the Paris Agreement, to enhance its capacity for climate action and secure funding for sustainable development initiatives.

However, policies and initiatives on the ground often focus on short-term survival strategies rather than long-term transformative solutions. Challenges such as lack of funding, limited financing options, and critical governance issues contribute to a reactive rather than proactive approach in running development programmes, which are often constrained by a five-year presidential cycle. This has led to extensive but incremental climate change across the country.

This chapter presents an overview of climate change impacts in Indonesia and the policies adopted to reduce and further mitigate its impacts. Currently, Indonesia is transitioning from merely seeking to survive the effects of climate change to actively transforming its societal structure and policies to build its resilience and sustainability. By integrating climate considerations into development strategies, the nation aspires to ensure a more secure and prosperous future for its population. Although research on coastal areas primarily focuses on adaptation strategies in response to climate change impacts, this chapter will also elaborate on mitigation measures as proactive steps to move forward into a sustainable future.

Climate Change Causes and Impacts in Indonesia: An Overview

Indonesia, particularly its major cities, continues to produce a significant volume of emissions due to increasing development and economic activities. Figure 2.1 shows the time-series data of GHG emissions in the twenty-first century as measured by the Ministry of Environment and Forestry (MoEF). The graph indicates that the forestry sector generates the highest emissions, contributing over 30% of the total emissions every year from 2000 to 2020. However, a notable aspect is the consistent growth of emissions from the energy sector, apart from slight decreases in 2020 (due to COVID-19) and in 2023 (as a result of the implementation of renewable energy aimed at achieving a 51.30-million-tonne reduction in carbon dioxide equivalent, or CO2e). According to the MoEF, over the 23-year period, cumulative CO2e emissions reached 27 million gigagrams (Gg; 1 Gg equals 1000 tonnes). This ongoing increase exposes Indonesia to greenhouse gases, including rising temperatures.

As a consequence of increasing emissions, Indonesia is coping with the inevitable climate change phenomenon. Climate change has had a range of negative impacts on human activities in Indonesia for decades (ADB, 2021); for example, sea levels rose at an average rate of 3.9 mm/year between 1992 and 2020 (NOAA, 2020). This situation, combined with land subsidence, has caused some areas to be permanently inundated and threatens communities living in low-lying coastal areas with eventual large-scale displacement (Nurhidayah & McIlgorm, 2019). This is

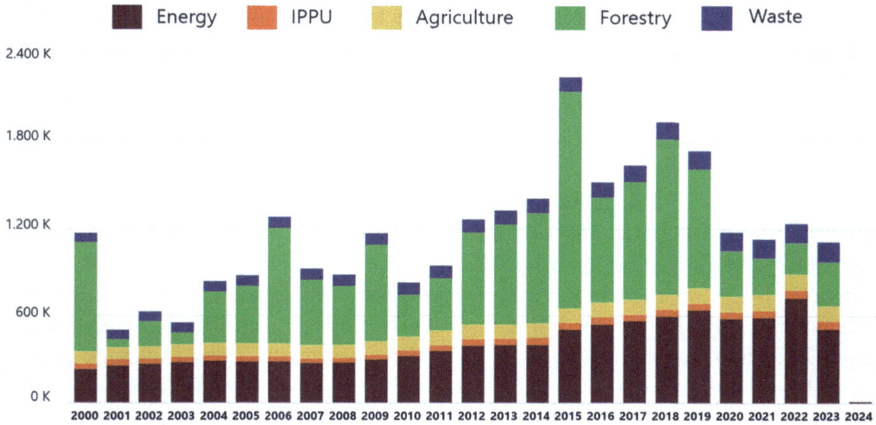

Fig. 2.1 Indonesia's greenhouse gas emissions 2000–2024. Source: Ministry of Environment and Forestry, 2024

just one example of how the impacts of climate change exacerbate communities' vulnerability.

In addition to rising sea levels, increasing temperatures have also caused disruption in civil development. Indonesia's Meteorology, Climatology, and Geophysics Agency (MCGA) has reported that the average temperature rise in Indonesia in the period 1981–2023 was 0.6 °C (MCGA, 2024a). This warming trend, along with increasing urban activity, generates surface urban heat island (SUHI) effects in densely populated areas, a condition in which the temperature in urban areas is significantly higher than that in rural areas. Using satellite-derived Land Surface Temperature and Enhanced Vegetation Index data for the 2001–2022 period, Ferdiansyah and Siswanto (2024) report that cities in Java Island (the most populated island in Indonesia) show the highest land surface temperature, with 23 cities experiencing daytime SUHI expansion and eight cities having night-time SUHI growth affecting surrounding municipalities. The increase in SUHI effects is mostly caused by rapid urban expansion. According to Sejati et al. (2019), in the Semarang Metropolitan Region, areas with intensive development due to industrial zones and transportation hubs, such as harbours and air terminals, experience a more significant SUHI expansion, with temperatures exceeding 30 °C.

This increased heat not only causes thermal discomfort by elevating the surface temperature within the cities but also interlinks with other climate risks that may cause harm to communities. Triana and Wahyudi (2020) predict, based on a scenario of a sea temperature rise by more than 2 °C, followed by an estimated thermosteric sea level rise of up to 40–80 cm by 2100, that the total sea level rise in Indonesia will reach 1–2 m by 2100. This situation will not only cause damage to infrastructure but also harm people and disrupt their livelihoods.

Rainfall patterns also reflect the ongoing climate change impacts in Indonesia. Based on an analysis of rainfall data from 1981 to 2023, MCGA (2024b) reports that over a 30-year period, Indonesia experienced precipitation changes, including annual increases of up to 2784 mm in some areas and, conversely, decreases of up to 750 mm in other areas. These significant changes can influence the occurrence of extreme weather events, causing more disasters, such as floods and droughts, posing severe risks to communities in crucial sectors, such as agriculture and water resource management.

Climate-related risks, such as extreme weather, not only cause damage to infrastructure but also significantly disrupt the primary sectors on which Indonesia is heavily dependent, including fisheries, agriculture, and forestry (Murniati, 2020; Wijaya, 2014). Romanello et al. (2022) report that climatic factors also significantly influence the spread of vector-borne diseases in Indonesia such as dengue, chikungunya, Zika, and malaria. These phenomena demonstrate that climate change effects extend beyond mere environmental degradation and affect multiple aspects of society through impacts such as economic losses and public health issues. Figure 2.2 portrays the most severe impacts of climate change across Indonesia, with the increased intensity and frequency of climatic events putting immense pressure on the country's resilience.

Urbanisation significantly exacerbates the impacts of climate change, particularly on Java Island. According to the Central Bureau of Statistics (CBS, 2021), 56.10% of Indonesia's population lives in Java, even though it covers only 7% of the country's total area. This concentration is largely due to rapid urban growth in the northern coastal areas of Java, which attracts in-migration due to their urban features (Handayani & Kumalasari, 2015). This extensive urban development creates pressure on these coastal regions, risking overcapacity as they try to accommodate the increasing population and associated activities. Consequently, northern coastal Java has become more vulnerable to climate change.

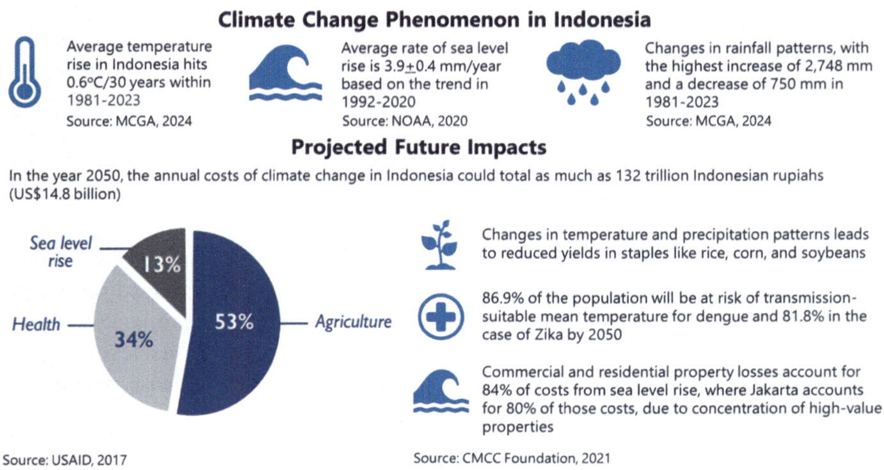

Fig. 2.2 Severe climate change impacts in Indonesia. Source: MCGA, 2024a; NOAA, 2020; USAID, 2016; CMCC Foundation, 2021 (modified)

Of the various climate-related disasters impacting northern Java, flooding is the most frequent threat, with the frequency of flood events in the northern coastal cities having been on the rise for decades. Sejati et al. (2024) note that, based on data from 1970 to 2020, flood occurrences are primarily concentrated in the northern coastal regions of West and Central Java. This frequent flooding is closely linked to the geographical characteristics of the area and is exacerbated by land subsidence. According to Bott et al. (2021), cities along Java's northern coast experience land subsidence rates of up to 15 cm per year. Figure 2.3 illustrates the climate-related risks faced by the northern coastal areas of Java.

The factors outlined above, along with the growing development pressures, significantly increase the vulnerability of cities in these areas. Handayani et al. (2020) note that rapid urbanisation in low-lying cities heightens exposure to flood risks, while climate change exacerbates this threat through rising sea levels and altered rainfall patterns. As a result, cities with dense populations and built environments are more likely to be susceptible to flooding.

Flooding in coastal areas has caused significant disruptions to both the environment and the livelihoods of local communities. Communities are experiencing issues such as freshwater contamination due to seawater intrusion, damage to housing and infrastructure, and the spread of diseases (Yuliastuti et al., 2023). Moreover, the negative impacts of flooding extend beyond daily community life, significantly affecting the fishery sector that is crucial for the livelihoods of people in these regions. Many residents rely on the fishery industry, working as fishermen, fish pond farmers, or employees in various fishing enterprises. Flooding damages fishing infrastructure, including harbours and fish ponds, thereby disrupting economic activity (Rudiarto et al., 2019). This, in turn, affects economic productivity, potentially leading to financial hardships for those employed in the fishery sector. Furthermore, this situation imposes additional burdens on women and children, as

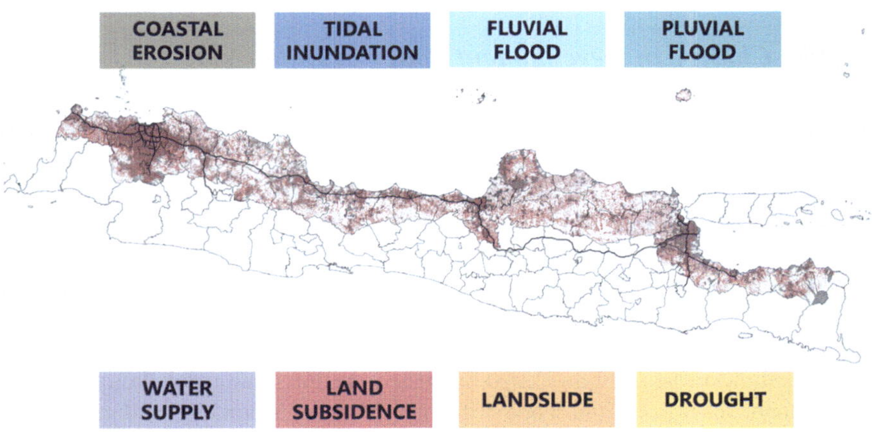

Fig. 2.3 Climate-related risks in northern coastal areas of Java

women—who are primarily responsible for domestic management in Indonesia—face decreasing household income and children's education disrupted by disasters.

The challenges caused by climate change involve not only excess water, such as flooding, but also water scarcity. Droughts have become a significant issue, with consequences for both the environment and communities. According to Rudiarto et al. (2018), drought is one of the most frequent disasters in the northern coastal region of Central Java, alongside floods and tidal floods, primarily due to decreased precipitation and population growth. This situation disrupts the ability to meet the water needs of various activities in coastal areas, from basic daily needs to economic activities. The impacts of drought extend beyond public health; they also affect the coastal environment by disrupting the salinity balance in aquatic ecosystems, which negatively impacts fish habitats and fishing activities (Xu et al., 2023). Collectively, these issues place significant strain on the region and exacerbate the vulnerabilities of the local population.

The points outlined above illustrate that the impacts of climate change are complex and multifaceted, with coastal regions being particularly vulnerable to escalating climate risks. Environmental degradation, infrastructure damage, and resource depletion pose significant challenges for residents, especially for the most vulnerable populations, such as women and children, adversely affecting their overall well-being. The economic losses resulting from climate change further exacerbate these vulnerabilities, impacting whole communities' livelihoods (Insani et al., 2022). Addressing these climate challenges is crucial for enhancing the resilience and security of affected populations. It is essential to formulate comprehensive policies that both mitigate and adapt to the effects of climate change. Integrating these policies into development planning is equally important to ensure effective implementation in urban development.

Mainstreaming Climate Change Into Development Policy

Climate Change in Development Policy in Indonesia

Indonesia demonstrated its commitment to tackling climate change by signing the United Nations Framework Convention on Climate Change (UNFCCC) in Rio de Janeiro, Brazil, in 1992, with this commitment later solidified through Law No. 6/1994. This commitment was followed by the ratification of the Kyoto Protocol—an international treaty to reduce GHG emissions by 2008–2012—through Law No. 17/2004 and by Indonesia's ratification of the Paris Agreement in 2015 during the 21st Conference of the Parties (COP21).

These commitments have been embedded in policies that focus on adaptation and mitigation measures. Law 32/2009 concerning Environmental Protection and Management accentuated the scope differences between adaptation and mitigation, with adaptation aiming to enhance resilience to reduce climate change impacts, while mitigation focuses on efforts to reduce emissions as a long-term objective. These concepts are translated into action plans, with a focus on adaptation outlined in the 2014 National Action Plan on Climate Change Adaptation and mitigation efforts detailed in the 2011 National Action Plan on Greenhouse Gas Emission Reduction.

Following the mandated contribution to the Paris Agreement, Indonesia's target was set out in its Nationally Determined Contribution (NDC). Indonesia also submitted its Forest Reference Emission Level (FREL) as the country's reference for measuring performance under the Reducing Emissions from Deforestation, Forest Degradation, Role of Conservation, Sustainable Management of Forests, and Enhancement of Forest Carbon Stocks (REDD+) framework. These commitments have driven Indonesia to release information systems on monitoring its achievement in climate change adaptation and mitigation, such as SIS REDD+,[1] SRN-PPI,[2] SIPONGI+,[3] and SIDIK,[4] as well as programmes at the local level—Programme *Kampung Iklim* (ProKlim).[5] These approaches, illustrated in Fig. 2.4, reflect Indonesia's commitment to aligning sustainable development with global climate goals.

Indonesia has also followed up on the ratification of various international agreements by incorporating them into development planning. Mogelgaard et al. (2018) note that mainstreaming mitigation and adaptation strategies into development policy is essential to ensure the resilience of development outcomes in the face of changing climates, promote the efficient use of resources and prevent maladaptive investments. This highlights that translating climate strategies into development policy means operationalising initiatives as tangible and actionable outcomes, as well as allocating resources for implementation, supporting their pivotal role in climate adaptation and mitigation.

In Indonesia, the integration of climate change into development policy reached a significant milestone with the inclusion of climate priorities in RPJMN 2020–2024, specifically under the sixth national priority (see Fig. 2.5). The integration of climate adaptation and mitigation initiatives into development is exemplified by the Climate-Resilient Development (CRD) and Low Carbon Development Indonesia (LCDI) programmes. These strategies are further detailed and operationalised in the

---

[1] Safeguard Information System (SIS) Reducing emissions from deforestation and forest degradation in developing countries (REDD). The '+' stands for additional activities related to forestry, including sustainable management of forests and the conservation and enhancement of forest carbon stocks.

[2] Sistem Registrasi Nasional – Pengelolaan Pengetahuan Inovasi, or Indonesia's National Registry System for Climate Change Mitigation and Adaptation Control, provided by the Ministry of Environment and Forestry.

[3] Forest and Land Fires Monitoring System, provided by the Ministry of Environment and Forestry.

[4] Sistem Informasi Data Indeks Kerentanan, or Indonesia's Vulnerability Index Data Information System, developed by the Ministry of Environment and Forestry.

[5] Indonesia's Climate Village Programme, which provides community-based climate action at the village or urban neighbourhood level, initiated by the Ministry of Environment and Forestry.

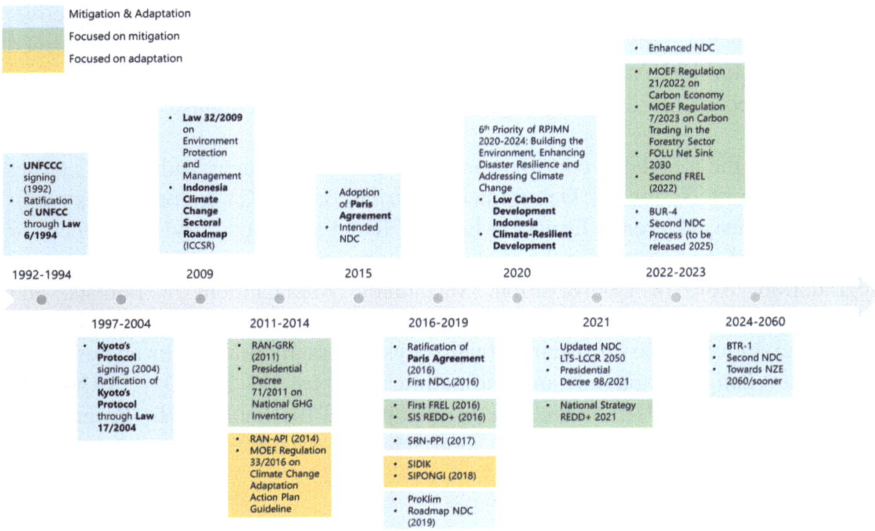

Fig. 2.4 Milestones of Climate Change Mitigation and Adaptation Policy in Indonesia. Source: MOEF, 2024 (*modified*)

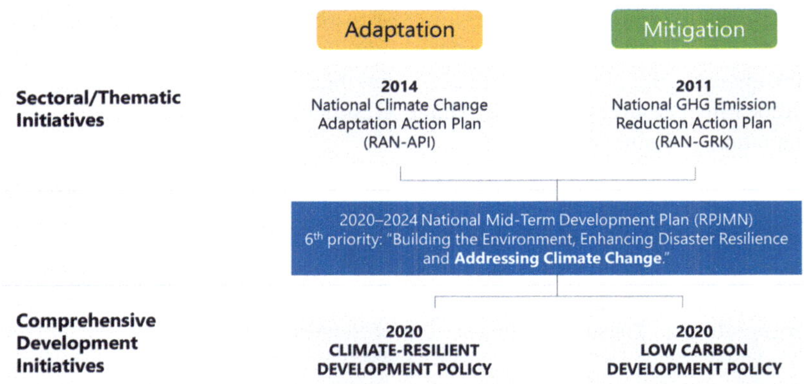

Fig. 2.5 The connectivity of adaptation and mitigation mainstreaming into development policy in Indonesia. Source: Government of Indonesia, 2021 (*modified*)

Long-Term Strategy for Low Carbon and Climate Resilience (LTS-LCCR). This strategy document includes a roadmap towards climate resilience and net zero emissions through adaptation initiatives directed towards climate-resilient development and mitigation concepts that have evolved from simple GHG emission reduction to a low-carbon development strategy aimed at enhancing the nation's capacity to withstand climate-related risks.

In the context of multilevel governance in Indonesia, the climate agenda has been implemented via integrative measures in national, regional/provincial, and local or city/regency-level development planning policies. Desdiani et al. (2021) highlight the importance of regional and local government involvement in climate change adaptation and mitigation, as such involvement enables more targeted approaches that account for different geographical and fiscal conditions and administrative capacities. This has been showcased in several cases in Indonesia, with several cities developing Regional Action Plans for Climate Change Adaptation (RAD-API), such as Semarang and Kendal, and Regional Action Plans for GHG Emission Reduction (RAD-GRK), such as Cilacap and Medan.

One notable programme that has been implemented at the regional level is ProKlim. The MoEF notes that Indonesia has established 10,113 climate villages in 2024. These require intensive coordination between national, provincial, and local governments and successfully promote active participation of the community in the implementation of adaptation and mitigation, allowing them to co-benefit from climate action (Wiati et al., 2022). The following sections provide a detailed explanation of adaptation and mitigation actions at the regional and local levels.

Climate Adaptation and Development Policy

Aligning climate change adaptation with development is crucial to ensure that adaptation actions create coherent and comprehensive climate change responses, supporting the objectives of sustainable development (Kalogiannidis et al., 2024). The implementation of climate change adaptation in Indonesia was first marked by the formulation of the Indonesian Climate Change Sectoral Roadmap (ICCSR) in 2009. This document contains climate change adaptation strategies for various sectors (e.g. water, marine and fisheries, agriculture, and health) and a roadmap for mainstreaming climate change into development planning, particularly via the National Long-Term Development Plan (RPJPN) 2005–2025. The ICCSR was subsequently adapted as the National Action Plan on Climate Change Adaptation (RAN-API) in 2014.

RAN-API focuses on climate adaptation efforts in four key sectors: marine and coastal, water, agriculture, and health. Strategies in these sectors are organised into four clusters: infrastructure, technology, capacity building, and governance. The document determines key programmes on climate change adaptation and also provides policy mechanisms and location priorities for implementation. RAN-API serves as a cross-sectoral plan within the national development planning, aiming to strengthen the strategic planning of multiple stakeholders involved in climate adaptation. It also serves as a guideline for regional and local governments in preparing their Regional Action Plans on Climate Change Adaptation (RAD-API).

MoEF Regulation 33/2016 on Guidelines for the Formulation of Climate Change Adaptation Actions outlines the steps in the formulation of RAN-API and RAD-API. The process begins with planning climate change adaptation actions by creating an inventory of climate change impacts, developing climate resilience baselines, setting targets, and establishing adaptation action plans. The formulation reflects priorities from the development policy document, such as the RPJMN. Following the action plan, the next step is implementing adaptation measures, which involve stakeholders at the national, provincial, district, and city levels, including the government, private sector, and civil society. Monitoring and evaluation are also carried out at every level with coordination by the MoEF, covering—at a minimum—the implementation of climate change adaptation policies, actions, and capacity-building for climate change resources. MoEF also develops similar initiatives on climate adaptation, such as ProKlim, which aims to foster a climate-friendly environment through community-based action at the neighbourhood level.

In accordance with changes in the Indonesian government's commitment following the ratification of the Paris Agreement, along with the continuous monitoring and evaluation process throughout the implementation of RAN-API, the government transformed RAN-API into Indonesia's Climate Resilience Development (CRD) approach. Bappenas—Indonesia's National Development Planning Agency—promoted CRD as part of the sixth National Priorities in RPJMN 2020–2024, along with Low Carbon Development for mitigation. Bappenas defines CRD as an anticipatory action to minimise potential losses caused by climate change hazards, vulnerabilities, impacts and risks, especially for communities in areas affected by climate change. Based on its operational definition, the CRD aligns with the four priority sectors outlined in RAN-API and focuses on its four key strategic clusters, as visualised in Fig. 2.6.

Following the establishment of Indonesia's NDC target in 2021 through Presidential Decree 98/2021, climate change adaptation is one of the required actions to achieve the target and build resilience to climate risks at national, regional, and community levels. This initiative is accommodated through the NDC document, which is later updated through the Enhanced National Determined

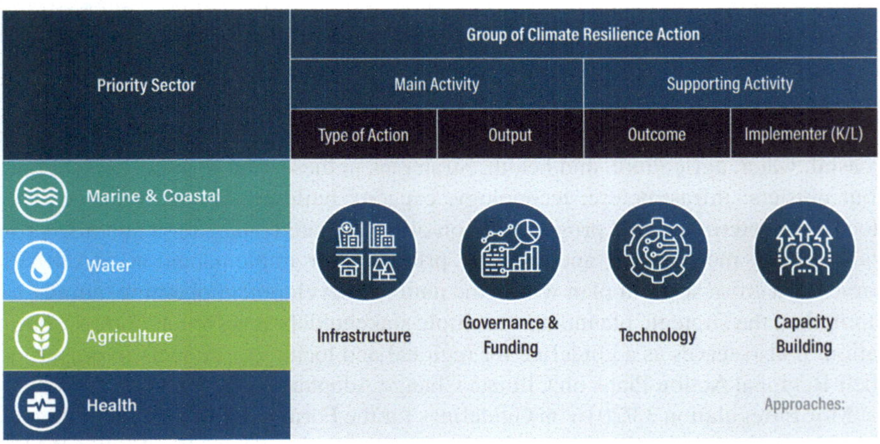

Fig. 2.6 CRD priority sectors and key strategies. Source: Ministry of National Development Planning/Bappenas (2021)

Contribution (ENDC) in 2022. The adaptation pathways under the ENDC are set to reduce climate change impacts on national GDP by 3.45% in 2050, with the main strategies focusing on four key sectors:

a. Marine and coastal: focuses on reducing pollution and optimising potential through upstream–downstream collaboration.
b. Water: emphasises securing water resources against droughts and floods through integrated systems such as water service infrastructure and water retention.
c. Agriculture: implements practices such as vertical farming to address productivity threats from rising temperatures and irregular rainfall.
d. Health: targets vector-borne diseases such as dengue fever exacerbated by densely populated neighbourhoods and lack of quality in sanitation.

The ENDC elaborates on three areas of resilience to be prioritised for operationalisation during the implementation of strategies across the four key sectors. These resilience aspects are required to ensure a holistic and comprehensive approach for addressing climate change.

a. Economic Resilience: ensuring potential climate risks do not significantly disrupt food production and distribution by transforming to a low-carbon economy and building food, water, and energy systems resilience.
b. Social and Livelihood Resilience: preventing further disparity caused by climate change-induced natural disasters, especially among Indonesia's most vulnerable populations.
c. Ecosystem and Landscape Resilience: providing environmental protection services such as watershed protection, carbon sequestration and conservation, and disaster risk reduction through an integrated and landscape-based approach in managing terrestrial, coastal, and marine ecosystems.

LTS-LCCR adapted the goals set in both RAN-API and ENDC to provide national, regional, and sectoral pathways for building climate resilience, as illustrated in Fig. 2.7. LTS-LCCR emphasises the principle of justice and sustainability in building climate resilience, utilising investment, human resources capacity development through green job opportunities, and the management of land and seascape. The efforts are carried out by modality mapping to strengthen capacities in natural resources, governance, culture, and local wisdom. In mapping modalities, LTS-LCCR groups the targets into various dimensions, such as literacy and adaptive technology, the proportion of renewable energy, ecosystem-based analysis units, and the utilisation of climate resources. These efforts are also translated to the regional level, in which the adaptation actions are determined based on the four key sectors combined with the vulnerability index data and climate hotspots within the region.

At the local level, climate change adaptation efforts have been implemented in various cities in Indonesia. The approaches, as illustrated in Fig. 2.8, are mostly related to flood management infrastructure since flooding remains the predominant threat. For example, in Semarang, the government has developed water retention ponds, drainage channels, and pump stations to manage urban flooding and

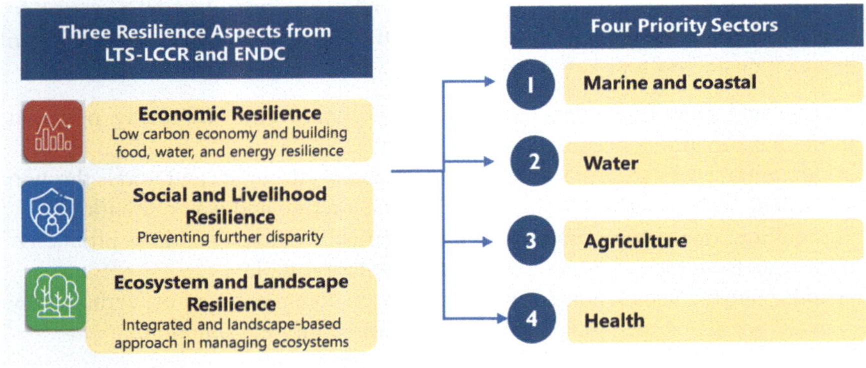

Fig. 2.7 LTS-LCCR and ENDC's three areas of resilience and four priority sectors in climate change adaptation. Source: Government of Indonesia (2021), Government of Indonesia (2022) (*modified*)

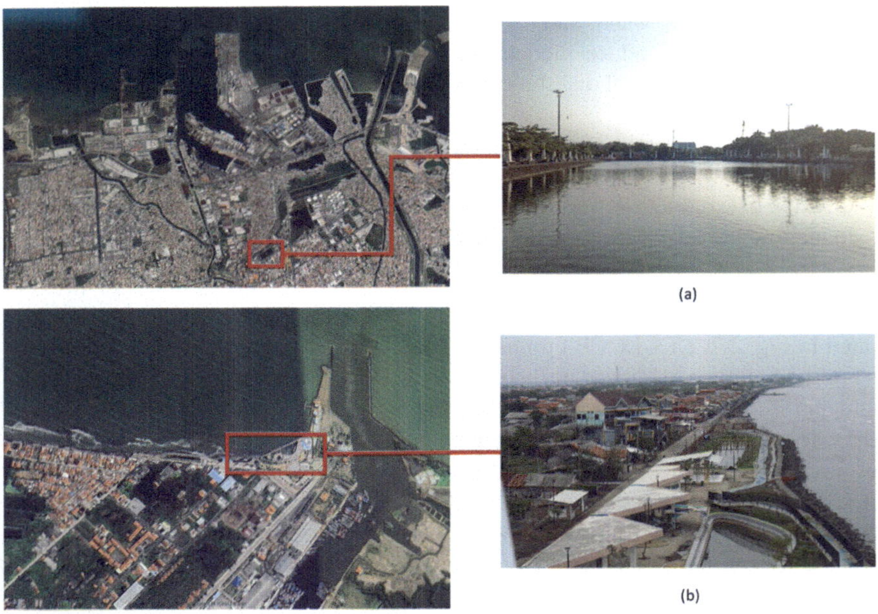

Fig. 2.8 Flood management infrastructure provision in Indonesia. (**a**) Retention Pond in Tawang, Semarang (**b**) Sea Wall in Panjangwetan, Pekalongan. Source: Google Earth, imagery © Airbus, 2025; Documentation, 2019; Documentation, 2024

inundation as part of polder systems in areas such as Tawang. Other coastal cities, such as Jakarta and Pekalongan, have constructed seawalls to reduce the impacts of tidal flooding and rising sea levels.

However, these adaptive strategies often focus on immediate solutions without addressing the underlying vulnerabilities in a transformative manner. Many urban areas continue to struggle with obtaining long-term funding, facilitating interdepartmental cooperation, and incorporating eco-friendly solutions into their infrastructure plans. For example, in Semarang, while the polder system has proven effective, ongoing maintenance and sustainable financing remain problematic. Similarly, in Jakarta, the sea wall alone is insufficient to fully tackle subsidence without comprehensive reforms in land use and water management. Nevertheless, the majority of strategies primarily rely on physical construction (i.e. grey infrastructure), which typically fails to enhance people's ability to cope with disasters and often leads to adverse environmental consequences in the long run (Onuma & Tsuge, 2018). Without adopting a more holistic and inclusive strategy that involves the community, these measures risk becoming temporary fixes rather than enduring solutions for climate resilience. This highlights the need for adaptive strategies that enhance people's capacity to respond to climate challenges, ensuring that solutions are implemented effectively from upstream to downstream.

Climate Change Mitigation and Development Policy

Indonesia's approach to climate change mitigation involves a combination of both international cooperation and national policy that is further elaborated into the National Action Plan for GHG Emission Reduction (RAN-GRK) stipulated under Presidential Decree 61/2011. The decree sets RAN-GRK as a national guideline for direct and indirect emission reduction to be conducted collectively by the national and local government, the private sector, and civil society. RAN-GRK covers emission reduction in several sectors: agriculture, forestry, and other land uses (AFOLU); energy and transportation; industry; and waste management. The modelling of the inventory of these sectors was developed into two stages, with AFOLU sectors modelled first and non-AFOLU sectors modelled separately.

Following these initiatives, MoEF also issued the Regulation of the Minister of Environment and Forestry (P.72/MENLHK/SETJEN/KUM.1/12/2017) on Guidelines for Implementation of Measurement, Reporting, and Verification of Action and Means of Implementation. Through this regulation, the Ministry launched the National Greenhouse Gas Inventory System (SIGN SMART), a GHG inventory tool. The GHG inventory serves as a measure of the emission baseline that will later be incorporated into the emission reduction action plan.

Indonesia's commitment under the Paris Agreement demanded a more integrated and comprehensive approach to climate change mitigation. Recognising this, the government transformed the GHG emission reduction strategy to the Low-Carbon Development Indonesia (LCDI) strategy in 2019. LCDI offers a broader strategy by

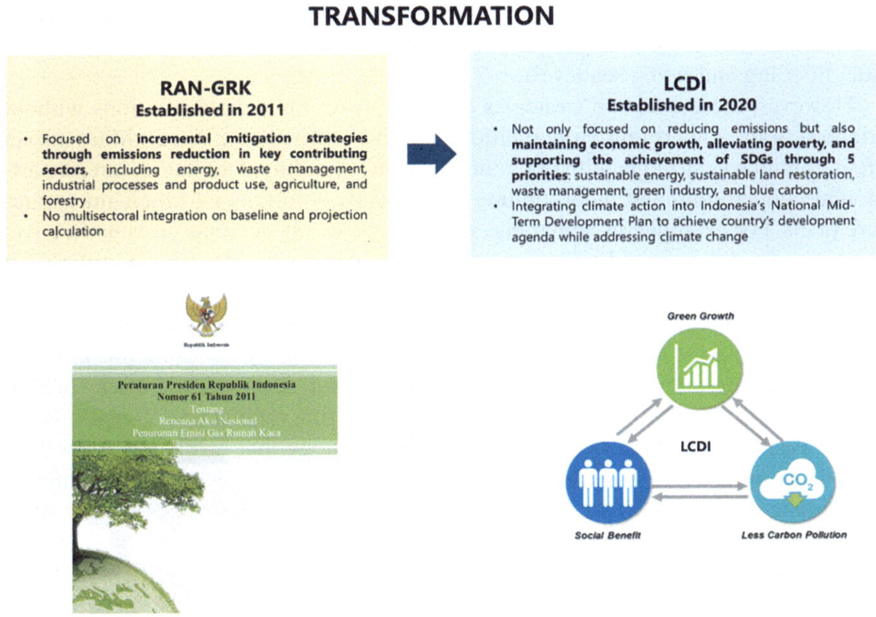

Fig. 2.9 Transformation from RAN-GRK to LCDI. Source: Ministry of National Development Planning/Bappenas, 2019 (*modified*)

integrating climate change mitigation and Indonesia's overall economic and social planning, as illustrated in Fig. 2.9. This initiative was later internalised into RPJMN 2020–2024 to ensure that the strategies are taken into account by and implemented within the development sectors.

LCDI focuses on five main strategies: increasing renewable energy, enhancing energy efficiency, sustainable land use, waste management, and innovative financing.

a. Increasing renewable energy involves transitioning from coal to renewable sources, with the latter scaling up from an 8% share of renewable energy from the national energy mix in 2015 to 23% by 2030 and 30% by 2045.
b. Enhancing energy efficiency, with the transition to renewable energy sources, is expected to reduce energy consumption per person by 3.5% by 2030 and 4.5% by 2045.
c. Sustainable land use to preserve 41.1 million hectares of primary forests, including nearly 15 million hectares of peatlands, by 2045 to enhance resilience and contribute to carbon emission reduction targets.
d. Waste management by ensuring effective policies on solid and industrial waste to reduce waste generation and emission factors.
e. Innovative financing by blending various sources, such as private capital (both domestic and foreign), to support investment in low-carbon transitions.

The implementation of these strategies involves integration into national and regional development plans, strengthening policy frameworks, and building capacity across sectors. The initiative also focuses on robust monitoring and reporting systems, public participation, and collaboration with international partners to ensure transparency and accountability.

As noted above, following the ratification of the Paris Agreement, Indonesia committed to reducing GHG emissions through the enactment of the NDC in Presidential Decree 98/2021 on the Implementation of Carbon Pricing for Achieving National Contribution Targets and Controlling Greenhouse Gas Emissions in National Development. Indonesia is committed to reducing GHG emissions by 29% below the Business as Usual (BaU) scenario as an unconditional target and 41% as a conditional target with international support by the year 2030. These targets were submitted by Indonesia to the UN Framework Convention on Climate Change (UNFCCC).

Presidential Decree 98/2021 stipulated the mechanism for achieving GHG emission reduction targets, which has three main steps. It begins with a GHG emissions inventory, with the government obligated to identify emissions sources, trends, and status, and which sectors produce the most emissions. The findings of this step serve as a basis for determining GHG emission reduction targets for the preparation of an emission reduction action plan or low-carbon development planning. This process is then followed by monitoring and evaluation, after which reporting is compiled to assess achievements in GHG emission reduction.

Indonesia also submitted its own long-term low-carbon plan to the UNFCCC, namely the LTS-LCCR, which is also integrated with the climate change adaptation strategy. The document incorporates NDC targets and states that Indonesia also aims to achieve net zero emissions (NZE)—a condition in which greenhouse gases entering the atmosphere are balanced with their removal from the atmosphere (IRID, 2022)—by the year 2060. LTS-LCCR includes scenarios such as the Current Policy Scenario (CPOS), Transition Scenario (TRNS), and the Low Carbon Scenario in compliance with the Paris Agreement Target (LCCP). These scenarios guide the country towards a sustainable future, emphasising resilience and global climate responsibility. Along with achieving the NZE target by the year 2060, Indonesia proposed updated targets in the ENDC in 2022. The new targets were increased to 32% for the unconditional target and 43% for the conditional target.

One of the key strategies for climate-change mitigation at the local level is the provision of public transportation. Jakarta has been a pioneer of public transport development since 2004, when it became the first city to develop a Bus Rapid Transit (BRT) network (named Transjakarta). Since then, Jakarta has continuously integrated public transport modes such as Mass Rapid Transit (MRT), Light Rapid Transit (LRT), commuter lines, and Jaklingko Microbuses.[6] Several other cities, such as Semarang and Yogyakarta, have also showcased their efforts in enhancing public transport networks, mostly related to BRT systems. These initiatives (see

[6] A public transport service system that operates integrated microbuses with fixed routes, standardised fares, and seamless connections to other modes such as TransJakarta, MRT, and KRL.

(a) (b)

Fig. 2.10 Public transportation provision in Indonesia. (**a**) MRT in Jakarta, (**b**) BRT in Semarang. Source: Ekonomi Bisnis, 2024; Documentation, 2016

Fig. 2.10) promote non-motorised transport that aligns with Indonesia's commitment to reduce GHG emissions.

However, challenges remain in ensuring the success and sustainability of these programmes. Significant barriers for these initiatives include infrastructure limitations, funding constraints, and resistance to behavioural change among commuters. For instance, some cities still struggle with inadequate connectivity and inconsistent policy implementation, which impose unsustainable maintenance burdens. These barriers need to be addressed effectively since these initiatives serve as a foundation for a long-term, impactful climate change solution. Walawalkar et al. (2022) argue that behavioural change plays a major role in enabling effective climate-related actions since it aims at large-scale changes, drawing on society's capacity to adapt to and tackle climate change.

Pathways to Integrating Climate Change into Development Policy in Indonesia

Over the years, Indonesia has developed climate change policies to align with international commitments and promote sustainable development. These efforts are reflected in integrated climate-related initiatives that encompass both adaptation and mitigation strategies across the country's multilevel and multisectoral development policies, aiming to reduce vulnerability to climate risks and GHG emissions to enable significant change. However, conventional and incremental strategies are not always sufficient to mitigate the severe impacts of climate change on communities and the environment. Fedele et al. (2019) observe that this type of approach may enable change through social or ecological modification, but it does not fundamentally change the core characteristics of the socio-ecological system, leaving communities vulnerable to climate change impacts. There is thus a need to alter the approach to climate change adaptation and mitigation, especially in Indonesia, where climate risks are imminent.

The ideal approach to addressing climate change must accommodate transformations that go beyond maintaining the status quo. It requires a comprehensive strategy to drive significant progress. Schreuder and Horlings (2022) suggest that climate action should be able to address the root causes of risks and vulnerabilities by shifting systems from unsustainable or undesirable trajectories, while Singh et al. (2024) argue for the importance of providing long-term interventions as a guide to prevent fragmented action in the short term. These views emphasise that climate action will be more effective if implemented over the long term, as it ensures sustainability beyond short-term measures and fosters the development of climate-resilient systems.

In response to these concerns, implementing a transformative approach to climate change adaptation and mitigation strategies is crucial for achieving long-term and sustainable outcomes. Chu et al. (2019) propose a transformative climate action approach that focuses on systemic changes in the development process, enabling quality-of-life improvement, promoting social and economic vibrancy, and ensuring more sustainable, resilient and inclusive urban development. This transformative approach can be applied in climate adaptation and mitigation actions to simultaneously address the problems posed by climate change and move cities on a path towards resilience.

A transformative approach means enabling significant changes within a system to challenge and alter the status quo. Deubelli and Mechler (2021) propose that a transformative approach to climate risk management would involve comprehensive, profound, and fundamental changes within the system by enhancing capacities, fostering broad stakeholder engagement, implementing continuous monitoring and evaluation, and exercising strategic leadership in decision-making. This implies that implementing a transformative approach to climate change adaptation and mitigation requires comprehensive initiatives in both strategy scope and execution within actions. In accordance with this, Fedele et al. (2019) outline six main characteristics to achieve a deep, long-term systemic change as desired from the transformative approach (see Fig. 2.11).

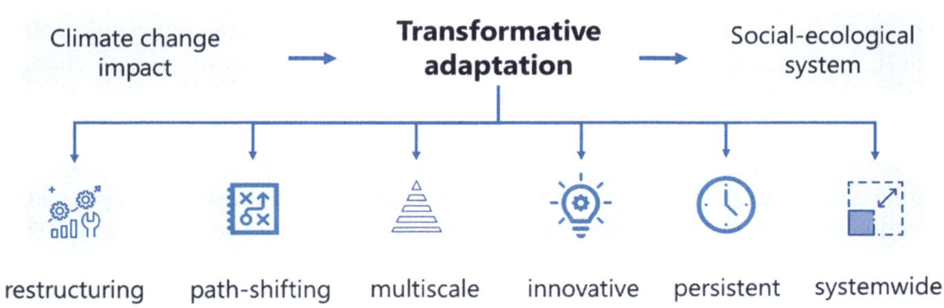

Fig. 2.11 Components of transformative climate actions. Source: Fedele et al. (2019) (*modified*)

a. Restructuring: enabling fundamental, functional, and interactional shifts in the socio-ecological systems.
b. Path shifting: altering current systems' trajectory towards an alternative direction to promote new options within the context of sustainable and resilient development.
c. Innovative: incorporating new knowledge, policies, and/or technologies learnt from long-term monitoring and evaluation to tackle climate issues effectively.
d. Multiscale: involving cross-sectoral partnerships and engaging with multilevel governance to spread transformative approaches and generate impacts on multiple scales.
e. Systemwide: enabling transformation to occur on a large scale through multi-stakeholder collaborations and participatory approaches to enable systemic changes.
f. Persistent: creating long-term (although not necessarily irreversible) impacts by integrating new practices into regulatory frameworks to secure political will and funding support to ensure long-term implementation sustainability.

Integrating transformative climate actions into policy is the next critical step in addressing climate challenges comprehensively. This integration is important to ensure that stakeholders' initiatives are harmonised and sustainable. Wamsler and Osberg (2022) describe climate change adaptation and mitigation as key prerequisites for achieving sustainable development, and they thus require systematic mainstreaming across all sectors that need to be accommodated within the policy. This highlights the importance of translating transformative climate actions into policies, including their integration into developmental policies. Such synergistic efforts, followed by more transformative initiatives, may lead to more climate-resilient development in the face of ongoing and imminent climate risks.

As coastal areas in Indonesia are at risk, adaptation must also embrace innovation to tackle the evolving challenges posed by climate change. This highlights the need for transformation in adaptation actions, particularly within communities, to enable fundamental changes in society. It is essential to invest in non-conventional adaptation strategies that promote behavioural changes within society, ultimately allowing for systemic transformations in development. Moreover, engaging vulnerable people (such as women and children) in the decision-making process is important to help them develop measures to cope with climate change impacts and ensure that the strategies are context-specific (Darjee et al., 2023). Involving vulnerable people who are experiencing (or will experience) the impacts of climate change allows for transformation within their adaptive capacity.

Similarly, mitigation actions should also incorporate a transformative approach, emphasising the importance of behavioural change in reducing emissions. Williamson et al. (2018) state that collective behavioural and lifestyle changes could effectively reduce global emissions by 19.9% to 36.8% between 2020 and 2050. This suggests the need to integrate transformative climate mitigation into policy, transitioning from upstream policies to their downstream implementation in daily life. Schreuder and Horlings (2022) provide examples of transformative practices that involve the shared ownership and management of renewable resources. One notable example is the local energy transition in Ilha da Culatra, Portugal, which utilises solar farms, desalinated water, an energy storage system, and off-grid energy distribution. These projects could serve as blueprints for future development policies.

Indonesia has started a transition from a survival-oriented approach to a more transformative strategy for addressing climate change, and these efforts have been implemented at the regional and local levels. However, both adaptation and mitigation efforts remain incremental and focus on short-term solutions. For instance, in terms of adaptation, flood management infrastructure provision has often focused on mitigating immediate disasters in sea walls, drainage, and polder systems. While these measures help reduce the impact of climate change, this approach cannot ensure long-term resilience. Likewise, for mitigation, public transportation networks represent positive steps in reducing GHG emissions. However, cross-city connectivity and long-term maintenance remain inadequate. Although these initiatives are substantial in scale, they often operate as standalone solutions rather than as components of a comprehensive, long-term solution. The primary challenge lies in integrating these measures across urban planning, policy, and community engagement to transition from reactive responses to systemic, forward-looking solutions.

In conclusion, promoting effective action in climate change adaptation and mitigation requires a comprehensive approach that includes innovative policies, various stakeholders' active engagement, and sustainable practices. Encouraging collaboration among governments, the private sector, and local communities can enhance collective efforts, ensuring that adaptation strategies are both culturally relevant and effective. By prioritising education and awareness, strengthening governance frameworks, and mobilising financial resources, societies can not only mitigate the impacts of climate change but also proactively adapt to it, fostering resilience and sustainability for future generations.

References

Asian Development Bank & World Bank. (2021). *Climate risk country profile: Indonesia*. Asian Development Bank. https://www.adb.org/sites/default/files/publication/700411/climate-risk-country-profile-indonesia.pdf

Bott, L.-M., Schöne, T., Illigner, J., Haghighi, M. H., Gisevius, K., & Braun, B. (2021). Land subsidence in Jakarta and Semarang Bay: The relationship between physical processes, risk perception, and household adaptation. *Ocean & Coastal Management, 211*, 105775. https://doi.org/10.1016/j.ocecoaman.2021.105775

CBS. (2021). *Results of the 2020 population census*. Statistics Indonesia. Retrieved from https://www.bps.go.id/id/pressrelease/2021/01/21/1854/hasil-sensus-penduduk%2D%2Dsp2020%2D%2Dpada-september-2020-mencatat-jumlah-penduduk-sebesar-270-20-juta-jiwa-.html

Chu, E., Brown, A., Michael, K., Du, J., Lwasa, S., & Mahendra, A. (2019). Unlocking the potential for transformative climate adaptation in cities. *Global Center on Adaptation*. https://gca.org/reports/unlocking-the-potential-for-transformative-climate-adaptation-in-cities/

CMCC Foundation. (2021). *Indonesia: G20 climate risk atlas*. Euro-Mediterranean Centre on Climate Change. Retrieved from https://files.cmcc.it/g20climaterisks/Indonesia.pdf

Darjee, K. B., Neupane, P. R., & Köhl, M. (2023). Proactive adaptation responses by vulnerable communities to climate change impacts. *Sustainability, 15*(14), 10952. https://doi.org/10.3390/su151410952

Desdiani, N. A., Afifi, F. A. R., Cesarina, A., Sabrina, S., Husna, M., Violeta, R. M., Adinegoro, A., & Halimatussadiah, A. (2021). *Climate and environmental financing at regional level: Amplifying and seizing the opportunities*. LPEM, Faculty of Economics and Business, University of Indonesia.

Deubelli, T. M., & Mechler, R. (2021). Perspectives on transformational change in climate risk management and adaptation. *Environmental Research Letters, 16*(5), 53002. https://doi.org/10.1088/1748-9326/abd42d

Fedele, G., Donatti, C. I., Harvey, C. A., Hannah, L., & Hole, D. G. (2019). Transformative adaptation to climate change for sustainable social-ecological systems. *Environmental Science & Policy, 101*, 116–125. https://doi.org/10.1016/j.envsci.2019.07.001

Ferdiansyah, M. R., & Siswanto, S. (2024). Assessment of surface urban heat island in Indonesia's municipal cities. In S. Lestari et al. (Eds.), *Proceedings of the international conference on radioscience, equatorial atmospheric science and environment and humanosphere science (INCREASE 2023)*. (Springer Proceedings in Physics, Vol. 305). Springer. https://doi.org/10.1007/978-981-97-0740-9_71

Government of Indonesia. (2021). *Long-Term Strategy for Low Carbon and Climate Resilience 2050 (Indonesia LTS-LCCR 2050)*. Jakarta: Government of Indonesia. https://unfccc.int/sites/default/files/NDC/2022-09/ENDC%20Indonesia.pdf

Government of Indonesia. (2022). *Enhanced Nationally Determined Contribution (ENDC) of the Republic of Indonesia*. Jakarta: Government of Indonesia. https://unfccc.int/sites/default/files/NDC/2022-09/ENDC%20Indonesia.pdf

Handayani, W., Chigbu, U. E., Rudiarto, I., & Putri, I. H. S. (2020). Urbanization and increasing flood risk in the northern coast of Central Java, Indonesia: An assessment towards better land use policy and flood management. *Land, 9*(10), 343. https://doi.org/10.3390/land9100343

Handayani, W., & Kumalasari, N. R. (2015). Migration as future adaptive capacity: The case of Java, Indonesia. In F. Hillmann, M. Pahl, B. Rafflenbeul, & H. Sterly (Eds.), *Environmental change, adaptation and migration*. Palgrave Macmillan. https://doi.org/10.1057/9781137538918_7

Insani, T. D., Rudiarto, I., Handayani, W., & Wijaya, H. B. (2022). Rural livelihood resilience on multiple dimensions: A case study from selected coastal areas in Central Java. *World Review of Science, Technology and Sustainable Development, 18*(2), 176–193. https://doi.org/10.1504/WRSTSD.2022.121303

Institute for Research and Innovation for Development (IRID). (2022). *Dasar-dasar net zero emission*. Retrieved from https://irid.or.id/wp-content/uploads/2022/07/2022.04.01-Dasar-Dasar-Net-Zero-Emission_SPREADS.pdf

Kalogiannidis, S., Kalfas, D., Papaevangelou, O., Chatzitheodoridis, F., Katsetsiadou, K. N., & Lekkas, E. (2024). Integration of climate change strategies into policy and planning for regional development: A case study of Greece. *Land, 13*(3), 268. https://doi.org/10.3390/land13030268

MCGA. (2024a). *Analysis of the annual average air temperature change rate*. Retrieved from https://www.bmkg.go.id/iklim/analisis-laju-perubahan-suhu-udara

MCGA. (2024b). *Analysis of the annual average rainfall change rate*. Retrieved from https://www.bmkg.go.id/iklim/analisis-laju-perubahan-curah-hujan

Ministry of Environment and Forestry of the Republic of Indonesia (MOEF). (2024). *National Registry System (SRN)*. https://srn.kemenlh.go.id/index.php?r=home%2Findex

Ministry of National Development Planning/Bappenas. (2019). *Low Carbon Development: A Paradigm Shift Towards a Green Economy in Indonesia* (Full Report). https://lcdi-indonesia.id/wp-content/uploads/2022/03/Low-Carbon-Development-A-Paradigm-Shift-Towards-a-Green-Economy-in-Indonesia-Full-Report-2019.pdf

Ministry of National Development Planning/Bappenas. (2021). *Executive Summary Climate Resilience Development Policy 2020–2045*. https://lcdi-indonesia.id/wp-content/uploads/2021/11/0_Executive-Summary.pdf

Mogelgaard, K., Dinshaw, A., Ginoya, N., Gutiérrez, M., Preethan, P., & Waslander, J. (2018). *From planning to action: Mainstreaming climate change adaptation into development* (Working paper). World Resources Institute. https://www.wri.org/publication/climate-planning-to-action

Murniati, K. (2020). The impact of climate change on the household food security of upland rice farmers in Sidomulyo, Lampung Province, Indonesia. *Biodiversitas: Journal of Biological Diversity, 21*(8). https://doi.org/10.13057/biodiv/d210809

NOAA. (2020). *Laboratory for satellite altimetry/sea level rise*. National Oceanic and Atmospheric Administration. Retrieved from https://www.star.nesdis.noaa.gov/socd/lsa/SeaLevelRise

Nurhidayah, L., & McIlgorm, A. (2019). Coastal adaptation laws and the social justice of policies to address sea level rise: An Indonesian insight. *Ocean & Coastal Management, 171*, 11–18. https://doi.org/10.1016/j.ocecoaman.2019.01.011

Onuma, A., & Tsuge, T. (2018). Comparing green infrastructure as ecosystem-based disaster risk reduction with gray infrastructure in terms of costs and benefits under uncertainty: A theoretical approach. *International Journal of Disaster Risk Reduction, 32*, 22–28. https://doi.org/10.1016/j.ijdrr.2018.01.025

Romanello, M., Di Napoli, C., Drummond, P., Green, C., Kennard, H., Lampard, P., Scamman, D., Arnell, N., Ayeb-Karlsson, S., Ford, L. B., Belesova, K., Bowen, K., Cai, W., Callaghan, M., Campbell-Lendrum, D., Chambers, J., van Daalen, K. R., Dalin, C., Dasandi, N., et al. (2022). The 2022 report of the Lancet Countdown on health and climate change: Health at the mercy of fossil fuels. *The Lancet, 400*(10363), 1619–1654. https://doi.org/10.1016/S0140-6736(22)01540-9

Rudiarto, I., Handayani, W., & Sih Setyono, J. (2018). A regional perspective on urbanization and climate-related disasters in the northern coastal region of Central Java, Indonesia. *Land, 7*(1), 34. https://doi.org/10.3390/land7010034

Rudiarto, I., Handayani, W., Wijaya, H. B., & Insani, T. D. (2019). Rural livelihood resilience: An assessment of social, economic, environment, and physical dimensions. *MATEC Web of Conferences, 280*. https://doi.org/10.1051/matecconf/201928001002

Schreuder, W., & Horlings, L. G. (2022). Transforming places together: Transformative community strategies responding to climate change and sustainability challenges. *Climate Action, 1*(1), 24. https://doi.org/10.1007/s44168-022-00024-3

Sejati, A. W., Buchori, I., Lakshita, N. M., Wiratmaja, I. A., & Ulfiana, D. (2024). The spatial analysis of urbanization dynamic impacts in a 50-year flood frequency in Java, Indonesia. *Natural Hazards, 120*(3), 2639–2659. https://doi.org/10.1007/s11069-023-06298-y

Sejati, A. W., Buchori, I., & Rudiarto, I. (2019). The spatio-temporal trends of urban growth and surface urban heat islands over two decades in the Semarang Metropolitan Region. *Sustainable Cities and Society, 46*, 101432. https://doi.org/10.1016/j.scs.2019.101432

Singh, A., Pope, F. D., Radcliffe, J., Luiu, C., Bakare, H., Bartington, S. E., Bonsu, N. O., Bryson, J. R., Cheeseman, N., Flowe, H., Krause, S., Newbigging, K., Nunan, F., Reardon, L., Rogers, C. D. F., Rowlingson, K., & Thomson, I. (2024). Delivering sustainable climate action: Reframing the sustainable development goals. *NPJ Climate Action, 3*(1), 110. https://doi.org/10.1038/s44168-024-00194-2

Triana, K., & Wahyudi, A. J. (2020). Sea level rise in Indonesia: The drivers and the combined impacts from land subsidence. *ASEAN Journal of Science and Technology for Development, 37*(3), 115–121. https://doi.org/10.29037/ajstd.627

USAID. (2016). *Indonesia costs of climate change: A 2050 policy brief*. United States Agency for International Development (USAID). https://www.climatelinks.org/sites/default/files/asset/document/Indonesia%20Costs%20of%20CC%202050%20Policy%20Brief.pdf

Walawalkar, T. P., Hermans, L. M., & Evers, J. (2022). Evaluating behavioural changes for climate adaptation planning. *Journal of Environmental Planning and Management, 66*(7), 1453–1471. https://doi.org/10.1080/09640568.2022.2028610

Wiati, C. B., Dharmawan, I. W. S., Sakuntaladewi, N., Ekawati, S., Wahyuni, T., Maharani, R., Hadiyan, Y., Naibaho, Y., Satria, W. I., Ngatiman, N., Abdurachman, A., Karmilasanti, K., Laksmita, A. N., Angi, E. M., & Khadka, C. (2022). Challenges to and strategies for the climate village program plus: A lesson learned from Indonesia. *Sustainability, 14*(9), 5530. https://doi.org/10.3390/su14095530

Wijaya, A. S. (2014). Climate change, global Warming and global inequity in developed and developing countries (Analytical perspective, issue, problem and solution). *IOP Conference Series: Earth and Environmental Science, 19*(1), 12008. https://doi.org/10.1088/1755-1315/19/1/012008

Xu, H., Zhu, Y., Yagci, A. L., Lü, H., Gou, Q., Wang, X., et al. (2023). Development of composite drought indices for the coastal areas of southeastern China: A case study of Jinjiang and Jiulongjiang River basins. *Journal of Hydrology, 626*, 130210. https://doi.org/10.1016/j.jhydrol.2023.130210

Yuliastuti, N., & Sariffudin, & Syafrudin. (2023). Social vulnerability level appraisal at tidal flood areas. *International Review for Spatial Planning and Sustainable Development, 11*(2), 99–113. https://doi.org/10.14246/irspsd.11.2_99

Wamsler, C., & Osberg, G. (2022). Transformative climate policy mainstreaming – engaging the political and the personal. *Global Sustainability, 5*, e13. https://doi.org/10.1017/sus.2022.11

Williamson, K., Satre-Meloy, A., Velasco, K., & Green, K. (2018). Climate change needs behavior change: Making the case for behavioral solutions to reduce global warming. Rare. https://rare.org/wp-content/uploads/2019/02/2018-CCNBC-Report.pdf

Part I
Climate Change

Chapter 3
Sea Level Rise and Environmental Migration: The Future of Climate Change Law in Indonesia

Laely Nurhidayah and Ana Borges Jelinic

Introduction

The Intergovernmental Panel on Climate Change (IPCC) states that sea level rise (SLR) is a present and future danger (Moore, 2019). The threat of submergence facing certain islands and coastal areas is recognised by scientists and politicians, with the Special Report on the Ocean and Cryosphere in a Changing Climate finding that global mean sea levels will most likely rise between 0.29 m and 1.1 m by the end of this century (Oppenheimer et al., 2019). The IPCC suggests that in the absence of adaptation, more intense and frequent extreme sea level events, together with trends in coastal development, will increase the expected annual flood damage by 2–3 orders of magnitude by 2100 (Oppenheimer et al., 2019). The Organization for Economic Co-operation and Development has predicted that, as a result, 40 million people living near coastal areas are threatened by submersion (Causse et al., 2016). The IPCC (2019) reports that climate change could increase population displacement, particularly in coastal areas and low-lying islands in developing countries, affecting the poorest communities. SLR and land subsidence potentially render coastal areas uninhabitable and unavailable for economic production, spurring migration by permanently destroying irrigated coastal agriculture and fresh drinking water supplies, disrupting vital human systems, reducing property values, and ultimately destroying property and infrastructure (Hauer et al., 2020).

The SLR threat to coastal communities in Northern Java, including Jakarta, Semarang, Pekalongan, and Demak, is already significant, and human mobility and

L. Nurhidayah (✉)
Research Center for Law, National Research and Innovation Agency (BRIN), Jakarta, Indonesia
e-mail: lael003@brin.go.id

A. B. Jelinic
Disrupting Violence Beacon, Griffith University, Brisbane, Australia

L. Nurhidayah et al. (eds.), *Climate Change, Labour and Migration in Indonesia*, Springer Climate, https://doi.org/10.1007/978-3-032-15334-0_3

migration due to SLR and land subsidence are already occurring, with Pekalongan and Demak the worst-affected areas. For example, Simonet Village in Pekalongan is sinking, and 96 households were relocated in 2020, with many people leaving their houses and properties. Likewise, the Mondoliko area in Bedono Village is experiencing subsidence, forcing the inhabitants to relocate to another village in 2022, while subsidence in Dukuh Rejosari, also in Bedono, forced the local community to relocate to a riverbank where they were unable to build permanent housing, as it was state land.

The inadequate climate change law in Indonesia has worsened the situation by delivering insufficient protection and slowing down responses to vulnerable groups affected by SLR and land subsidence. Disaster Management Law No. 24/2007 does not categorise SLR as a disaster. In addition, the legislation is silent on providing protection and relocation for those who have lost their land and housing as a result of SLR and coastal inundation. As there is no guideline in this law on how to address the issue of people displaced by climate change, policy responses to forced migration are not uniform between regions, an issue that is explored further in this chapter. Disaster Management Law provides only temporary protection for victims affected by flooding, with inadequate protection regarding employment (particularly in informal sectors in disaster-prone areas) and no provision for social security. Strict regulations around the workplace and business responses contain important gaps.

International law on solutions to address forced environmental migration has produced neither an internationally agreed-upon definition of environmental migration nor common assistance and protection arrangements (Scissa, 2021). The legal and human rights impacts of the sea level rise phenomenon are wide-ranging, requiring innovative legal and practical solutions to protect human rights—particularly as vulnerable groups, such as women, children, the elderly, and indigenous populations, are likely to suffer the most.

This chapter presents reflections on the specific issues of forced migration and climate change that have emerged as part of a larger research project in the regions of Pekalongan and Demak. Chapter 17 also engages with some of the findings. These reflections aim to examine the existing laws and their gaps. In addition, this chapter highlights the case studies of Pekalongan and Demak and suggests a way for climate law in Indonesia to address climate change-driven displacement and forced migration.

Forced Migration Due to Climate Change and International Protection

At the global level, climate change has forced many people to migrate, whether within or across regions. Millions of people are displaced every year due to climate change. According to the Internal Displacement Monitoring Centre, 17.7 million people were internally displaced in 2018 by natural disasters, many driven or exacerbated by climate change, while in 2020, disasters resulted in 40.5 million people being internally displaced across 149 countries (IDMC, 2020). People displaced due to climate change have faced difficulties accessing housing, education, income, food, and even health services.

Currently, 'climate refugee' is not a category under international law. Therefore, protection for people displaced due to climate change varies internationally and even internally. To bridge this gap in international law, states can use human rights-based approaches, as most have ratified international human rights treaties. Protection includes human rights before, during, and after migration. The UNHCR has advocated for the inclusion of displaced people in local and global climate policy, finance, and action (UNHCR, 2025) and supports legislative and policy reforms that strengthen protection and solutions for internally displaced people (UNHCR, 2025).

An important but often overlooked component of disaster risk management, particularly in terms of managing IDPs, is legal preparedness. Legal preparedness is important, particularly when a planned relocation is being contemplated. Many countries—for example, Fiji and Malawi (IFRC, 2021)—have been revising their disaster risk management legislation to include planned relocation as a solution to IDPs; however, the current domestic legislation in Indonesia is lacking protection in cases of displacement as a result of climate change.

Loss and damage have become a topic of discussion under the United Nations Framework Convention on Climate Change (UNFCCC) Conference of the Parties (COP) agenda. COP19 (2013) in Warsaw established the Warsaw International Mechanism for loss and damage associated with Climate Change Impacts, which included the impacts of slow-onset disasters (such as SLR) through Decision 2/CP 19. This decision recognised that loss and damage associated with the adverse effects of climate change can be reduced through adaptation. In addition, the COP encourages states to establish institutional arrangements, such as an international mechanism, including functions and modalities, to address the loss and damage associated with climate change impacts in developing countries that are particularly vulnerable to the adverse effects of climate change. At COP29 (2024), the Raul Wallenberg Institute of Human Rights and Humanitarian Law advocated that states need to consider 'rights-based approaches to planned relocation which can help reduce potential economic and non-economic loss and damage by promoting inclusive decision-making and consideration of holistic economic, social, cultural, civil, and political rights at all stages of the process' (Platform on Disaster Displacement, 2024).

In relation to climate change adaptation and loss and damage, the two pillars of the UNFCCC are the consensus on the importance of an institutional framework and the role of sustainable development in reducing the loss and damage arising from slow-onset disasters. In addition, governments should consider inclusive decision-making, with vulnerable groups participating in decision-making to address relocation and their livelihoods. The 2011 Cancún Adaptation Framework (paragraph 14(f)) invites states to 'enhance action on adaptation' by undertaking '[m]easures to enhance understanding, coordination and cooperation with regard to climate change induced displacement, migration and planned relocation, where appropriate, at national, regional and international level'.

The International Law Commission (ILC) has been actively working on the legal implications of sea level rise in relation to international law since 2018. According to its final report in 2025, the international legal framework for protecting people affected by sea level rise is fragmented and general in nature (ILC, 2025). The current protections are scattered across various branches of international law, including Disaster Risk Reduction Law, Human Rights Law, and Climate Change Law. Importantly, the ILC emphasises that these legal regimes do not provide a comprehensive or coordinated approach to addressing the complex human impacts of sea level rise, especially with respect to displacement and migration. The report urges states to uphold the principle of **human dignity** when responding to the needs of individuals and communities affected by sea level rise (ILC, 2025). This includes ensuring the rights and protection of persons who are forced to move due to environmental degradation and rising seas, and calls for states to integrate human-centred approaches into both national and international responses.

Protection in Case of Forced Migration: Indonesian Legislation

Indonesian domestic legislation is silent on the protection of displaced people as a result of climate change impacts. There are three main pieces of legislation that will be discussed, as they provide potential future protection in the case of forced migration, although all three are currently inadequate. They are Human Rights Law No. 39/1999, Disaster Management Law No. 39/1999, and Local Government Law No. 23/2014.

Human Rights Law of the Republic of Indonesia No. 39/1999

Recognition of the right to an adequate and healthy environment can be found in Human Rights Law No. 39/1999 (Article 9). In addition, Article 9(2) states that every person has the right to tranquillity, safety, peace, happiness, and physical and mental prosperity. SLR and land subsidence have brought suffering to some coastal communities due to coastal inundation and the loss of their land and property, and SLR is thus considered a threat to the right to life and a healthy environment. Despite the government's obligation to protect the right to life and the healthy environment of its citizens, no case based on human rights violations has been brought against the government by members of communities affected by SLR, subsidence, and forced migration. To meet their obligations, states should facilitate migration with dignity and address the specific human rights protection needs of migrants. These include the provision of food and clean water; access to adequate housing, healthcare and social security, education, and decent work opportunities; and safeguarding the principles of non-refoulement and the prohibition of collective expulsion, as well as the rights to liberty, personal integrity, and family unity (OHCR, 2024). In addition, states have an obligation to address the causes of forced migration in the context of climate change and should seek to protect rights, strengthen social protection systems, reduce disaster risk and exposure, and increase adaptive capacity (OHCR, 2024). While there is no provision in human rights law governing forced migration as a result of climate change, Article 27 of Law No. 39/1999 does state that every citizen has the right to freely move, transmigrate, and settle in the territory of the Republic of Indonesia.

Disaster Management Law Republic of Indonesia No. 24/2007

Disaster Management Law of the Republic of Indonesia (Law No. 24 of 2007). This is the legal basis for disaster management in Indonesia. It covers all disasters, natural and man-made, including tsunamis, earthquakes, and flooding. The legislation also covers a holistic approach to disaster management, including pre-disaster, during disaster, and post-disaster. However, it is silent on responding to SLR. SLR and coastal inundation are thus not categorised as disasters. Article 1 of Law No. 24/2007 states that natural disasters mean an event, or a series of events, caused by nature, such as an earthquake, tsunami, volcanic eruption, flood, drought, typhoon, or landslide. From this definition, the SLR and subsidence are notably missing. In addition, based on interviews with officials at the local government level, it is apparent that SLR is not categorised as a disaster, resulting in SLR being treated as a low priority. Even some of the local people affected consider SLR not as a disaster but, rather, as a day-to-day situation, one to which they are accustomed and to which they adapt. Officials from the National Disaster Management Agency (BNPB) acknowledge during interviews the need for a revision of Law No. 24/2007 to explicitly include SLR as a category of disaster.

The current priority in disaster management is focused on rapid-onset events rather than slow-onset events. Contingency plans prioritise efforts to address earthquakes, tsunamis, flooding, and other rapidly occurring disasters, and SLR and land subsidence are missing in contingency plans prepared by the National Disaster Management Agency (BNPB). During an interview, the BNPB advised that to categorise land subsidence as a disaster, it would be necessary to establish new institutions to identify and address the issue. The Information and Geospatial Agency (BIG) is currently responsible for monitoring SLR in Indonesia and is the closest to a potentially responsible agency.

Law No. 24/2007 is also silent on relocation as a result of forced migration due to climate change. Instead, it governs how to manage rehabilitation and reconstruction in situ, omitting the arrangements and mechanisms of relocation or resettlement. As a result, there is no uniformity when the local government is required to address this forced migration. The absence of legal instruments can give rise to various problems and injustices in the implementation of migration, refugee, and relocation programs and policies for communities affected by climate change. While at the local government level, there are several policies, regulations, and programs to respond to the impacts of human mobility and climate change; these are inadequate to provide just relocation for displaced persons.

Local Government Law Republic of Indonesia No. 23/2014

In the future, cities will become the centre of human mobility due to climate change (Roderick et al., 2021). The climate crisis is either accelerating the inclusion of migrants and displaced people or exacerbating marginalisation and exposure to inequality and risk (Roderick et al., 2021). Local governments play important roles in addressing the issue of climate change and migration. Decentralised systems could perhaps be best placed to manage the first phases of the migration crisis, as they are closer to the communities. The problem, however, is that not all local governments and cities have sufficient resources and budgets to cope with relocation and migration. One problem with migration is a lack of vacant land and housing for relocation. Under the Local Government Law, housing comes under the authority of local governments, with one of the mandatory responsibilities being to provide housing. However, in cases where the local government does not have a sufficient budget for housing in relocation areas, it can request assistance from the central government. The local government then needs only to locate and designate a suitable land for relocation.

Currently, there is no priority given in the Local Government Law to policy addressing the slow-onset disaster of SLR and **land subsidence**; instead, the government favours protection and accommodation approaches, a reactive approach. As a result, local governments do not undertake planning for relocation. In addition, there is no uniform approach to this issue at the local level and no road map for planned relocations. Planned relocation tends to occur only years after the community has been resettled in temporary housing.

There is an urgent need to synchronise efforts and develop collaboration between central and local governments in terms of who has the authority to provide housing for displaced persons. According to interviews with the central government, the local government has the authority to provide housing in cases of disaster; however, based on interviews with local government representatives, they do not have adequate funding to address the issue and provide housing for many households affected by climate change. A possible solution is to escalate the response to the housing problem for those affected by subsidence by categorising their situation as a national disaster. Such a classification could generate collaboration between central and local governments, with funding coming from both levels.

Pekalongan and Demak Case Studies

Pekalongan and Demak suffer from coastal inundation and subsidence in several suburbs. In Pekalongan, residents experienced coastal inundation and sinking of their village in 2020, which required relocation. Residents of Simonet Village were forced to move to Semut Village and occupy temporary housing in a location on the village land (tanah bengkok). In fact, the current relocation policy prevents them from obtaining adequate permanent housing and access to livelihood. In Pekalongan, the initiative for relocation came from the village leader, who designated a site for relocation despite restrictions on building permanent residential buildings on the land in question. The relocation program, which should be planned carefully and in detail, still uses a house renovation program scheme. This problem illustrates the lack of clarity and guarantee of rights for relocated communities. Only recently, after four years of living in temporary housing, are the residents of Simonet planning to move to a new relocation provided by the government. Lack of income and permanent housing have become serious issues due to the lack of laws on how to address forced migration in Indonesia.

A similar situation also occurred in the Demak Regency in the wake of coastal inundation and subsidence in several suburbs in 2005, which affected Rejosari and Senik residents. The government issued several programs and policies, including relocating several affected residents to outside Java, but not all relocation programs have been successful, as the community preferred to be relocated near their existing village. The relocation program for the Rejosari suburb and Bedono Village is an example. In 2005, the regional government relocated some residents—around 206 households—to Gemulak Village. However, problems arose when the relocation destination area turned out to be an irrigation area or a riverbank managed by the National River Basin Agency (BBWS), which is vulnerable to flooding. Additionally, the residents' land ownership status is unclear; residents are only given occupancy permits for a certain period, and there are allegations of discriminatory treatment towards new arrivals from long-term residents. In 2022, another relocation initiative was implemented in Bedono Village in the Demak Regency. The aim of the initiative was to relocate all residents from Dukuh Mondoliko to Dombo Village. This was a major undertaking involving the transfer of 130 households. According to interviews with community members, the local government facilitated this relocation effort by providing substantial material assistance valued at IDR 50 million per household. In addition, the head of Bedono Village played a pivotal role in aiding the acquisition of suitable land for the relocation process and facilitating flexible payment options, thus allowing residents to build their new homes in Dombo Village.

The Future of Climate Change Law in Indonesia: Addressing Displacement

Indonesia does not have a special climate change law. Instead, policy on adapting to and mitigating climate change is mainstreamed into sector-specific legislation, action plans, and policy. There are several legislative instruments that regulate coastal areas, including planning law, environmental protection and management law, and disaster management law, which contribute to climate change adaptation. Although there is no single piece of legislation that mentions SLR, the current policy and legislation still indicate a commitment to reducing the vulnerability associated with climate change in coastal areas. However, based on observations and interviews with local government representatives in the Pekalongan and Demak Regencies, the existing law is not adequate to address and protect displaced persons as a result of SLR and land subsidence. The protection approach that became foundational under the current legislation cannot cope with the rapid impacts of SLR and land subsidence. The government also needs to adopt retreat approaches—plans to relocate people, buildings, and infrastructure away from vulnerable coastal areas—in future legislation for adapting to climate change, where there is an increasing intensity, frequency, and extent of flooding, SLR, and coastal erosion that increase the risk of local communities. As the challenges imposed by climate change become more complex and all-encompassing, definitions in law need to be updated. The slow pace of law reform is a particular challenge given the increasing pace and increased complexity of climate change, but it is a challenge that must be overcome.

Indonesia ratified the United Nations Framework Convention on Climate Change (UNFCC) and the Paris Agreement and thereby agreed to abide by the commitment to implement mitigation and adaptation strategies. However, current adaptation action is mainstreamed under sector-specific legislation. In the Enhanced National Determined Contribution submitted to the UNFCCC secretariat, Indonesia stated its commitment to more ambitious adaptation measures as elaborated in programs, strategies, and actions to achieve economic, social, and landscape resilience, including the protection of communities' livelihoods and ecosystems.

To address SLR and land subsidence, the priority policies of the current government focus on protection and accommodation rather than a retreat policy. This is reflected in the choice to establish a 700 km "Giant Sea Wall" from Banten to Gresik, East Java. There are four key elements that should be taken into consideration in establishing the protection of displaced persons under a retreat policy. These are outlined below.

Human Rights and Human Security-Based Approach

As revealed by the case studies in Pekalongan and Demak, the law in Indonesia is behind in addressing the needs of people displaced by climate change-related disasters. A new law addressing displacement is needed and should focus on a human rights and human security-based approach, acknowledging that different people are affected by disasters in different ways and recognising their different capacities to adapt. It should also highlight the varied exposure and vulnerability of diverse people. Marginalised groups, women, those with a disability, and low-income groups are more vulnerable to climate change impacts and thus need more protection than others. In addition, the law should create cultural, economic, environmental, political, and social systems and structures that provide people with equitable rights and opportunities to live with dignity (QCEA, 2025). Resettlement and relocation should include consultation, participation, and equality of information to those affected. In addition, relocation should also consider their economic needs. Most local people affected by displacement in Pekalongan and Demak have been forced to change their source of income, such as moving from farming to aquaculture farming or from fishing to labouring in factories. According to our interviews with local communities, they preferred to adapt in situ to disaster-prone areas rather than being relocated; relocation is the last choice when there are no other options available. While relocation as a response to climate change impacts (or as an adaptation strategy) is now becoming increasingly acceptable, it remains a last resort and is not part of communities' adaptive strategies or cultures. People in Java are emotionally attached to place, in contrast to Pacific island communities, where retreating from coastal areas in response to changing environmental conditions has long been a part of adaptive strategies, culture, and practices (Gharbaoui & Blocher, 2016). According to the interviews, people fear being relocated, as they need to adjust to a new environment, a new livelihood, and a new neighbourhood.

The policy adopted in future legislation should be based on 'moving with dignity'. Moving with dignity is a phrase that encompasses the idea of respecting the value of the rights of individuals in the context of migration, care, and social justice. Gray (2022) refers to the 'moving with dignity' approach as a strengths-based approach that looks at what people can do for themselves and how aids and adaptations can be used to ensure that a sufficient level of care is provided. It focuses on rethinking policy approaches to make them more human-centred and rights-based. This principle is based on the Universal Declaration of Human Rights, which recognises the inherent dignity and the equal and inalienable rights of all members of the human family as the foundation of freedom, justice, and peace in the world.

Meaningful Public Participation in Decision-Making Concerning the Place of Relocation

In addition, based on the analysis of the existing law, there is a lack of guidance in the legislation on how to involve the community in decision-making, especially on challenges such as climate change adaptation measures. Community participation is guaranteed by Law No. 32/2009 on Environmental Protection and Management in conducting an Environmental Impact Assessment (EIA) and the development process. However, in practice, the community is rarely consulted about development implementation and its impact on the environment (Nurhidayah & Mcilgorm, 2019).

The current law and policy are more focused on protection or accommodation in situ rather than relocation or retreat options for displaced people. The government launched the Giant Sea Wall project to protect Jakarta Province and other coastal cities in Semarang and Demak; however, this project has generated controversy with claims that it benefits industry and the wealthy rather than the local people. For example, plans to construct a sea wall in the shape of a Garuda in Jakarta were abandoned after academics and the community protested due to the project's environmental and social implications for the local fishing industry. There is a need to bring all parties to the table and provide greater visibility and transparency when evaluating the social, economic, and environmental impacts through the EIA process (Nurhidayah et al., 2022).

From the case studies in Tambak Lorok village in Semarang and Bedono, it appears a lengthy process of local participation and negotiation was necessary before the community agreed to the relocation plans and areas selected, as the community insisted on sites near their sources of livelihood—for example, near coastal areas for fishermen.

Adequate Funding for Relocation

Future legislation addressing forced migration should provide clarity to assist the public and private sectors through funded adaptation initiatives. The availability of funding mechanisms would also help offset the disproportionate economic burden on disadvantaged individuals or groups. In the case of Demak and Pekalongan Regency, for example, the Java regional government has developed policies to provide finances for housing if the local communities provide land. However, the money provided is limited to IDR 50 million, whereas the cost of building new housing is usually more than IDR 80 million. Therefore, according to interviews with local communities, the funding provided by the local government is not sufficient to cover the cost of building a new house. Local communities should be provided with additional funding to top up and cover all the expenses of relocation. In addition, the availability of land remains a crucial issue. Not all displaced communities have sufficient money to buy land, which is necessary to receive financial assistance for housing from the government. In extreme cases, local communities live on riverbanks and in other public places that are not suitable for housing. However, they lack other options. In Demak Regency, for example, the head of the village bought land from other villages for the community. In return, the community members are paying him back monthly, without interest. Clarity around land ownership and sources of funding is crucial in relocation policies.

In addition, there should be social security measures that particularly target people affected by climate change, including displaced people. According to interviews with staff from the Ministry of Social Affairs, such funding would need to be proposed by the coordinating ministry at the central level or by the local government.

Multi-Sectoral Approach and Coordination

New climate laws are needed, either in the form of a special law or as a revision of the existing law. As climate change-related forced migration is a multifaceted issue, the responses from the government must involve multiple layers of government—central government, provincial and regency governments, and city and village level authorities. Relocation requires the involvement and coordination of a range of institutions, including the Disaster Management Agency at national and local levels (BNPB and BPBD), Public Works and Housing (PUPR), the Development and Planning Agency at national and local levels (Bappenas and Bappeda), and others. In addition, funding for relocation comes from the central and local governments. Generally, local governments should provide the available land with a clear legal status for relocation. For example, in the case of Simonet village in Pekalongan, the local government provided one hectare of land to accommodate 66 households affected by the loss of homes and properties. The government allocated a housing size of 64 square metres to each household, ensured they were equipped with public facilities, and provided certificates of ownership for the land and building. However, the scarcity of available land is a major hurdle for relocation, and examples such as this are rare. The community must be consulted first about the relocation place to increase the likelihood of finding suitable areas for relocation, thereby increasing buy-in and reducing the chance of resistance or return to the former location due to lack of work in the new region. Relocation requires more inclusive and proactive approaches rather than reactive approaches.

Conclusion

Sea level rise is a major environmental problem in coastal Java, with coastal communities in northern Java suffering from coastal inundation and the subsidence of their village. Many households have been displaced and relocated to new places, making forced environmental migration a pressing human security problem that nonetheless still lacks serious attention and protection under existing law and policy at both the international and local levels. Local government responses to this problem need to be integrated and harmonised with new legislation that addresses displacement and forced migration due to environmental disasters and climate change impacts.

The current legislation addressing environmental migration as a result of climate change is not adequate. There is an urgent need to revise the existing Disaster Management Law No. 24/2007 to address the management of displacement and mobility as a result of coastal inundation, SLR, and land subsidence. The absence of any law specifically addressing such displacement and migration results in slow and ineffective responses from the government in addressing this issue. There is also no uniformity across governments in managing such situations.

The important features that need to be incorporated in future climate law or laws on disaster management are human rights-based approaches, under which different people have their specific needs addressed when considering the impact of climate change, recognising that they have varied exposure and vulnerabilities, and adaptive capacities. Marginalised groups, the poor, women, and children are more vulnerable and need more protection against climate change impacts. In addition, public participation in decision-making regarding the relocation site is crucial to ensure relocation to a place that suits the affected community. In addition, adequate funding for relocation is essential, as the relocated communities need new housing urgently. The slow response of the government, due to inadequate regulation, is a threat to human rights and the obligation of the state to protect its citizens. Finally, a multi-stakeholder and multi-sectoral approach to addressing this issue is essential. There is a need for clarity of the roles and responsibilities of different institutions and jurisdictions in disaster management law or special laws addressing climate change.

References

Causse C., Mokhnacheva, D., & Camus, G. (2016). *Scientific Notes*. https://environmentalmigration.iom.int/sites/g/files/tmzbdl1411/files/ocean-environment-climate-changemigration_ScientificNotes_Oct2016_BD_ppp-12.pdf

Gharbaoui, D., & Blocher, J. (2016). The reason land matter: Relocation as adaptation to climate change in Fiji islands, *Migration, Risk Management and Climate Change: Evidence and Policy Responses*, 149.

Gray, M. (2022) *Moving with Dignity Approach Impact Assessment, powys*. https://powys.moderngov.co.uk/documents/s67239/IA35%20Moving%20with%20dignity.pdf

Hauer, M. E., Fussell, E., Mueller, V., Burkett, M., Call, M., Abel, K., McLeman, R., & Wrathall, D. (2020). Sea-level rise and human migration. *Nature Review Earth Environment, 1*, 28–39. https://doi.org/10.1038/s43017-019-0002-9

IDMC, International Displacement Monitoring Centre. (2020). *Global Report on Internal Displacement*. https://www.internal-displacement.org/global-report/grid2020/

IFRC, International Federation of Red Cross and Red Crescent Societies. (2021). *Displacement in Changing Climate, Localized Humanitarian Action at the Forefront of the Climate Crisis,*

Geneva, https://www.ifrc.org/sites/default/files/2021-11/2021-Climate-Displacement-Report-Final.pdf

International Law Commission (ILC). (2025). *Sea Level rise in relation to International Law, Seventy sixth session, Geneva, 2025*, https://documents.un.org/doc/undoc/gen/g25/015/61/pdf/g2501561.pdf

International Organization for Migration (IOM). (2019). *Glossary on Migration*. https://publications.iom.int/system/files/pdf/iml_34_glossary.pdf

Moore, R. (2019). *IPCC Report: Sea Level Rise Is a Present and Future Danger*, September 25, 2019. https://www.nrdc.org/bio/rob-moore/ipcc-report-sea-level-rise-present-and-future-danger

Nurhidayah, L., Davies, P., Alam, S., Saintilan, N., & Triyanti, A., (2022), Responding to sea level rise: Challenges and opportunities to govern coastal adaptation strategies in Indonesia, *Maritime Studies*, 350.

Nurhidayah, L., & Mcilgorm, A. (2019). Coastal adaptation laws and the social justice of policies to address sea level rise: An Indonesian insight. *Ocean and Coastal Management, 171*, 11–18. https://doi.org/10.1016/j.ocecoaman.2019.01.011

OHCR–The Office of the High Commissioner for Human Rights, Integrating Human Rights at the UNFCCC. (2024). https://www.ohchr.org/sites/default/files/Documents/Issues/ClimateChange/materials/KMMigration.pdf

Oppenheimer, M., Glavovic, B. C., Hinkel, J., van de Wal, R., Magnan, A. K., Abd-Elgawad, A., Cai, R., Cifuentes-Jara, M., DeConto, R. M., Ghosh, T., Hay, J., Isla, F., Marzeion, B., Meyssignac, B., & Sebesvari, Z. (2019). Sea Level Rise and Implications for Low-Lying Islands, Coasts and Communities. In H.-O. Pörtner, D. C. Roberts, V. Masson-Delmotte, P. Zhai, M. Tignor, E. Poloczanska, K. Mintenbeck, A. Alegría, M. Nicolai, A. Okem, J. Petzold, B. Rama, N. M. Weyer (Eds.) *IPCC Special Report on the Ocean and Cryosphere in a Changing Climate* (pp. 321–445). Cambridge University Press. https://doi.org/10.1017/9781009157964.006

Platform on Disaster Displacement. (2024). *Human Rights-based Approaches to Planned Relocation in the Context of Disasters and Climate Change – Side Event at COP29*. https://disasterdisplacement.org/news-events/human-rights-based-approaches-to-planned-relocation-in-the-context-of-disasters-and-climate-change/

Quaker Council for Eurpean Affairs. (2025). *Moving with Dignity A positive peace approach to migration*. A handbook. https://www.qcea.org/wp-content/uploads/2025/04/Migration-Handbook-Full-Interactive.pdf

Roderick ,W., Gary, S., Morel, L., Brick, K., & Power, M. (2021) *Cities, climate and Migration: The role of cities at the climate migration Nexus*. C40 Cities, https://climatechange.issuelab.org/resources/43166/43166.pdf

Scissa. (2021). Recognition and Protection of Environmental Migrants in International Law. In *Dignity in Movement: Borders, Bodies and Rights*. https://www.e-ir.info/2021/06/24/recognition-and-protection-of-environmental-migrants-in-international-law/#google_vignette

UNHCR. (2025). *Climate Change Protecting the Rights of Migrants*. https://www.ohchr.org/sites/default/files/Documents/Issues/ClimateChange/materials/2PMigrationLight.pdf

Chapter 4
Gender-Responsive Adaptation for Climate and Disaster Risk Mitigation: Recommendations for Effective Implementation from Case Studies

Landung Esariti, Dina Auliya Kharisma, and Celeste Harris

Introduction

Climate change is a global phenomenon that has attracted public attention in recent decades, with its impact increasingly experienced across the globe. Climate change phenomena cause temperature increases and extreme weather changes, triggering an increase in the frequency and intensity of climate-related disaster events (Ghazali et al., 2018). Climate change disaster events can have devastating impacts on infrastructure, the environment, and human life. However, the impacts of disasters are not the same for all groups or in all locations. Human vulnerability and capacity to respond to disasters vary depending on socioeconomic factors, including gender (Eastin, 2018). Previous studies have identified the persistence of gender inequality in societal responses to disasters and the significant differences between the experiences of men and women from the catastrophic impacts of climate change.

Many people in the Global South experience high vulnerability to climate change, as well as gender inequality (Ngcamu, 2023). These conditions are exacerbated by the unstable economies of countries in the Global South and high dependence on natural resources, such that when climate change occurs, it can rapidly disrupt livelihood stability. Gender inequality is a crucial factor that amplifies the negative impact of disasters, especially for women (Verduci et al., 2024). Robles and Benavidez

L. Esariti (✉) · D. A. Kharisma
Department of Urban and Regional Planning, Faculty of Engineering, Universitas Diponegoro, Semarang, Indonesia
e-mail: landungesariti@lecturer.undip.ac.id

C. Harris
Disrupting Violence Beacon, Griffith University, Logan, QLD, Australia

ARC Centre of Excellence for the Elimination of Violence Against Women (CEVAW), Melbourne, VIC, Australia

L. Nurhidayah et al. (eds.), *Climate Change, Labour and Migration in Indonesia*, Springer Climate, https://doi.org/10.1007/978-3-032-15334-0_4

(2020) focused on experiences in Low- and Middle-Income Countries (LMICs) and highlighted that women have been disproportionately affected by disaster events and are at a higher risk of being adversely affected. These include experiences of gender-based violence, health risks, livelihoods, education, financial security, and reduced life expectancy. These adverse impacts can exacerbate pre-existing gender inequality in societies and communities. Limited access to resources, along with women's multiple roles, can impact their psychological well-being. Vulnerability factors that cause gender inequality for women include power differences, differences in economic and employment conditions, social and cultural roles and norms, and differences in biological conditions (Björnberg & Hansson, 2013).

Culturally, Indonesian society mainly considers men as the head of the family; their role is to earn a living for the family, organise, and make key family decisions. Women often have the primary responsibility of managing the household, raising children, and taking care of the family. The demands for different roles and responsibilities create conditions of inequality, whereby women often bear a greater burden of caring responsibilities and managing household affairs, especially in times of disaster (Ashraf & Azad, 2015). Research related to gender inequality conducted in Indonesia (Handayani et al., 2025) found that in the cities of Jakarta, Semarang, and Pekalongan, there were inequalities in roles, access, and control between men and women, especially in the context of urban planning and disaster resilience. In Jakarta, women faced structural barriers in accessing basic services and participating in decision-making. In Semarang and Pekalongan, women were often not involved in deliberative and decision-making forums related to climate change adaptation, even though they played a major role in households and communities in the event of disasters. Inequality is therefore exacerbated by the low recognition of women's role in disaster risk adaptation and reduction, as well as the lack of gender-disaggregated data that can support inclusive policies.

In Indonesia and across the globe, there is an increasing need for communities and governments to pay attention to gender perspectives in making policy decisions, particularly in terms of disaster management, to increase gender equality in a way that benefits society as a whole. The goal from a gender perspective is to ensure that women enjoy equal rights, justice, and equality in terms of access, participation, benefits, and control (Dhuli et al., 2023). Research has highlighted the unique strengths and capacities of women in responding to and adapting to climate change and the associated disaster risk. The roles of women as leaders and change agents are increasingly recognised in disaster risk reduction and climate adaptation, acknowledging that a sole focus on vulnerability can undermine this important work (Azad & Pritchard, 2023). Women are often agents of disaster recovery through their contributions to society, but their involvement is often invisible or unacknowledged (McNamara et al., 2021). Women build their communities as social networks that increase their adaptive capacity, often experiencing increased safety and contribution to decision-making in the communities of other women (Jerin et al., 2023). Although women often have a lower economic status and higher social vulnerability than men, particularly in the Global South, in the context of adaptive capacity, women may have an advantage in surviving and responding to disasters. Women's primary role in managing household affairs fosters a strong connection with local resources and the surrounding environmental conditions. This close familiarity enables them to better anticipate, respond to, and manage disasters by utilising available local resources more effectively (Lee et al., 2022).

As one of the countries heavily affected by climate change, threatening national development, the Government of Indonesia has developed several key policies related to disaster risk reduction and climate change adaptation. Two major national-level policies are the National Action Plan for Climate Change Adaptation (RAN-API) and the Disaster Management Master Plan (RIPB) (Kawanishi et al., 2016; Putra et al., 2021). These policies aim to strengthen community adaptation capacity and reduce the vulnerability of populations in disaster-prone areas, in line with the United Nations Sustainable Development Goals (SDGs), which emphasise inclusive development and the improvement of community welfare. Although these national policies generally support the principle of inclusivity, a Gender, Disability, and Social Inclusion (GEDSI) approach has not been explicitly and comprehensively integrated into the core framework of disaster adaptation policies. GEDSI principles, which ensure fair protection and meaningful participation for vulnerable groups such as women, persons with disabilities, children, and the elderly, are mentioned in several sectoral policy documents. However, GEDSI is not a key element in policy implementation across levels and tends to appear only in guidelines as a consistently mandated approach across sectors. Integrating GEDSI into disaster adaptation policies is a strategic and urgent necessity, given that vulnerable groups often have higher levels of exposure and susceptibility to disasters, yet limited access to resources, information, and decision-making processes. For instance, women often bear dual responsibilities in the household and community, yet are rarely involved in planning or evacuation planning. People with disabilities frequently face physical and informational barriers during emergencies. Children and the elderly require specific protection and services during and after disasters. Without a GEDSI approach, adaptation policies tend to be generalised and fail to address the specific needs of these vulnerable groups. Moreover, mainstreaming GEDSI could enhance the effectiveness and sustainability of adaptation programs. When all segments of society, including the most vulnerable, are actively involved in planning and implementation, the resulting solutions become more targeted, inclusive, and capable of strengthening overall social resilience. Social inclusion in disaster adaptation is not only a matter of justice but is also essential for creating communities that are more resilient and responsive to increasingly complex risks driven by climate change. However, the implementation of GEDSI at the local level remains limited. Many local governments lack the technical capacity, understanding, or resources to fully integrate GEDSI into their adaptation programs (Rimbawan & Nurhaeni, 2024). As a result, this approach has not yet been considered a primary priority. Key challenges include weak interagency coordination, limited funding, and insufficient institutional support. Therefore, efforts to strengthen human resource capacity, regulatory support, and cross-sector collaboration are necessary so that GEDSI principles can become a truly integral part of every stage of disaster adaptation policy, both at the national and local levels.

This chapter underscores the need for governments and community stakeholders to maintain mainstream gender in policy and implementation at all stages of disaster management, from mitigation to recovery. A descriptive review was conducted by analysing three selected case study articles focusing on gender-responsive disaster adaptation in Pakistan (Ahmad et al., 2021), Kenya (Ngigi et al., 2017), and Bangladesh (Naz et al., 2018). These case studies were supported by a review of the relevant reports referenced within each article. These countries were selected due to their socioeconomic and cultural similarities with Indonesia, particularly in terms of gender inequality and disaster vulnerability. The analysis focused on manually identifying success factors from research studies conducted in each country, focusing on how gender considerations were integrated into disaster risk management. This approach sought to identify key factors that can address the gaps caused by social norms, economic inequality, and policies that are not responsive to the specific needs of women. By considering the gender perspective, disaster management agencies can better identify needs that are often overlooked and develop more contextually relevant solutions. Women's involvement in decision-making can result in a more holistic approach to the needs of society as a whole. In this way, disaster management not only becomes fairer but also more effective in protecting all aspects of people's lives.

Gender-Responsive Disaster Adaptation as Part of Climate Change Adaptive Policies

Adaptation and mitigation are the two main climate change strategies, with different but complementary actions. Mitigation aims to reduce the causes of climate change, such as greenhouse gas emissions, while adaptation focuses on adjustments to climate change impacts (Ackerl et al., 2023; Landauer et al., 2019). Mitigation and adaptation should be carried out simultaneously to achieve maximum results in environmental risk management (Berrang-Ford et al., 2021). Mitigation has a long-term risk prevention and reduction focus on reducing the intensity of future phenomena (Pinto, 2014). Examples of mitigation measures in disaster reduction involve efforts such as building embankments, spatial planning that considers disaster risks, and strengthening building structures to withstand earthquakes or floods (Huq et al., 2018). Adaptation, in contrast, is an effort to increase community resilience to the impacts of climate change and disasters. These adjustments can be in the form of physical and behavioural changes in response to changing environmental conditions (Heltra Pradana et al., 2022)—for example, raising the floors of houses and building stilt houses in flood-prone areas (physical change) or participating in disaster training (behavioural change). The success of adaptation is strongly influenced by the adaptive capacity of the community, which includes the level of vulnerability, exposure, social sensitivity, and institutional support (Aksa & Afrian, 2022).

This paper emphasises adaptation strategies, especially those that are gender-responsive. Gender-responsive disaster adaptation is an approach to disaster risk reduction and climate change that considers the different roles, needs, and capacities of men and women, as well as other vulnerable groups. Greve (2016) notes that women in disaster conditions often face double vulnerabilities due to limited access to information, resources, and decision-making. Therefore, gender-responsive adaptation emphasises the importance of women's active participation in the planning and implementation of disaster management strategies, the provision of safe and women-friendly facilities, and the strengthening of adaptive capacity through training and education that is sensitive to their specific needs (Sagala et al., 2024). This strategy also encourages changes in social norms that limit the role of women in creating an inclusive and equitable adaptation system.

In adaptive climate change policies, the integration of gender perspectives is carried out in various ways, one of which is the application of gender mainstreaming principles in all policy stages from formulation to evaluation (Liu, 2021; Moore, 2018). These adaptive and inclusive policies allow women to play an important role in building community resilience to climate change—for example, through agricultural adaptation training programs for women farmers or strengthening the right to water and land access in disaster-prone areas (Kasim et al., 2024; Pundir et al., 2025). Collaboration between governments, communities, and the private sector plays an important role in strengthening gender-responsive climate change adaptation policies (Klein et al., 2017). The government plays a key role in creating a regulatory framework that is inclusive and responsive to the needs of vulnerable groups, especially women. It provides policy and funding platforms to ensure that gender perspectives are included in strategic documents. Communities are the key actors in the real-world implementation of gender-responsive adaptation strategies. The involvement of communities, particularly women's groups, in local planning and decision-making processes can ensure that adaptation strategies are informed by gender-specific and local contextual needs (Abakah, 2018). Additionally, Pauw and Pegels (2013) explain the important role of the private sector as a strategic partner in supporting innovation, financing, and adaptation of technology.

Why Is a Gender-Responsive Adaptation Strategy Important?

Gender-responsive adaptation strategies are an important approach to climate change and disaster risk because they take into account the differences in roles, responsibilities, risk perceptions, and needs between men and women (Ngigi et al., 2017). This approach not only encourages fairness in decision-making but also allows for more targeted and effective adaptation strategies. The goal is to reduce vulnerability and inclusively increase community resilience. In the context of disaster risk reduction, understanding gender perspectives is key to formulating equitable adaptation policies and measures. Gender mainstreaming is also important to help meet the intersecting needs of marginalised groups such as women, children, the elderly, and people with disabilities so that they are not left behind in facing disaster threats (Ashari & Nurhafifa, 2024).

These marginalised groups can face limited access to information, resources, and decision-making and can be at high risk when disasters occur. Their capacity for preparedness is influenced by inequitable sociocultural factors, education levels, experience in dealing with disasters, and knowledge of risks (Chang et al., 2023). Gender-responsive adaptation strategies should be designed to strengthen the capacity of marginalised groups through education, empowerment, and active engagement (Aziz & Anjum, 2024), requiring effective cross-sectoral collaboration.

Figure 4.1 highlights that vulnerability is an outcome of climate change that impacts the well-being of individuals, households, and society at large (Adger et al., 2009). Bryan and Behrman (2013) explain that this diagram uses the Sustainable Livelihoods (SL) and Institutional Analysis and Development (IAD) frameworks to represent the context of vulnerability in humans to climate change, which is influenced by various components such as user characteristics, biophysical conditions, information and technology, and institutional settings. The emergence of this vulnerability encourages adaptive actions at the individual, household, and community levels. The purpose of such actions is to improve welfare by strengthening the adaptive capacity of actors to their vulnerability. Thus, the diagram emphasises the existence of a reciprocal relationship between the level of vulnerability and adaptation strategy, which is influenced by the level of individual, family, and societal well-being.

This model continues to be relevant and useful for articulating the complex interactions between vulnerability, adaptation, and well-being, particularly when analysed through a gender lens. It is important to note that more recent gender-climate adaptation models are limited, and no widely accepted comprehensive framework has

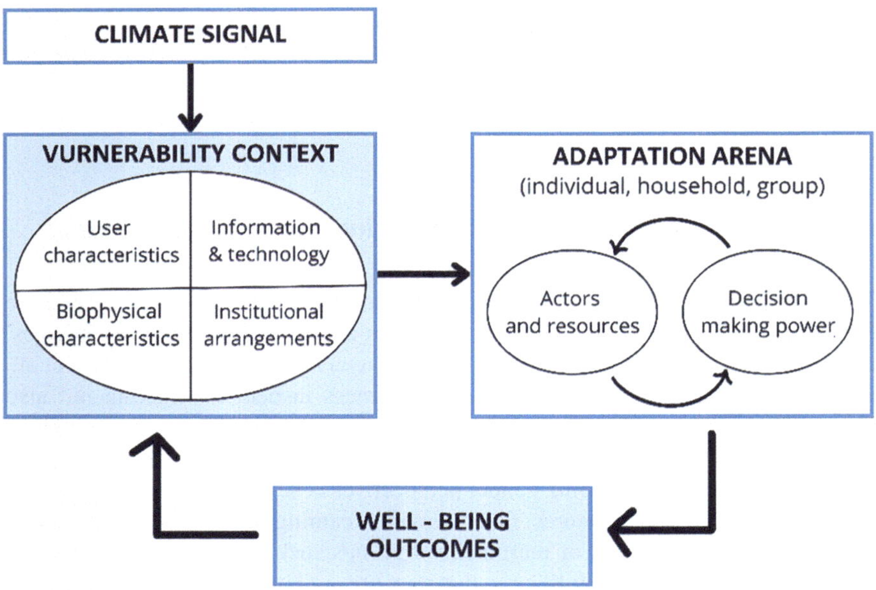

Fig. 4.1 Conceptual gender-climate change framework. *Source: Bryan and Behrman,* 2013

emerged to entirely replace this model. The conceptual framework of this study, therefore, builds upon this foundation by focusing specifically on gender-differentiated responses to climate variability. Within this framework, gender perspectives play an important role in exploring the dynamics of vulnerability and adaptation (Holmes et al., 2011). Gender not only affects the extent to which individuals or groups are affected by climate change but also determines their access to resources, participation in decision-making, and ability to adapt. Thus, gender mainstreaming is a crucial aspect in designing a fair and effective adaptation strategy.

Case Studies of Gender-Responsive Adaptation Strategies

This paper reviewed illustrative case studies of gender-responsive adaptation in Pakistan, Kenya, and Bangladesh, three countries that face challenges similar to Indonesia in terms of climate vulnerability and gender inequality. In Bangladesh, gender inequality is evident in socioeconomic disparities, such as limited access to income-generating opportunities, education, and participation in social activities, which are reinforced by traditional gender roles (Naz et al., 2018). Nevertheless, Bangladesh is considered a pioneer in integrating gender perspectives into disaster risk reduction (DRR) through the active involvement of political actors and women in policy formulation (Naz et al., 2018; Roy et al., 2022). In Pakistan and Kenya, studies have shown that the absence of gender perspectives in disaster response policies resulted in women's needs being overlooked, while gender-neutral approaches exacerbated inequalities by ignoring specific vulnerabilities, such as the need for reproductive health services and safety in shelters (Ali et al., 2022; Gocio & Kulkarni, 2016). Both countries have begun to implement gender-responsive adaptation, with promising efforts centred on community-based participatory approaches and economic empowerment to reduce gender gaps (Ahmad et al., 2021; Ngigi et al., 2017). These cases demonstrate that, despite persistent challenges, effective gender-based adaptation strategies are possible and offer valuable lessons for similarly vulnerable countries such as Indonesia.

This review explores key success factors and lessons relevant to the Indonesian context from the case studies, with a focus on gendered aspects of adaptation strategies documented in these communities. The authors analysed a case study from each of the selected countries—Pakistan (Ahmad et al., 2021), Bangladesh (Naz et al., 2018), and Kenya (Ngigi et al., 2017)—to examine real-world adaptation strategies adopted across genders and their success in reducing vulnerability to disaster risk and climate change. Each study investigates multiple locations—specifically, three villages in Pakistan, three villages in Bangladesh, and approximately 70 villages across three different agro-ecological zones in Kenya. Articles were reviewed manually to identify key factors for successful implementation and lessons learned, as summarised in Table 4.1. Within this analysis, some shared factors for gender-responsive disaster adaptation emerged and are summarised below, including increasing women's participation and access to resources, promoting education and training for women, and the important role of institutions and policy.

Table 4.1 Key success factors from the implementation of gender-responsive adaptation strategies in Pakistan, Kenya, and Bangladesh

Community profile and location	Review of the implementation of gender-responsive adaptation strategies
Ngigi et al. (2017) *Profile and Location:* Climate Change Affects Crops in Rural Kenya *Number of Respondents:* 156 pairs *Method:* Individual and intra-household interviews	*Key Factors for Successful Implementation* • **Recognition of gendered roles and responsibilities.** Options for adapting to climate change closely interplayed with the different roles and responsibilities of men and women in households, impacting the coping strategies and access to resources, which in turn influenced adaptive behaviour. Policy interventions should reflect on-the-ground gender realities to amplify men's and women's specific abilities to manage risks. • **Integrating climate risk perceptions.** Risk perceptions of weather changes, such as rainfall and signs of disasters, shape different adaptation responses between men and women. This study found that women are more responsive and sensitive to weather changes and signs of disasters because they have experience and sensitivity through the perception of high risk. This influences them to make decisions in determining the adaptation that suits their needs. • **Addressing gender-specific access to resources and information.** Women and men had different ways of accessing resources and information, allowing households to have a varied perspective on adaptation options. Men sought more information on increasing agricultural production, while women sought information on weather forecasts and advice on appropriate adaptation options. There were also different levels of trust in information sources between men and women. Women's access to and control over assets, particularly consumer durable assets, positively influenced their adoption of climate-smart practices. There is a need for gender-sensitive information delivery approaches that utilise the preferred channels and meet the different climate information needs of men and women, and a need to address gender disparities in asset ownership and control • **Leveraging gendered participation in community-based social capital.** Men and women in the same household belonged to different social groups/communities in which they shared information and resources. Participation in groups benefited men and women differently; men acquired more climate information, adaptation ideas, and access to farm inputs through social groups, while women benefited from livelihood diversification and risk management. Women's groups also offered pathways for empowerment and inclusion in broader decision-making. Community groups play an important role in adaptation, and it is important to understand and leverage gendered differences in participation and benefits • **Promoting educational empowerment for adaptive capacity.** Education positively influenced the adoption of adaptation strategies for both men and women, with a more pronounced effect among women. Women with more years of schooling were better positioned to understand and implement technically demanding practices. Supporting women's education and integrating climate literacy into educational programs can significantly enhance adaptive capacity and promote gender-responsive adaptation at the household level • **Formulation and support of gender-sensitive adaptation policies and frameworks.** Gender-sensitive policy and program formulation are critical for effective adaptation. The study showed that adaptation strategies should be designed taking into account women's needs and perspectives, ensuring that interventions are inclusive and equitable. The study also highlighted the need to embed gender perspectives in government across all levels, designing institutional mechanisms that consider gender as a key factor
	Lessons learned It is important to understand gender perspectives to consider the different roles, responsibilities, and access to resources of men and women. Adaptation strategies should take into account gendered realities on the ground, ensuring equitable access to information and assets, education, and leveraging different forms of community participation. Gender-sensitive communication and inclusive policy frameworks are essential to empower both women and men, enabling more comprehensive risk management and fostering resilience at the household and community levels

(continued)

Table 4.1 (continued)

Community profile and location	Review of the implementation of gender-responsive adaptation strategies
Naz et al. (2018) *Profile and Location:* *Charcoal farmer households in* Zanjira, Bangladesh, are affected by floods *Number of Respondents:* 115 men and 128 women *Method* • Key informant interviews • Focus group discussion • In-depth interviews with farmers	*Key Factors for Successful Implementation* • **Enhancing women's access to land.** Land access was found to be critical to enhance adaptation practices, and increasing women's ability to access land is essential for their adaptive capacity but was limited by gendered social and cultural barriers. Improving land tenure security can empower women and increase their resilience to climate impacts • **Empowering women through education and literacy.** These were significant determinants for adaptation strategies, especially for women. Literacy increased women's ability to determine, receive support for, and adopt adaptation strategies. Providing information and inclusive training programs on adaptation increased their knowledge and skills to adapt to the risks of floods and other climate impacts, and it is important to increase adaptive capacity • **Ensuring women's participation and influence in adaptation processes.** Involving women in adaptation decision-making can help ensure their needs are recognised and integrated, as well as provide a more equitable perspective for all communities • **Providing gender-specific support mechanisms.** Recognising the unique challenges faced by women is critical to successful implementation. The study showed that women often needed different supports compared to men, such as help accessing training programs and credit • **Addressing social and cultural norms.** The study found that some religious and social norms prevented women from working outside the home and limited their participation in other programs. It is important to address these underlying social and cultural barriers to ensure that women can actively participate in and benefit from adaptation initiatives • **Recognition of farming experience and age.** In this study sample, some male and female participants came from the same household, such as husbands and wives. They agreed that men had longer farming experience and were older than women, which resulted in men more often taking the lead in decision-making processes. Although this finding may not fully represent the general population, the gap highlights that women were less involved and underrepresented in decision-making. Therefore, it was important to empower women by sharing knowledge and skills to help reduce this disparity • **Supportive political policies and gender-responsive institutions.** Supportive policies that promote gender equality in both development and adaptation strategies were identified as critical. The study emphasised the importance of policy interventions that focus on improving women's access to vital resources and training, using fair and gender-equitable processes. They also highlighted the need to have on-the-ground female field staff to directly support women with adaptation efforts
	Lessons learned A gender-based approach to implementing adaptation measures can enhance women's resilience to climate change. Improving access to land, education, and training can strengthen their adaptive capacity to access information and make adaptation choices. In addition, women's participation in decision-making and providing tailored support mechanisms addresses their specific needs and promotes equity. Overcoming social and cultural norms and implementing gender-responsive policies is also crucial for fostering inclusive adaptation and strengthening women's adaptive capacity

(continued)

Table 4.2 (continued)

Community profile and location	Review of the implementation of gender-responsive adaptation strategies
Ahmad et al. (2021) *Profile and Location:* Flood hazard-affected communities in Punjab, Pakistan *Number of Respondents* 139 men and 154 women *Method:* • Focus group discussion • In-depth interviews with key informants	*Key Factors for Successful Implementation* • **Increased access to education, training, and land.** The study found that for females, access to schooling and training on adaptation strategies were important factors in adopting flood adaptation strategies. Training helped to increase knowledge and skills to determine appropriate adaptation actions. In addition, increasing access to resources in the form of land ownership and credit helped women invest in the agricultural sector and increased their household economic resilience. It is important to ensure access to schooling, training, and land for women to improve adaptation to climate disasters • **Strong Social Networks.** The study found that women rely on community-based social networks to obtain information, training, and adaptive resources. Through daily interactions and gender solidarity, women can improve their household resilience, especially via income diversification actions such as animal husbandry or small businesses. This is an effective strategy in increasing women's adaptive capacity to disasters • **Farming experience and age.** The study found that both older male and female farmers had a higher tendency to adopt experience-based adaptation strategies. In particular, experienced female farmers had used their farming knowledge to diversify household economies and to enhance both social and economic resilience to floods • **Supportive and inclusive institutions.** The study highlighted that it was important for institutions to actively engage with people on the ground, particularly women. They recommended developing interactions between institutions and the community and increasing the number of female staff within institutions to enhance women's access to support and training opportunities
	Lessons learned It is important to recognise the critical role of gender in shaping adaptation capacities, as women and men often have different levels of access to education, resources, and decision-making opportunities. Successful adaptation strategies must account for these differences by promoting equitable access to training, land, and credit, particularly for women. Community-based social networks serve as vital platforms for women to exchange information, build solidarity, and diversify household income. Therefore, inclusive and responsive institutions, especially those with active female representation, are essential to ensure meaningful engagement and support. A gender-sensitive approach that integrates these components strengthens resilience and enables more effective disaster risk management

Based on a review of the successful implementation of gender-responsive adaptation strategies from case studies in Pakistan, Kenya, and Bangladesh, seven success factors for implementing gender-responsive disaster adaptation strategies can be identified. From these studies, the successful implementation of gender-based adaptation strategies is influenced by the presence and role of the following factors.

Recognition of Gender Roles and Perspectives in Perceptions of Adaptation Risk

Risk perception influenced the selection of adaptations by considering the level of threat based on varied experiences and needs. Case studies have demonstrated that the perception of climate change risks can be gendered and influenced by different lived experiences and sensitivities. The research from Kenya (Ngigi et al., 2017) indicated that in study sites, husbands and wives within the same household can have different perceptions of changes in climate, such as variations in rainfall patterns. While the villages studied in Pakistan (Ahmad et al., 2021) and Bangladesh (Naz et al., 2018) did not explicitly compare climate change risk perception between genders, their separate analyses of men's and women's adaptation strategies suggest that climate-related hazards, such as floods, can affect genders differently, potentially influencing their underlying risk perceptions. Gender-specific perceptions of climate change risks are important factors to consider when designing and implementing adaptation strategies that are responsive to the needs of both men and women.

Research by Enarson et al. (2018) and Reyes and Lu (2016) reinforced the finding of this study that in emergencies, men and women tend to have different roles and responsibilities. Men, as heads of households, were often involved in matters related to the physical protection of the family and the technical needs of the household, while women focused more on protecting household assets and managing logistics, such as food, water, and other necessities. These differences stemmed from experiences shaped by perceptions of disaster risk. Understanding these differences is useful for designing appropriate and inclusive adaptation strategies.

Equal Access to Information and Resources for Adaptation

Access to information and resources was found to be important in gender-responsive adaptation, as it increased knowledge and supported each gender's preferences for accessing disaster-related information, while also enabling the implementation of appropriate strategies (Neuman, 2016). However, the case studies demonstrated that women often faced barriers in accessing critical resources, such as land ownership, savings, and credit, which negatively impacted their adaptive capacity. These structural limitations reduced their ability to make autonomous decisions related to disaster preparedness and livelihood strategies. In all three case studies, women showed a higher level of trust in information that they accessed directly through their networks, such as from fellow women farmers, community groups, or local facilitators. When women were provided access to relevant and reliable information, they effectively used it to implement adaptation strategies, including income diversification and improved agricultural practices (Ahmad et al., 2021; Naz et al., 2018; Ngigi et al., 2017). However, their ability to act on this information was often constrained by limited access to land and productive assets, which restricted their control over the key resources needed to implement adaptation measures. Across the case studies, land ownership and access to capital remained largely male-dominated, further reinforcing gender disparities in resilience-building efforts.

Improving Education and Literacy as Knowledge for Gender-Responsive Adaptation

Education and literacy levels varied between men and women and affected the selection of adaptations, impacting understandings of disaster-related risks and solutions. Lower levels of education tended to act as a limitation in understanding information related to adaptation. The case studies in Kenya, Pakistan, and Bangladesh demonstrated that education and literacy were important factors for women to enable better adaptation outcomes (Naz et al., 2018). Women with higher levels of education also tended to be more respected and involved in the decision-making process and thus better able to advocate for their needs (Ahmad et al., 2021).

Strengthening Adaptive Capacity Through Gender-Responsive Training

Engagement with training influenced the selection of gender-responsive adaptations, as it improved skills and enabled more diverse disaster responses, particularly among women. Training was typically provided by both government and non-government agencies within communities. In these case studies, training was often focused on agriculture, since farming was the main source of livelihood. In Bangladesh and Pakistan, training has emphasised the adoption of disaster-resilient agricultural practices, such as cultivating flood- or drought-resistant crop varieties (Ahmad et al., 2021; Naz et al., 2018). In Kenya, group-based training initiatives have focused on entrepreneurship and agribusiness to strengthen women's economic capacity (Ngigi et al., 2017). These training programs played a crucial role in enhancing women farmers' adaptive capacity by expanding their access to resources and risk management strategies. They also aimed to boost women's confidence and decision-making, which helped overcome social barriers and enabled greater control over agricultural production and income diversification in the face of climate risks.

Strengthening Social Capital Through Inclusive Community Involvement

Social capital created through a group-based approach was able to help individuals, households, and communities share information, knowledge, and support. Utilising social capital, strong social bonds and trust can be created in the community (Wagner & Fernandez-Gimenez, 2009). In addition, community-based approaches encourage women to participate more actively in decision-making related to adaptive selection (Po & Hickey, 2020). Women tended to prefer communities of other women, while men were more involved in mixed communities (Ngigi et al., 2017). In case studies from Pakistan, Kenya, and Bangladesh, women played an important role in community networks that could increase social capital (Ahmad et al., 2021; Naz et al., 2018; Ngigi et al., 2017). These networks provided support for them to collaborate and access resources by sharing knowledge and experiences related to climate adaptation so that interventions could be more effective and tailored to their specific needs.

The Influence of Farming Experience and Age on Adaptive Capacity

Farming experience and age influenced the choice of gender-based adaptation strategies by contributing to knowledge, skills, and risk tolerance. Zhang et al. (2020) note that the greater the experience and age, the stronger the preference for more practical adaptations with a lower risk of adaptation failure. In Pakistan, older male and female farmers had a stronger tendency to adopt experience-based adaptations, with experienced women using their knowledge to diversify household economies and strengthen resilience to floods (Ahmad et al., 2021). Older women typically relied on traditional practices, while younger women were more open to innovation. In Bangladesh, men had more farming experience and were generally older, while women were younger and less experienced, which limited their role in decision-making (Naz et al., 2018). Empowering women through knowledge sharing has been considered crucial for bridging this gap. Although experience supports livelihood diversification and resilience, age also creates physical limitations and encourages reliance on non-technological strategies (Tessema et al., 2018).

Policy and Institutional Support for Gender-Responsive

The existence of institutional and policy support for gender-responsive adaptation strategies can promote gender equality in accessing resources and information, thereby helping increase women's adaptive capacity (Sumarni & Rudiatin, 2020). When women receive full support from public institutions, they are more likely to actively participate in effective adaptation strategies for their families and communities (Jannat & Kusakabe, 2024). All the case studies pointed to the importance of these structural factors. In Bangladesh, strong political commitments and gender-equitable institutions have improved women's access to resources and training (Naz et al., 2018). However, Kenya and Pakistan still face challenges in gender-responsive policies and require interventions that better support women in adaptation (Ahmad et al., 2021; Ngigi et al., 2017).

Conclusion: Recommendations for Gender-Responsive Adaptation Strategies

This chapter identifies several key factors for implementing gender-responsive adaptation strategies, as well as factors that influence people to choose adaptation approaches according to their adaptive capacity. While these factors are valuable as broad considerations, strategies must be contextualised to meet the needs of different marginalised groups, including women, children, the elderly, and people with disabilities in the location of climate disasters. From this perspective, the authors recommend the following considerations, informed not only by insights derived from the reviewed case studies but also by the authors' professional experience in the field.

Firstly, the case study of villages in Pakistan (Ahmad et al., 2021) demonstrated that increasing women's adaptive capacity may require expanding women's access to resources such as land and education while removing patriarchal social and cultural barriers that limit women's rights. The establishment of women's communities and community-based approaches can encourage women's involvement in decision-making, helping integrate their needs into disaster adaptation. Age and farming experience influence the application of practical, experience-based knowledge to strengthen household resilience. Policies that support income diversification for women are also needed, such as small business and livestock training programs to increase their economic resilience. The Kenyan case (Ngigi et al., 2017) highlights the use of group-based approaches and improved gender-sensitive access to critical information to support effective adaptation strategies for all gender groups. Information is helpful in climate adaptation and practices if it is accessible to both men and women. Increased access to information can be provided by appropriate organisations through social networks in the form of groups/communities. In addition, social networks can encourage women to actively participate in decision-making related to disaster adaptation and agricultural practices that are gender-specific. Education and training can improve women's skills and strengthen their adaptive capacity, as well as help diversify their income, which can improve family finances and resilience. Adaptation policies need to be gender-sensitive so that men and women have equal access to information and resources. Finally, the case of Bangladesh (Naz et al., 2018) revealed that increased access to education and training for women opened up more employment opportunities outside the agricultural sector and highlighted how other sectors can provide alternative sources of income. Government enforcement of regulations, particularly those ensuring fair and secure land ownership rights for women, has played a crucial role in promoting equitable access to resources. Legal recognition of women's land rights not only improved their economic security but also increased their ability to participate in livelihood decisions and adaptation strategies. In addition, institutional support for women must be strengthened to improve engagement with women and advance gender equity.

The case studies point to the role of networks at the neighbourhood level, which can enable community members, including vulnerable groups such as women, to mobilise their capacity to become more visible and vocal. In this context, gender-responsive adaptation strategies are particularly relevant, as women often have local knowledge and an important role in disaster risk management that has not yet been fully recognised. The understanding that sustainable financial support is not always available may encourage communities to create alternative forms of collaboration, such as crowdfunding, that can support the implementation of adaptation strategies. This could also open up space for women to be actively involved in decision-making, adaptation planning, and resource management. This is an important consideration for local governments, as recognising and empowering marginalised groups, including women, can strengthen the resilience of the communities and society as a whole. This indicates the need to move from gender-responsive strategies that acknowledge and address gender differences and inequalities towards gender-transformative approaches that challenge the systemic and social structures that cause gender inequality. This is an important consideration for future research and policy development in this area, particularly in the Global South, and aligns with global trends in gender equality and climate adaptation (Resurrección et al., 2019). Ultimately, integrating gender perspectives into adaptation policies and programs is not only important but also vital for interventions to be truly inclusive, effective, and equitable.

References

Abakah, E. (2018). Participation without a voice? Rural women's experiences and participation in local governance activities in the Abura–Asebu–Kwamankese district of Ghana. *Cogent Social Sciences, 4*(1), 1–19. https://doi.org/10.1080/23311886.2018.1549768

Ackerl, T., Weldemariam, L. F., Nyasimi, M., & Ayanlade, A. (2023). Climate change risk, resilience, and adaptation among rural farmers in East Africa: A literature review. *Regional Sustainability, 4*(2), 185–193. https://doi.org/10.1016/j.regsus.2023.05.004

Adger, W. N., Dessai, S., Goulden, M., Hulme, M., Lorenzoni, I., Nelson, D. R., Naess, L. O., Wolf, J., & Wreford, A. (2009). Are there social limits to adaptation to climate change? *Climatic Change, 93*(3–4), 335–354. https://doi.org/10.1007/s10584-008-9520-z

Ahmad, D., Afzal, M., & Rauf, A. (2021). Flood hazards adaptation strategies: A gender-based disaggregated analysis of farm-dependent Bait community in Punjab, Pakistan. *Environment, Development and Sustainability, 23*(1), 865–886. https://doi.org/10.1007/s10668-020-00612-5

Aksa, F. I., & Afrian, R. (2022). Community adaptation strategies toward tidal flood: A case study in Langsa, Indonesia. *Jamba: Journal of Disaster Risk Studies, 14*(1). https://doi.org/10.4102/JAMBA.V14I1.1258

Ali, T. S., Ali, S. S., Nadeem, S., Memon, Z., Soofi, S., Madhani, F., Karim, Y., Mohammad, S., & Bhutta, Z. A. (2022). Perpetuation of gender discrimination in Pakistani society: Results from a scoping review and qualitative study conducted in three provinces of Pakistan. *BMC Women's Health, 22*(1). https://doi.org/10.1186/s12905-022-02011-6

Ashari, M. R., & Nurhafifa, S. (2024). Faktor yang Berhubungan dengan Kesiapsiagaan Bencana Banjir Pada Pelompok Rentan di Desa Beka Kabupaten Sigi. *Preventif: Jurnal Kesehatan Masyarakat, 14*(2). https://doi.org/10.22487/preventif.v14i2.680

Ashraf, M. A., & Azad, M. A. K. (2015). Gender issues in Disaster: Understanding the relationships of vulnerability, preparedness and capacity. *Environment and Ecology Research, 3*(5), 136–142. https://doi.org/10.13189/eer.2015.030504

Azad, M. J., & Pritchard, B. (2023). The importance of women's roles in adaptive capacity and resilience to flooding in rural Bangladesh. *International Journal of Disaster Risk Reduction, 90*, 103660. https://doi.org/10.1016/j.ijdrr.2023.103660

Aziz, M., & Anjum, G. (2024). Transformative strategies for enhancing women's resilience to climate change: A policy perspective for low- and middle-income countries. *Women's Health, 20*. https://doi.org/10.1177/17455057241302032

Berrang-Ford, L., Siders, A. R., Lesnikowski, A., Fischer, A. P., Callaghan, M. W., Haddaway, N. R., Mach, K. J., Araos, M., Shah, M. A. R., Wannewitz, M., Doshi, D., Leiter, T., Matavel, C., Musah-Surugu, J. I., Wong-Parodi, G., Antwi-Agyei, P., Ajibade, I., Chauhan, N., Kakenmaster, W., et al. (2021). A systematic global stocktake of evidence on human adaptation to climate change. *Nature Climate Change, 11*(11), 989–1000. https://doi.org/10.1038/s41558-021-01170-y

Björnberg, K. E., & Hansson, S. O. (2013). Gendering local climate adaptation. *Local Environment, 18*(2), 217–232. https://doi.org/10.1080/13549839.2012.729571

Bryan, E., & Behrman, J. A. (2013). *Community-based adaptation to climate change: A theoretical framework, overview of key issues and discussion of gender differentiated priorities and participation*. International Food Policy Research Institute (Issue 109). https://doi.org/10.2499/CAPRiWP109

Chang, K. J., Villeneuve, M., Crawford, T., Yen, I., Dominey-Howes, D., & Llewellyn, G. (2023). Disaster preparedness, capabilities, and support needs: The lived experience perspectives of people with disability. *Disabilities, 3*(4), 648–665. https://doi.org/10.3390/disabilities3040042

Dhuli, B., Dhamo, A., & Dhamo, I. (2023). Gender equality as a necessary approach for the country's development process and for gender integration. *Migration Letters, 21*(2), 1045–1054. https://doi.org/10.59670/ml.v21i2.6433

Eastin, J. (2018). Climate change and gender equality in developing states. *World Development, 107*, 289–305. https://doi.org/10.1016/j.worlddev.2018.02.021

Enarson, E., Fothergill, A., & Peek, L. (2018). *Gender and disaster: Foundations and new directions for research and practice* (pp. 205–223). Springer. https://doi.org/10.1007/978-3-319-63254-4_11

Ghazali, D. A., Guericolas, M., Thys, F., Sarasin, F., Arcos González, P., & Casalino, E. (2018). Climate change impacts on disaster and emergency medicine focusing on mitigation disruptive effects: An International Perspective. *International Journal of Environmental Research and Public Health, 15*(7), 1379. https://doi.org/10.3390/ijerph15071379

Gocio, D., & Kulkarni, K. G. (2016). Women's empowerment and economic development: The cases of Bangladesh and Kenya. *International Journal of Education Economics and Development, 7*(3–4), 113–132. https://doi.org/10.1504/IJEED.2016.082171

Greve, A. I. (2016). Sustainable development, climate change adaptation and disaster management. In *Sustainable development and disaster risk reduction* (pp. 13–36). Springer. https://doi.org/10.1007/978-4-431-55078-5_2

Handayani, W., Tatius, C., Hillmann, F., Esariti, L., Young, A., & Nurhidayah, L. (2025). (Im)mobility and environmental change in the coastal sinking cities of Java, Indonesia. *Cosmopolitan Civil Societies: An Interdisciplinary Journal, 17*(1), 64–80. https://doi.org/10.5130/ccs.v17.i1.9414

Heltra Pradana, A., Ayu Fai'zati, E., Dwi Astuti, G., Satya Grahai, I. M., & Julasry Tolan, R. (2022). Adaptasi rumah warga sebagai bentuk mitigasi bencana di Desa Wonolelo, Kabupaten Bantul. *Prosiding SEMSINA, 3*(2), 291–298. https://doi.org/10.36040/semsina.v3i2.5131

Holmes, R., Jones, N., Mannan, F., Vargas, R., Tafere, Y., & Woldehanna, T. (2011). Addressing gendered risks and vulnerabilities through social protection: Examples of good practice from Bangladesh, Ethiopia, and Peru. *Gender and Development, 19*(2), 255–270. https://doi.org/10.1080/13552074.2011.592637

Huq, S., Mahid, Y., & Suliman, N. (2018). Evolution of climate change adaptation policy and negotiation. In *Resilience* (pp. 63–75). Elsevier. https://doi.org/10.1016/B978-0-12-811891-7.00005-0
Jannat, G., & Kusakabe, K. (2024). Women's participation in community-based adaptation to climate change (CBACC): An intersectional analysis. *Social Sciences, 13*(7), 354. https://doi.org/10.3390/socsci13070354
Jerin, T., Azad, M. A. K., & Khan, M. N. (2023). Climate change-triggered vulnerability assessment of the flood-prone communities in Bangladesh: A gender perspective. *International Journal of Disaster Risk Reduction, 95*(August 2022), 103851. https://doi.org/10.1016/j.ijdrr.2023.103851
Kasim, S. S., Hos, J., Taridala, S. A. A., & Moita, S. (2024). The influence of transforming the role of women farmers on access to resources: A gender sociology perspective in agriculture. *Journal of Global Innovations in Agricultural Sciences, 12*(4), 1083–1092. https://doi.org/10.22194/JGIAS/24.1462
Kawanishi, M., Preston, B. L., & Ridwan, N. A. (2016). Evaluation of national adaptation planning: A case study in Indonesia. In *Climate change policies and challenges in Indonesia* (pp. 85–107). Springer Japan. https://doi.org/10.1007/978-4-431-55994-8_4
Klein, J., Juhola, S., & Landauer, M. (2017). Local authorities and the engagement of private actors in climate change adaptation. *Environment and Planning C: Politics and Space, 35*(6), 1055–1074. https://doi.org/10.1177/0263774X16680819
Landauer, M., Juhola, S., & Klein, J. (2019). The role of scale in integrating climate change adaptation and mitigation in cities. *Journal of Environmental Planning and Management, 62*(5), 741–765. https://doi.org/10.1080/09640568.2018.1430022
Lee, C. C., Huang, K. C., Kuo, S. Y., Lin, Y. J., Ke, K. Y., Pan, T. Y., Tai, L. L., Cheng, C. K., Shih, Y. L., Lai, H. T., & Ke, B. H. (2022). Gender matters: The role of women in community-based disaster risk management in Taiwan. *International Journal of Disaster Risk Reduction, 77*(February), 103046. https://doi.org/10.1016/j.ijdrr.2022.103046
Liu, L. (2021). Mainstreaming gender in actions on climate change. *Climate Change Research, 17*(5), 548–558. https://doi.org/10.12006/j.issn.1673-1719.2020.298
McNamara, K. E., Clissold, R., & Westoby, R. (2021). Women's capabilities in disaster recovery and resilience must be acknowledged, utilized and supported. *Journal of Gender Studies, 30*(1), 119–125. https://doi.org/10.1080/09589236.2020.1801397
Moore, S. (2018). Sustainable energy transformations, power, and politics. In *Global development policy in the 21st century: New challenges*. Routledge. https://doi.org/10.4324/9780429508011
Naz, F., Doneys, P., & Saqib, S. E. (2018). Adaptation strategies to floods: A gender-based analysis of the farming-dependent char community in the Padma floodplain, Bangladesh. *International Journal of Disaster Risk Reduction, 28*(December 2017), 519–530. https://doi.org/10.1016/j.ijdrr.2017.12.016
Neuman, L. (2016). The right of access to information: Exploring gender inequities. *IDS Bulletin, 47*(1), 83–98. https://doi.org/10.19088/1968-2016.108
Ngcamu, B. S. (2023). Climate change effects on vulnerable populations in the Global South: a systematic review. *Natural Hazards, 118*(2), 977–991. https://doi.org/10.1007/s11069-023-06070-2
Ngigi, M. W., Mueller, U., & Birner, R. (2017). Gender differences in climate change adaptation strategies and participation in group-based approaches: An intra-household analysis from rural Kenya. *Ecological Economics, 138*, 99–108. https://doi.org/10.1016/j.ecolecon.2017.03.019
Pauw, P., & Pegels, A. (2013). Private sector engagement in climate change adaptation in least developed countries: an exploration. *Climate and Development, 5*(4), 257–267. https://doi.org/10.1080/17565529.2013.826130
Pinto, F. (2014). Urban planning and climate change: Adaptation and mitigation strategies. *TeMA-Journal of Land Use, Mobility and Environment*.
Po, J. Y. T., & Hickey, G. M. (2020). Cross-scale relationships between social capital and women's participation in decision-making on the farm: A multilevel study in semi-arid Kenya. *Journal of Rural Studies, 78*, 333–349. https://doi.org/10.1016/j.jrurstud.2020.04.024

Pundir, S., Padaria, R. N., Muralikrishnan, L., Aiswarya, S., Mukherjee, S., Shravani, K., & Ghosh, B. (2025). Cluster analysis-based discernment of farmers' typologies and climate change adaptation strategies among rural women. *Indian Journal of Extension Education, 61*(1), 78–82. https://doi.org/10.48165/IJEE.2025.61114

Putra, A., Dewata, I., & Gusman, M. (2021). Literature reviews: Hydrometeorological disasters and climate change adaptation efforts. *Sumatra Journal of Disaster, Geography and Geography Education, 5*(1), 7–12. https://doi.org/10.24036/sjdgge.v5i1.363

Resurrección, B. P., Bee, B. A., Dankelman, I., Park, C. M. Y., Haldar, M., & McMullen, C. P. (2019). Gender-transformative climate change adaptation: Advancing social equity. *Paper Commissioned by the Global Commission on Adaptation (GCA).*

Reyes, D. D., & Lu, J. L. (2016). Gender dimension in disaster situations: A case study of flood prone women in malabon city, metro manila. *International Journal of Disaster Risk Reduction, 15*, 162–168. https://doi.org/10.1016/j.ijdrr.2015.11.001

Rimbawan, I. P. D., & Nurhaeni, A. (2024). Gender equality, disability and social inclusion approach to disaster management policy: The case of the Bali disaster response authority. *JISPO Jurnal Ilmu Sosial Dan Ilmu Politik, 13*(2), 169–192. https://doi.org/10.15575/jispo.v13i2.28396

Robles, C. P. Q., & Benavidez, M. J. (2020). *Gender equality and women's empowerment in disaster recovery*. https://coilink.org/20.500.12592/7mdswc

Roy, S., Mallick, B., & Bhattarai, U. (2022). Gender dimensions in disaster risks reduction policy: Insights from Bangladesh. *Gender and the Politics of Disaster Recovery: Dealing with the Aftermath, July*, 206–221. https://doi.org/10.4324/9781003290148-12

Sagala, S., Azhari, D., Nailah, N., Vicri, R. N., Paramitasari, D., Herdiansyah, A. R., & Patricia, C. (2024). Transformative participation in inclusive disaster risk reduction. In *Oxford Research Encyclopedia of Natural Hazard Science*. Oxford University Press. https://doi.org/10.1093/acrefore/9780199389407.013.522

Sumarni, L., & Rudiatin, E. (2020). Women's role in mitigation and disaster management. *BASKARA: Journal of Business and Entrepreneurship, 3*(1), 1–12. https://doi.org/10.54268/baskara.v3i1.8513

Tessema, Y. A., Joerin, J., & Patt, A. (2018). Factors affecting smallholder farmers' adaptation to climate change through non-technological adjustments. *Environmental Development, 25*, 33–42. https://doi.org/10.1016/j.envdev.2017.11.001

Verduci, L., Kaur, S., Venganai, H., & Amu, H. (2024). How natural disasters affect women and girls in the global south. In *SDG Talks*. Cassyni. https://doi.org/10.52843/cassyni.jthfnj

Wagner, C. L., & Fernandez-Gimenez, M. E. (2009). Effects of community-based collaborative group characteristics on social capital. *Environmental Management, 44*(4), 632–645. https://doi.org/10.1007/s00267-009-9347-z

Zhang, T., Yan, T., & Qiu, T. (2020). Effects of age on farmers' adoption of intertemporal green agricultural technology. *Resources Science, 42*(6), 1123–1134. https://doi.org/10.18402/resci.2020.06.10

Chapter 5
Epistemic Injustice: The Impact of Climate Change on Educational Facilities and Learning Processes

Dewi Analis Indriyani, Josefhin Mareta, and Nawawi

Introduction

The ongoing melting of ice in polar regions is causing sea levels to rise, leading to global impacts, including erosion and flooding in coastal areas (Harper, 2018). The north coast of Java has experienced climate change impacts, including erosion (Safuan, 2019; Iqbal, 2020; Sushmita, 2021; Suripto, 2022; Aryantie et al., 2023; Utama, 2024), which is exacerbated in several areas by land subsidence due to excessive use of groundwater (Putri, 2021). Several studies have highlighted the problems along the north coast of Java with extensive coastal degradation (Solihuddin et al., 2020), erosion, and accretion (Mahendra et al., n.d.; Solihuddin et al., 2020; Sulaiman, 2021), inundation (Solihuddin et al., 2020), land subsidence (Andreas et al., 2019; Sarah & Soebowo, 2018; Sidiq et al., 2021), coastline changes (Yulianto et al., 2019), and tidal flooding (Salim & Wibowo, 2024; Sauda et al., 2009; Solihuddin et al., 2020).

Computer simulations conducted at the Bandung Technology Institute (ITB) predict that by 2035, 80% of the Pekalongan area on the north coast of Java will be below sea level (Agustina, 2021). Pekalongan has been among the worst affected by climate change impacts, with frequent occurrences of erosion and tidal flooding (Iskandar et al., 2020; Kasbullah & Marfai, 2014; Mahendra et al., n.d.; Marfai et al., 2013; Salim & Siswanto, 2018). An analysis of coastline dynamics from 1978 to 2021 using remote sensing and GIS techniques in Pekalongan confirmed that changes in the coastline are one of the causes of tidal flooding (Chulafak et al., 2022; Yulianto et al., 2019). In addition, Pekalongan also experiences one of the

D. A. Indriyani (✉) · J. Mareta
Research Center for Law, National Research and Innovation Agency (BRIN), Jakarta, Indonesia
e-mail: dewi043@brin.go.id

Nawawi
Research Center for Population, National Research and Innovation Agency, Jakarta, Indonesia

L. Nurhidayah et al. (eds.), *Climate Change, Labour and Migration in Indonesia*, Springer Climate, https://doi.org/10.1007/978-3-032-15334-0_5

highest rates of land subsidence, up to 20 cm/year, compared to other cities, which are still recording changes of less than 10 cm/year (Nashrullah et al., 2013; Pratami et al., 2022).

Blacanan, Depok, Semut, Pecakaran, Jeruksari, Bandengan, and Kandang Panjang are the areas of Pekalongan most severely impacted by erosion, with a reach up to roughly 232 m^2. Along with erosion, tidal flooding is also a problem in the Bandengan, Jeruksari, and Kandang Panjang regions (Chulafak et al., 2022). Pekalongan's constant tidal flood conditions have an indisputable impact on virtually all community activities, leading to economic losses, causing health problems, and interfering with education (Ismanto et al., 2021). Some local schools continue to conduct teaching and learning activities despite the tidal flooding surrounding the school area (Sukarno, 2018), assessing that flooding below 30 cm is still within a safe range. Others have relocated classes to private homes in areas less affected by tidal flooding (Purnomo, 2022). Both options cause serious inconvenience and interfere with classroom instruction. Moreover, in the longer term, tidal flooding will undoubtedly destroy a number of school buildings (Riyatmoko et al., 2022).

Thus, persistent flooding disrupts educational activities, threatens the safety of students and teachers, and results in extended disruptions to the education system. Experts assert that such interruptions impede academic advancement and intensify socio-economic disparities, as educational institutions in flood-prone regions frequently lack the necessary resources for effective adaptation. This crisis underscores profound concerns over equity, consistent with Miranda Fricker's notion of epistemic injustice (Fricker, 2007), which addresses the way in which systemic inequalities marginalise specific groups within the domain of knowledge. Students, teachers, and communities in impacted regions experience 'testimonial injustice', as their lived experiences and perspectives on the issue are frequently devalued or disregarded in policy formulation and planning. Furthermore, 'hermeneutical injustice' occurs when marginalised groups lack the resources or platforms to express their challenges within the existing frameworks of understanding. Addressing tidal flooding in Pekalongan necessitates both technical and physical measures, as well as a transition towards inclusive and equitable epistemic practices that emphasise the perspectives of people immediately affected.

Based on the background of this research, this chapter will concentrate on the issues of epistemic injustice among marginalised groups in the educational environment, which arise due to the impact of climate change on educational facilities and the learning process.

Methodology

We conducted this research qualitatively using a doctrinal research approach (Dobinson & Jhons, 2007), which applies Fricker's epistemic injustice theory (Fricker, 2003, 2007) to analyse and qualitatively assess field data, observations, and interview results. This represents a perspective on the inequitable treatment of those who know and are directly impacted by a circumstance. This research concentrated on two aspects: (1) the vulnerability of the educational infrastructure and (2) the learning processes affected by climate change. The data collection process relied on literature reviews and in-depth interviews with key stakeholders to delve into the perspectives of the research subjects (Silverman, 2006). These research subjects include those directly involved in teaching and learning activities in schools—representatives of the education office, principals, teachers, and students—and those involved indirectly, such as the parents of students. The research location was Pekalongan Regency, Central Java Province, which, as mentioned, is one of the areas on the northern coastline of Java that has suffered severe erosion, resulting in significant damage and loss.

The interview participants were connected with schools that have been directly impacted by tidal floods in three sub-districts: Wonokerto, Tirto, and Siwalan. These schools were recommended by the Pekalongan District Education Office, and the research was conducted at six elementary schools and two junior high schools.

Result and Discussion

Rights are moral entitlements inherent in individuals who possess self-respect and dignity. The conventional interpretation of 'right' is that it pertains to what one should obtain, receive, or have acknowledged, whereas 'duty' or 'obligation' refers to actions one is compelled to undertake (Hayati et al., 2024; Norman, 1977). In this chapter, we will discuss how epistemic injustice arises in efforts to fulfil the right to education for children in flood disaster situations due to climate change.

The right to education is an essential entitlement that must be upheld for everyone. The right to education for children is established in international law, particularly in the United Nations Convention on the Rights of the Child (UNCRC) (CRC, 2023). Similarly, the Constitution of the Republic of Indonesia declares that every citizen has the basic right to education, which must be available to all without discrimination. Nonetheless, climate change presents considerable hurdles to fulfilling this right, especially when educational facilities are impacted by occurrences such as tidal floods.

The UNCRC emphasises the child's right to education and stipulates that states should ensure that their responsibilities regarding children's right to be heard are integrated into international environmental decision-making processes, encompassing discussions and the execution of international environmental law instruments (Article 28). States must also respect and protect children's rights, including ensuring a secure and supportive environment, along with a legal and institutional framework that enables children to effectively exercise their rights (Article 29).

Notwithstanding these requirements under the UNCRC, there is a recognised gap in international law on the educational necessities of children displaced by climate-related circumstances (UNESCO, 2021). The United Nations Educational, Scientific, and Cultural Organization (UNESCO) acknowledges the scarcity of research exploring the relationship between climate-induced displacement and the right to education. Its own evaluation reports that climate disasters exacerbate poverty, thereby imposing more barriers on education. Moreover, UNESCO notes that such disasters directly damage or destroy educational institutions and resources (UNESCO, 2021). States must therefore devise and execute substantive strategies, including risk mitigation, for disaster-prone regions, especially in areas identified through comprehensive research as susceptible to recurring disasters.

Pekalongan, both a city and a regency, is frequently utilised as a research site for studies across numerous disciplines on the impacts of tidal flooding. Nonetheless, the response from both the central and regional governments lacks the seriousness required for effective long-term planning. Various studies indicate that the government appears to have normalised tidal flooding and typically offers emergency assistance only when such flooding intensifies (Siagian et al., 2024), although even in these cases, the aid provided is not uniformly distributed across all inundated areas.

Marginalised Groups' Knowledge of Climate Change

This paper aligns with Fricker's concept that within the realm of justice—particularly epistemic justice—certain injustices are perceived as normative (Fricker, 2007). Consequently, the sole method to elucidate the components of epistemic injustice is to examine the negative space surrounding it. This study elucidates the epistemic injustice experienced by stakeholders in the education sector, including school principals, teachers, students, and parents.

On several occasions, we purposefully engaged directly within classrooms at schools impacted by tidal floods in the Pekalongan Regency. We observed classroom interactions for Year 7 students in an elementary school and Year 9 students in a junior high school at two schools, both located in the Wonokerto sub-district. We found that all students lacked awareness of the concept of climate change, including global warming. When we inquired about the condition of the school environment and local residences recently affected by flooding, the students noted that it had been tidal flooding but could not elucidate the significance of the tidal flood or the factors contributing to its occurrence.

Moreover, in an in-depth interview conducted in the Siwalan and Tirto sub-districts with principals, teacher representatives, student representatives, and parent representatives, only two of the six principals could clearly explain the concept of climate change, whereas four of the five teacher representatives offered a reasonably satisfactory explanation. However, the principals and teachers appeared uncertain about the responses they gave.

We asked student and parent representatives about their knowledge regarding tidal floods rather than the broader concept of climate change. The students' responses varied; some asserted that tidal floods result from excessive rainfall, others that coastal water intrudes into the mainland, while some indicated that tidal floods occur when the embankment fails. The parent representatives all defined tidal floods in similar terms, as seawater encroaches on the mainland; only one of the six representatives could elucidate the mechanisms and processes underlying tidal floods.

Interactions and interviews with school principals, teacher representatives, student representatives, and parent representatives reveal that individuals impacted by climate change and tidal flooding affecting their schools and homes lack a clear comprehension of the phenomenon they are encountering; this exemplifies one of the forms of epistemic injustice that we will discuss.

Epistemic Injustice Framework in the Knowledge and Experiences of Marginalised Groups

Any assertion of injustice must depend on collective ethical intuition; nevertheless, we can attain a more lucid understanding of what defines an injustice by analysing the nature of the wrong inflicted. Fricker's (2007) framework encapsulates a uniquely epistemic form of injustice. A variety of phenomena may be categorised under the overarching concept of epistemic injustice. Considering conventional philosophical perspectives on justice, the concept of epistemic injustice primarily evokes considerations of distributive inequity concerning epistemic resources such as information or education.

Fricker identifies two categories of epistemic injustice: testimonial and hermeneutical. Testimonial injustice refers to when an individual's or a group's views, based on their knowledge or experience, are rejected because they are deemed not credible; this damages their identity as individuals and as a group. Negative prejudices can manifest in a variety of ways and encompass multiple phenomena. Hermeneutical injustice is conceptualised as an injustice that arises from a gap in collective hermeneutical resources, where the gap creates a cognitive disadvantage that has an uneven impact on different social groups.

This article aligns with the research conducted by the Center for Child Protection and Welfare Studies (Pusat Kajian dan Advokasi Perlindungan dan Kesejahteraan Anak, PUSKAPA) at Universitas Indonesia, which examines the effects and injustices of climate change on children and adolescents aged 13–24 in North Jakarta, Pontianak, and Pekalongan City. These findings indicate that the comprehension of climate change among these children remains confined to basic biophysical processes and anthropogenic factors such as waste disposal (Siagian et al., 2024).

The experiences of children and adolescents in Pekalongan City are largely similar to those observed in the eight schools in the Pekalongan Regency where our research was conducted. The effects of climate change on principals, teachers, students, and parents are palpable; however, there is a prevailing perception that these experiences are normal rather than indicative of recurring and permanent harm or damage. The government prioritises short-term demands over the development of long-term strategies to mitigate the risk of recurrent harm from tidal floods. Every application submitted to the Pekalongan Regency Education Office failed to align with the needs of the schools, leading to a misaligned allocation of support.

The injustice that is encountered pertains not only to unmet demands for and fundamental rights to safe and healthy living but also to injustice in the realisation of educational rights, which include the right to a secure, pleasant, conducive, and joyful learning experience. The uneven dissemination of climate change knowledge raises epistemic injustice concerns, with numerous individuals being oblivious to the prospective adverse effects of tidal flooding in their locality.

Epistemic Injustice in the Learning Process in Affected Schools Due to the Impacts of Climate

Normally, educational activities conducted by schools in Pekalongan Regency occur in person within classrooms; however, recurrent tidal floods have significantly impeded the learning process. This section delineates the forms of epistemic injustice that adversely affect the coherence of the learning process, encompassing resources, attendance, concentration, and the academic performance of students.

Inadequate Resources

Tidal floods occur frequently and recurrently, and they can be anticipated at specific times, such as at the full moon, periods of significant rainfall, and the onset of the rainy season. Currently, however, the local community is unable to anticipate when tidal floods may submerge roads and buildings in impacted regions, including the access road to the school and the school building itself, due to climate change. According to the testimonies of the principal, teachers, students, and parents, instruction continues in classrooms despite water rising high enough to submerge students' feet and school furnishings. This is based on assessments of whether the tidal flood remains within the "safe" classification, meaning that the water level is no higher than halfway up an adult's calf. This practice persists despite acknowledgement of the adverse effects of exposure to even this level of flooding, which can cause or exacerbate various health issues, including pruritus, pyrexia, tinea pedis, and asthma.

Where tidal flood rises above this level, policies vary between schools: some replace classroom teaching with online learning and allow students to return home, while others migrate classes to elevated facilities. Usually, these are village halls, mosques, or residences in proximity to the school or the homes of students' parents on a rotational basis; some even relocate their students to other schools within the same zoning area that are not adversely impacted by tidal flooding. The educational experience at off-site locations was reported to be highly unfavourable due to (1) the large number of students in contrast to the capacity of the relocation venue; (2) a scarcity of instructional resources required by teachers to effectively convey the material; and (3) the learning conducted not being comprehensive, lacking the full duration of standard instructional hours that students are entitled to receive.

At the same time, the online learning process had numerous deficiencies, specifically: (1) not all students had personal devices (handphones/tablets/laptops/computers) due to many hailing from lower-middle economic backgrounds, resulting in students being dependent on access to parental devices (if available) or borrowing devices from relatives or neighbours; (2) assigning schoolwork—mostly reading tasks—through WhatsApp groups posed challenges as not all students possessed the necessary reading materials; (3) while some teachers attempted to innovate by recording lessons and uploading them to YouTube, enabling students to engage with the instruction, this approach requires internet access both for teachers to upload and students to view the material, which not all parents can afford to provide; and (4) parents confirmed that when online learning was implemented, children studied less preferring to play. On the other hand, parents also expressed concerns about their children being required to attend school during tidal floods, given the hazardous conditions affecting their homes, streets, and school environment.

Almost all principals, teachers, and parents appeared confused by the situation and did not understand what they were actually experiencing. This scenario is an example of hermeneutical injustice.

Attendance of Teachers and Students

Recent tidal flooding near the school resulted in numerous student absences, with some students being unable to attend due to their residential areas also being inundated, while others saw the floods as an opportunity to help their fathers with fishing work to meet family needs. The number of students present at either the school or the temporary relocation sites thus constantly fluctuates. Nonetheless, principals and teachers noted that regardless of the flood level (provided it remained within the 'safe' parameters), they were compelled to be present at school as long as even one student was in attendance in order to complete the attendance roll. This was the case even if the class was moved to an alternative location or if the teaching hours did not meet the required standards. During flooding, this typically involved reaching the school with a small wooden boat, which posed a significant challenge due to the expense, particularly for those whose salaries are modest. It is important to note that not all teachers are civil servants; some are employed under contract, and their remuneration usually depends on the school's income (school operational cost funds (Dana BOS), based on the number of students, with funds disbursed every three months). In addition, one of the schools temporarily relocated its classes a considerable distance from the flood-affected area inundated by tidal floods, requiring the principal and teachers to walk several kilometres between the temporary site and the school solely to record attendance. Several of the teachers expressed concerns about their safety when traversing flood-affected areas, as they are unable to discern the road and must keep watch for aquatic creatures, including snakes.

Impact on Concentration

Extensive tidal flooding that submerges roadways, residences, and educational institutions causes substantial pressure for principals, teachers, and students alike. Frequently, when the school experiences tidal floods, their residences are likewise inundated, impacting their economic situation and daily activities. Despite this, both students and teachers are frequently obliged to attend school. Under these adverse conditions, students reported experiencing pruritus, particularly in the lower extremities, as well as abdominal discomfort while at school, leading them to favour recreational activities over academic pursuits.

This was also conveyed by the teachers, who noted that students frequently arrived at school or temporary classrooms with their clothing soaked up to the waist. Consequently, they are unable to focus on the lessons being presented, and teachers often observe them daydreaming or gazing blankly. There were also reports of other students distracted by enjoying the tidal deluge and intentionally soaking themselves. These circumstances resulted in mental fatigue among the teachers, bearing the responsibility of being their family's economic mainstay while also attempting to manage the physical and emotional exhaustion of their students. At the same time, they are expected to maintain their role as an educator, inspiring enthusiasm in both colleagues and students.

Academic Performance

The stress, both physical and mental, occasioned by tidal flooding events—at home, on the streets, and at school—unquestionably impedes the learning process. The principals and teachers constantly strive to maintain enthusiasm and support each other to ensure that adverse conditions do not impair their effectiveness in engaging with students. Nonetheless, student development has decreased significantly across attitude, mentality, and academic performance.

The situation reported by the principals, teachers, and parents in this research was corroborated by a statement from the head of the basic education development division of the Pekalongan District Education Office, who noted that during periods of significant rainfall, particularly during the full moon, tidal inundation is inevitable. The consequence is that the educational process is temporarily transferred to a local mosque situated above the tidal flood level. Officials have also noted substantial deterioration in the quality of education, affecting students and teachers. He highlighted that coastal regions exhibit deficiencies in academic performance, as evidenced by their achievements in academic competitions. Nonetheless, schools in these areas perform well in non-academic competitions, producing numerous athletes in sports and martial arts.

Epistemic Injustice in the Vulnerability of Education Infrastructure Due to the Impacts of Climate Change

This section addresses epistemic injustice concerning educational infrastructure. In Indonesia, the requirement for standardised educational facilities and infrastructure is regulated by the Minister of Education Regulation No. 22 of 2023 (MER No. 22 of 2023). MER No. 22 of 2023 delineates two categories of educational facilities: (1) tools and equipment utilised to attain learning objectives and (2) infrastructure, which encompasses essential facilities required for the operational functions of educational units, including school land, building standards, and various room types.

Inadequate Educational Facilities

As reported in the previous section, the experiences of schools impacted by tidal floods hamper the learning process—which should occur in a safe and comfortable environment—and cause damage to school property. Consequently, educational activities take place in temporary locations (such as residences, mosques, or village halls) or via online platforms that cannot be executed effectively by either teachers or students.

According to the principals and teachers in this research, numerous educational resources are damaged and unusable, necessitating their storage in warehouses or classrooms. Schools are also unable to significantly assist with online learning resources, such as providing students with Internet quotas or access to devices. School office supplies and stationery are insufficient and frequently run out, and there is a lack of computers and laptop facilities. Schools utilise operational cost funds to purchase office supplies and stationery, but these funds are sometimes insufficient, as the calculation of operational cost income is predicated on student statistics. Given that the schools impacted by the tidal flood are state institutions with a limited student population, the operational funding they receive from the education office is limited. Consequently, there are schools whose teachers contribute their own funds to procure school supplies or to assist in financing student competitions. Funds from the educators will be reimbursed following the disbursement of the school's operational expenses (Dana BOS), typically allocated every three months.

Inadequate and Unsafe Infrastructure

MER No. 22 of 2023 (Article 8) requires that school land must be constructed in a safe environment, devoid of health and safety hazards, and must provide access to emergency rescue services. The eight schools we visited were situated in unsafe and potentially hazardous settings, particularly in the event of tidal flooding. Five of the eight schools are located near rice fields that become muddy ponds due to inundation by tidal floods. In addition, the school area is deficient in water absorption zones, which causes water stagnation during rainfall and impacts the duration of water evaporation.

The construction requirements outlined in MER No. 22 of 2023 (Article 9) specify that constructed buildings shall adhere to provisions concerning safety, health, security, and comfort. Our observations indicate that all of these schools have been inundated by tidal floods, resulting in seawater exposure. Prolonged submersion in seawater leads to structural corrosion (Irawan et al., 2023), posing significant risks (Syahril et al., 2023; Wedhanto, 2017) to the safety of individuals in the school environment. The sole support provided by the Education Office for reconstruction necessitated by tidal flooding is the elevation of the building, indicating that the bottom portion of the supports for these structures remains constructed from the same material. Numerous schools that we visited exhibited inadequate structural conditions, with the walls already exhibiting swelling due to deterioration.

In all the schools, the condition of renovated buildings was sub-optimal, with issues including disorganised electrical installations, lack of adequate lighting, forcing children to study in darkness, lack of ceilings, and uncovered floors with exposed clay, resulting in dust when dry and mud when wet. The Education Office confirmed that the majority of the reconstructed buildings lacked ceilings and that most floors were merely covered with cement due to the persistent threat of tidal floods necessitating the recurrent elevation of the school structure.

Alongside the substandard infrastructure, the majority of schools visited lacked toilet facilities, forcing students to use nearby residences, mosques, or local village halls. In addition, the absence of laboratory space, health rooms, and prayer rooms was also noted. There were library spaces, although they were not operating effectively. Some libraries are merely structures, empty of materials; some remain perpetually closed due to ongoing tidal flooding despite the presence of shelves, books, and educational resources; and others have yet to be constructed.

The head of the Facilities and Infrastructure Section at the Pekalongan Regency Education Office suggested a possible solution in the form of 'floating schools'. Having an architectural background, he came up with the idea of a floating school after conducting a comparative study in a European nation, where he observed how the structures remained secure even when the land was below sea level. This concept is not new to Indonesia; floating schools have already been established on the Serayu River (Central Java), Kapuas River (Central Kalimantan), Lake Toba (North Sumatra), Musi River (South Sumatra), Suku Bajo, Buton (Southeast Sulawesi), and other river- or lakeside and coastal locations.

Regional Policies Not Optimally Implemented

There are no specific rules in the scope of educational content in Indonesia that directly relate to efforts to minimise or eliminate epistemic injustice; for example, there are no special lessons on the environment, or more specifically, climate change, to ensure that principals and teachers can transfer knowledge to students.

Local content (mulok) lessons should include environmental knowledge to ensure principals, teachers, and especially students understand what they are actually experiencing in their environment and, rather than normalising tidal flooding as something natural, recognise it as a warning that the area in which they live or attend school is not safe.

Other regional policies dealing with tidal flooding in Pekalongan Regency can be found in a number of instruments: (1) (Pekalongan Regency Regulation No 1 of 2016 concerning Disaster Management; (2) Pekalongan Regency Regulation No 8 of 2017 concerning the Pekalongan Regency Medium-Term Development Plan 2016–2021; (3) Pekalongan Regency Regulation No 5 of 2021 concerning the Pekalongan Regency Medium-Term Development Plan 2016–2021; (4) Regent Regulation No 180.18/65 of 2023 concerning the Determination of the Extension of the Emergency Alert Status for Wind, Flood, and Landslide Disasters in Pekalongan Regency; (5) Joint Agreement Number 614.11/9/MoU 2017 concerning Control of Tidal Flooding in Pekalongan Regency and City; and (6) (Pekalongan Regency Regulation No 3 of 2020 concerning the Pekalongan Regency Regional Spatial Planning Plan for 2020–2040.

Given these various regulations, the Pekalongan district government is well aware of the community's disaster mitigation needs, one of which is the construction of a flood control network system in the form of a water damage control system with embankments in Siwalan District, Wonokerto District, and Tirto District. However, as of the time of writing, the embankment has not entirely encompassed the three districts. As a result, the community, mindful of the need for security, collaborated to construct a temporary embankment but encountered difficulties in operating the water pump equipment, causing the embankment to collapse.

The Regional Disaster Management Agency (Badan Penanggulangan Bencana Daerah, BPBD) has limited capabilities. According to our interviewees, BPBD is always cost-constrained, and the only activities that can be carried out are non-physical ones, such as offering disaster preparedness training to village representatives, to which not all villages may be invited, and which takes place only twice a year.

Conclusion

Based on the range of insights and experiences concerning tidal flood conditions in Pekalongan, as conveyed by local students, teachers, principals, and parents, it is evident that there is a gap in informational resources pertaining to climate change and its repercussions. The impacts of climate change necessitate alterations in the learning environment, often resulting in a transition to new facilities or online education, provisional remedies accompanied by additional challenges such as inadequate capacity in relocation facilities for the student population, inequitable access to devices, and difficulties in acquiring internet packages. Tidal flooding adversely affects school infrastructure, resulting in damage to buildings, furniture, teaching materials, books, and other school properties.

The deficiency of collective hermeneutic resources creates cognitive losses that have uneven impacts. The Pekalongan Regency government, particularly the Education Office, must address distributive injustice in accessing epistemic resources, specifically information and education regarding climate change, within the education sector in Pekalongan Regency. The experiences of children, teachers, principals, and parents concerning the learning process and infrastructure requirements must be examined, and specific regulations established regarding these two aspects to ensure that the government upholds children's rights to quality education under all circumstances. In addition to the lack of a long-term plan to address tidal flooding, the expertise and insights of the Head of Facilities and Infrastructure at the Pekalongan Education Office regarding the construction of floating schools could serve as a valuable recommendation, ensuring the optimal utilisation of funds allocated for school building repairs.

References

Agustina, R. (2021, March 15). Ahli: Akibat pengambilan air tanah berlebihan, tahun 2035 wilayah Pekalongan bisa tenggelam. *Tribunews.Com*. https://www.tribunnews.com/regional/2021/03/15/ahli-akibat-pengambilan-air-tanah-berlebihan-tahun-2035-wilayah-pekalongan-bisa-tenggelam

Andreas, H., Abidin, H. Z., Sarsito, D. A., Meilano, I., & Susilo. (2019). *Investigating the tectonic influence to the anthropogenic subsidence along northern coast of Java Island Indonesia using GNSS data sets. 94*. https://doi.org/10.1051/e3sconf/20199404005

Aryantie, M. H., Winanti, W. S., & Amru, K. (2023, September 20). Pantai utara jawa dan dampak nyata perubahan iklim. *Mongabay*. https://www.mongabay.co.id/2023/09/20/pantai-utara-jawa-dan-dampak-nyata-perubahan-iklim/

Chulafak, G. A., Khomarudin, M. R., Ardha, M., Pranowo, W. S., Prayudha, B., & Mujio. (2022). Thirty years of change of pekalongan coastline based on landsat imagery. *Earth and Environmental Science, 1109*, 012056, 1–14. https://doi.org/10.1088/1755-1315/1109/1/012056

CRC. (2023, August 22). *Convention on the Rights of the Child—General comment No. 26 (2023) on children's rights and the environment, with a special focus on climate change*. https://docstore.ohchr.org/SelfServices/FilesHandler.ashx?enc=6QkG1d%2FPPRiCAqhKb7yhsqIkirKQZLK2M58RF%2F5F0vHrWghmhzPL092j0u3MJAYhyUPAX9o0tJ4tFwwX4frsfflPka9cgF%2FBur8eYD%2BEeDmuoVnVOpjkzwB9eiDayjZA&

Dobinson, I., & Jhons, F. (2007). Qualitative legal research. In *Researchs methods for law*. Edinburgh University Press.

Fricker, M. (2003). Epistemic injustice and a role for virtue in the politics of knowing. *Methaphilosophy, 34*(1/2), 154–173.

Fricker, M. (2007). *Epistemic injustice: Poser & ethics of knowing*. Oxford University Press.

Harper, L. (2018). Glaciers, ice sheets, and more: A primer on the different types of polar ice. *News from the Columbia Climate School*. https://news.climate.columbia.edu/2018/02/05/glaciers-ice-sheets-polar-ice/

Hayati, A. N., Indriyani, D. A., Firmanditya, N., & Citrawan, H. (2024). The vocabulary of right under the Indonesian constitution: A hohfeldian analysis. *Constitutional Review, 10*(1), 234–265. https://doi.org/10.31078/consrev1018

Iqbal, D. (2020, April 29). Pantura jawa terancam karam. *Mongabay*. https://www.mongabay.co.id/2020/04/29/pantura-jawa-terancam-karam/

Irawan, D., Niken, C., Zaenudin, A., & Helmi, M. (2023). Pengaruh perendaman air laut terhadap kuat tekan dan permeabilitas beton. *JRSDD, 11*(4), 789–800.

Iskandar, S., Helmi, M., Muslim, Widada, S., & Rochaddi, B. (2020). Analisis geospasial area genangan banjir rob dan dampaknya pada penggunaan lahan tahun 2020—2025 di kota pekalongan provinsi jawa tengah. *Indonesian Journal of Oceanography, 2*(3), 271–282.

Ismanto, K., Pratikwo, S., Madusari, B. D., & Christianto, P. A. (2021). Analisis kebutuhan masyarakat terdampak banjir rob: Studi kasus kota pekalongan. *Jurnal Litbang Kota Pekalongan, 19*(1), 20–28.

Kasbullah, A., & Marfai, M. (2014). Pemodelan spasial genangan banjir rob dan penilaian potensi kerugian pada lahan pertanian sawah padi studi kasus wilayah pesisir kabupaten pekalongan jawa tengah. *Geoedukasi, III*(2), 83–91.

Mahendra, I. W. W. Y., Maulana, E., Wulan, T. R., Rahmadana, A. D. W., & Putra, A. S. (n.d.). Pemetaan kawasan rawan abrasi di provinsi jawa tengah bagian utara. In *Bunga Rampai Kepesisiran dan Kemaritiman Jawa Tengah* (Vol. *II*, pp. 93–105). Parangtritis Geomaritime Science Park.

Marfai, M. A., Mardiatno, D., Cahyadi, A., Nucifera, F., & Prihatno, H. (2013). Pemodelan spasial bahaya banjir rob berdasarkan skenario perubahan iklim dan dampaknya di pesisir pekalongan. *Jurnal Bumi Lestari, 13*(2), 244–256.

Nashrullah, S., Aprijanto, Junita, M. P., Hazarika, M. K., & Samarakoon, L. (2013). Study on flood inundation In pekalongan, central java. *International Journal of Remote Sensing and Earth Sciences, 10*(2), 76–83.

Norman, E. B. (1977). Taking rights seriously by Ronald Dworkin. Massachusetts: Harvard University press. 1977. pp. 563. *Catholic University Law Review, 26*(4), 908–923.

Pekalongan Regency Regional Regulation Number 1 of 2016 Concerning Disaster Management.

Pekalongan Regency Regional Regulation Number 3 of 2020 Concerning the Pekalongan Regency Regional Spatial Planning Plan for 2020-2040.

Pekalongan Regency Regional Regulation Number 5 of 2021 Concerning the Pekalongan Regency Medium-Term Development Plan 2016-2021.

Pekalongan Regency Regional Regulation Number 8 of 2017 Concerning the Pekalongan Regency Medium-Term Development Plan 2016-2021.

Pratami, B., Larasati, R., Ratna Intan, S., & Kamaluddin, I. (2022). Status hukum tanah musnah berdasarkan permen ATR/BPN no. 17 tahun 2021. *Officium Notarium, 1*(2), 218–229. https://doi.org/10.20885/JON.vol1.iss2.art2

Purnomo, I. D. (2022, November 16). Gedung SDN di pekalongan, terendam banjir rob, siswa terpaksa belajar di rumah warga selama 3 pekan. *Tribun Banyumas.Com*. https://banyumas.tribunnews.com/2022/11/16/gedung-sdn-di-pekalongan-terendam-banjir-rob-siswa-terpaksa-belajar-di-rumah-warga-selama-3-pekan?page=2

Putri, G. S. (2021, September 17). Muka tanah pekalongan turun hingga 20cm per tahun, Ini penyebab dan dampaknya. *Kompas.Com*. https://www.kompas.com/sains/read/2021/09/17/173200423/muka-tanah-pekalongan-turun-hingga-20-cm-per-tahun-ini-penyebab-dan?page=all

Regent Regulation Number 180.18/65 of 2023 Concerning the Determination of the Extension of the Emergency Alert Status for Wind, Flood, and Landslide Disasters in Pekalongan Regency.

Regulation of the Minister of Education, Culture, Research, and Technology of the Republic of Indonesia Number 22 of 2023 Concerning Standards of Facilities and Infrastructure in Early Childhood Education, Elementary Education Level, and Secondary Education Level, Pub. L. No. 22 (2023).

Riyatmoko, A., Sanjoto, T., & Juhadi. (2022). Impact of the rob flood in north pekalongan. In *4th IGEOS: International Geography Seminar 2020, 1089 (2022) 012016*, 1–8. https://doi.org/10.1088/1755-1315/1089/1/012016

Safuan, A. (2019, December). Ribuan hektare daratan di pantura jateng hilang akibat abrasi. *Media Indonesia*. https://mediaindonesia.com/nusantara/280474/ribuan-hektare-daratan-di-pantura-jateng-hilang-akibat-abrasi

Salim, M. A., & Siswanto, A. B. (2018). Penanganan banjir dan rob di wilayah pekalongan. *Jurnal Teknik Sipil, 11*, 1–8.

Salim, M. A., & Wibowo, K. (2024). Rob flood control on the north coast of java (study on coastal areas of pekalongann and semarang). In *IOP Conf. Series: Earth and Environmental Science, 1321 (2024) 012026*, 1–10. https://doi.org/10.1088/1755-1315/1321/1/012026

Sarah, D., & Soebowo, E. (2018). *Land subsidence threats and its management in the north coast of java. 118.*

Sauda, R. H., Nugraha, A. L., & Hani'ah. (2009). Kajian pemetaan kerentanan banjir rob di kabupaten pekalongan. *Jurnal Geodesi Undip, 8*(1), 466–474. https://doi.org/10.14710/jgundip.2019.22777

Siagian, C., Sari, W. L., Rahmi, M. A., Kartaadipoetra, F. W., Lisan, I. H., Tieken, S., Robby, M. B., Matahelemual, G. J., Ufaira, R., Nisa, S. A., & Swanarta, A. (2024). *Bergerak menuju tata kelola iklim yang adil: Memahami perubahan iklim dari sudut pandang anak dan remaja yang terpinggirkan di Indonesia.* PUSKAPA-ANU.

Sidiq, T. P., Gumilar, I., Meilano, I., Abidin, H. Z., Andreas, H., & Permana, A. (2021). *Land Subsidence of Java North Coast Observed by SAR Interferometry. 873.*

Silverman, D. (2006). *Qualitative research: Theory method and practice* (2nd ed.). SAGE Publications.

Solihuddin, T., Husrin, S., & Salim, H. L. (2020). Morfodinamika pantura jawa. In *Penilaian Hasil (Outcome Assessment) Rehabilitasi Pantura Jawa.* AMAFRAD Press-BRSDMKP.

Sukarno, S. (2018, January 17). Ratusan siswa di pekalongan belajar dengan ruang kelas terendam air. *Inews.Id.* https://jateng.inews.id/berita/ratusan-siswa-di-pekalongan-belajar-dengan-ruang-kelas-terendam-banjir

Sulaiman, D. M. (2021). *Teknologi pegar untuk penanggulangan erosi dan abrasi pantai.* Deepublish.

Suripto, I. (2022, October 21). Pesisir pantura diprediksi hilang 2030, abrasi kini gerus 2 ribu Ha di brebes. *Detik.Com.* https://www.detik.com/jateng/berita/d-6361802/pesisir-pantura-diprediksi-hilang-2030-abrasi-kini-gerus-2-ribu-ha-di-brebes

Sushmita, C. I. (2021, Mei). *Abrasi pantura menelan jawa tengah.* Solopos Media Group. https://regional.espos.id/abrasi-pantura-menelan-jawa-tengah-1127648

Syahril, Mujiman, R. D., Riandi, M. R., Uli, P., Artanti, L. D., & Shaputra, M. R. E. (2023). Pengaruh penambahan vermikulit terhadap kuat lentur beton mutu rendah dengan simulasi banjir air laut. *Wahana TEKNIK SIPIL, 28*(2), 241–248.

UNESCO. (2021). *How climate change and displacement affect the right to education.* The Right to Education. https://www.unesco.org/en/right-education/climate-change-displacement?

Utama, D. A. (2024, July 29). Empat wilayah di pantura alami abrasi, garis pantai mundur sampai 5 kilometer. *Merdeka.Com.* https://www.merdeka.com/peristiwa/empat-wilayah-di-pantura-alami-abrasi-garis-pantai-mundur-sampai-5-kilometer-171704-mvk.html

Wedhanto, S. (2017). Pengaruh air laut terhadap kekuatan tekan beton yang terbuat dari berbagai merk semen yang ada di kota malang. *Jurnal Bangunan, 22*(2), 21–30.

Yulianto, F., Suwarsono, M., & T., & Khomarudin, M. R. (2019). The dynamics of shoreline change analysis based on the integration of remote sensing and geographic information system (GIS) techniques in Pekalongan coastal area, Central Jawa, Indonesia. *Journal of Degrades and Mining Lands Management, 6*(3), 1789–1802. https://doi.org/10.15243/jdmlm.2019.063.1789

Chapter 6
Capabilities, Empowerment, and Citizenship Among Women in Coastal Areas of North Jakarta: Response to Climate Change

Tuti Alawiyah, Nurrahman Aji Utomo, Vera Bararah Barid, and Ahmad Fathony

Introduction

Climate change is a global phenomenon with interconnected impacts on ecosystems, biodiversity, and human systems, with these impacts varying across regions, sectors, and communities (IPCC, 2022). While climate change impacts differ globally, those who reside in developing and under-resourced countries will be more severely affected due to having fewer resources and capabilities to tackle climate risks (Chinowsky et al., 2011; Eastin, 2018; Letta et al., 2022). Domains that are negatively impacted by climate change range across economics, food, and water scarcity (Stringer et al., 2021; Nhemachena et al., 2020), health and well-being (Ebi et al., 2021), and mental health (Palinkas & Wong, 2020), with flow-on effects in relation to gender inequalities and gender-based violence (Van Daalen et al., 2022).

Climate change has and will continue to have a disproportionate impact on women in many situations, since gender roles and inequalities influence resource availability and the mobility necessary to adapt to its consequences (Smith, 2020). Previous studies confirm that women face a greater climate change impact than men (Ahmed & Elkund, 2021; Djoudi et al., 2016; Eastin, 2018; Tickamyer et al., 2014). A cross-country study found that climate change impacted women and reinforced gender inequalities due to increased disaster exposure, reduced availability of food and water (with women globally bearing the burden for water collection: Carr et al.,

T. Alawiyah (✉)
Faculty of Social and Political Science, University of Muhammadiyah Jakarta, Banten, Indonesia
e-mail: tuti.alawiyah@umj.ac.id

N. A. Utomo · V. Bararah Barid · A. Fathony
Research Center for Law, National Research and Innovation Agency (BRIN), Jakarta, Indonesia

L. Nurhidayah et al. (eds.), *Climate Change, Labour and Migration in Indonesia*, Springer Climate, https://doi.org/10.1007/978-3-032-15334-0_6

2024), lack of ownership of assets, and limitations on women's ability to access the labour market, join civil society organisations, or mobilise for political change (Eastin, 2018). In Indonesia, one study on Pramuka Island showed that climate change impacts women more severely, with reduced fish catches reducing families' incomes and economic positions (Saputro & Kurian, 2023). Social norms and differences in gender roles and expectations also affect how men and women react and cope when exposed to climate risks (Alston, 2014; Chindarkar, 2012).

With women facing greater impacts from climate change, they—particularly at the community level—are well placed to offer essential information for responding to climate change (Wanjiru, 2012). Despite this, gender and intersectionality perspectives in climate change research and policies are noticeably lacking (Ahmed & Elkund, 2021; Alston, 2014; Djoudi et al., 2016; MacGregor, 2009). In their review of 41 papers on climate adaptation strategies, for example, Djoudi et al. (2016) found that most climate studies did not use an intersectionality lens, and there was little to no attention paid to power, social, and political relations. This chapter, therefore, highlights the gender lens, utilising Kabeer's (1999) three elements of women's agency—capabilities, empowerment, and citizenship—to understand women's climate adaptation strategies in a coastal area of Jakarta.

Climate Change Adaptation and Women's Agencies

The Paris Agreement's global goal of adaptation (Article 7 (1)) calls for measures to enhance 'adaptive capacity, strengthening resilience and reducing vulnerability to climate change, with a view to contributing to sustainable development and ensuring an adequate adaptation response in the context of the temperature goal referred to in Article 2' (UNFCCC, 2015, p. 28). Based on other clauses in the Agreement, adaptation planning should aim to protect people, livelihoods, and ecosystems, especially in developing country contexts, utilising a gender-responsive perspective and respecting indigenous people and local knowledge systems. Thus, the adaptation planning processes must identify adaptation practices, needs, and priorities using a local perspective. One of the steps in adaptation planning is assessing climate change's impact on vulnerable communities, including women and children. Adaptation strategies include economic diversification and sustainable management of natural resources to build local community resilience. Additionally, the UAE Framework for Global Climate Resilience calls for adaptation planning to reduce climate impacts, risks, and vulnerabilities and enhance adaptation action and support (UNFCCC, 2023).

Adaptation is thus defined as 'the process of adjustment to actual or expected climate and its effects in order to either lessen or avoid harm or exploit beneficial opportunities' (IPCC, 2022), while resilience—which is only mentioned in the 6th Assessment Report of the IPCC (Kikstra et al., 2022)—is defined as 'the capacity of social, economic, and environmental systems to cope with a hazardous event or trend or disturbance, responding or reorganising in ways that maintain their essential function, identity, and structure, while also maintaining the capacity for adaptation, learning, and transformation' (Ackerl et al., 2023, p. 186).

Women Agencies: Capabilities, Empowerment, and Citizenship

Women's agency has been defined as 'their ability to define and act on goals, make decisions that matter to them, and participate in the economy and public life' (Chang et al., 2020, p. 24). Donald et al. (2017) describes agency as 'a fundamental part of empowerment [that] implies exercising choices in ways that challenge power relations'. This definition is related to women's right to self-determination, and thus, agency can be varied across different contexts, and improvement in women's agency is instrumental to women's rights and well-being (Donald et al., 2017). To further develop the conceptual framework used in this chapter, we turn to Kabeer's (2021) three dimensions of women's agency: capabilities, empowerment, and citizenship.

Capabilities are 'people's ability to achieve various ways of 'being and doing', and 'reflect the interaction between the resources at their disposal and their ability to translate these resources into valued goals' (Kabeer, 2021, p. 1). Kabeer also notes that 'patriarchal structures, including in the form of gender inequalities, gender division of labour, and lack of opportunities to participate in the economy and political life, and gender hegemonic ideologies serve to constrain women's capabilities relative to those of men' (Kabeer, 2021, p. 1).

Empowerment 'builds on the capability approach but makes the interaction between agency and power structures more explicit' (Kabeer, 2021, p. 2). Kabeer defines 'power as the capacity to make choices', and it is 'inescapably bound up with the condition of disempowerment and refers to the processes by which those who have been denied the capacity to make choices gain this capacity'. Empowerment is about gaining the power 'from the perspective of the dominant groups ("the power over")', 'the power to transform the self ("the power within")', 'to achieve valued goals ("the power to")', and 'the new forms of power that come into existence through collectively act with others ("the power with")' (Kabeer, 2021, p. 2).

Citizenship and collective action encompass 'collective forms of agency that set out to tackle the structures of gender injustice'. Citizenship can be defined in passive terms as 'personhood in a society' or in active terms as 'the social practices', and 'citizenship as practice places the question of agency at the heart of contestations around citizenship' (Kabeer, 2021, p. 4).

Women and Climate Adaptation in the Coastal Area of North Jakarta

Indonesia ratified the Paris Agreement in 2016 and has pledged to achieve a 29% reduction in emissions with domestic resources and 42% with international funding by 2030, with this reduction occurring across the forestry, renewable energy, and peatland sectors (Republic of Indonesia, 2022). Indonesian regulations and policies governing climate change measures include the Indonesian Constitution of 1945 (which guarantees Indonesian citizens the right to a healthy environment), Law No. 31/2009 on the environment (some clauses of which were amended by the Cipta Kerja Law No. 11/2020 on Job Creation), and the latest Presidential Decree No. 98/2021 on carbon pricing.

As part of working towards Indonesia's nationally determined contributions (NDCs) to tackle climate change issues, the Jakarta Provincial Government (Jakarta Province) issued Jakarta Province Governor Regulation No. 90/2021 on the Regional Low Carbon Development Plan (RPRKD). With this regulation, Jakarta Province aims to reduce greenhouse gas emissions by 30–50% by 2030 and achieve net zero emissions by 2050. This regulation also aims to expedite climate change mitigation and improve the community's adaptive capacity in dealing with climate risks.

Previous studies on the impact of climate change in the North Jakarta coastal areas have shown that women are more severely impacted, as they often face a double burden as both breadwinners and family caretakers. As outlined by Wahyuni (2022), women in fishery-dependent communities were predominantly responsible for domestic work as well as earning additional income for their families, as their husbands were often absent for long periods. Fishermen spent extended periods at sea, which imposed greater domestic hardships on women, a fact often overlooked in policy. Prior to their husbands' departure, the women performed important tasks, including helping with preparation and crafting nets. Upon the husbands' return with the catch, the women engaged in cleaning, peeling fish and shellfish, and selling fish in the market. Thus, women's adaptation strategies should consider the interplay between economic, physical, psychological, and social dimensions (Yoga Putra et al., 2019) and the spatial context that could influence women's successful adaptation and livelihoods as a result of climate mitigation and strategies imposed by the government, including relocations (Mar'iyah, 2019).

Women's agency in climate change adaptations in the Jakarta area was investigated by Dewi et al. (2022) in their study of five women's NGOs working on sustainability and climate change. The authors found that women were important actors in climate adaptation, including initiating climate change programs to address issues such as food security, clean and inclusive energy, forest resource management, community resilience, and agrarian justice. Additionally, looking at the role of social capital, socio-cultural diversity, and local vulnerabilities, Wannewitz and Garschagen (2024) found differing local adaptation capacities in the coastal areas of Jakarta that were dependent on three key social identities: families and close neighbourhoods, traditional institutionalised local groups, and cross-neighbourhood Civil

Society Organization (CSO)-supported networks. According to Wannewitz and Garschagen (2024), these identities played a key role in long-term collective adaptation and influenced how communities could obtain knowledge and resources to adapt to climate risks.

Given that climate change has greater impacts on women in the North Jakarta coastal areas, it is important to understand how these women are adapting and what successful strategies they are using to mitigate climate risks. This chapter looks at women's adaptation strategies and agencies (including capabilities, empowerment, and citizenship) to tackle climate change risks.

Methodology

The study used qualitative research and Kabeer's (2021) women's agency perspective to understand how communities and women adapt to climate change impacts. A coastal area in North Jakarta was chosen as the research location, as it is an urban area that is vulnerable to climate change phenomena, such as flooding and sea level rise (Wannewitz & Garschagen, 2024; Prana et al., 2024). The research discussed in this chapter was part of the larger 'Forced Labour and Climate Change: Keeping a Focus on Women and Children' study led by researchers from Griffith University and the Indonesian National Research and Innovation Agency (Badan Riset dan Inovasi Nasional, BRIN) and funded by the KONEKSI Research Project.

Using snowball sampling to collect data, we recruited 18 respondents for the study; the sample comprised fishermen, government officials, community leaders, and women leaders from three sub-districts in North Jakarta (Muara Angke, Cilincing, and Kalibaru) and one sub-district in Bekasi (Muara Gembong). The Muara Gembong sub-district was included due to its proximity to the North Jakarta area, and it was considered as one location along the coastal area of North Jakarta. Details of the study participants are included in the Results section. Data collection involved in-depth interviews and focus group discussions across the study sites in January, July, and September 2024. The interview and focus group data were transcribed, coded, and analysed using Creswell's (2014) process, turning data into categories and themes. This chapter provides insights into the effects of climate change, general adaptation strategies, and resilience of women's agency in addressing these impacts in the coastal areas of North Jakarta.

Results

Table 6.1 outlines the demographic information of the participants. There were 12 male and seven female participants. Most of the participants were fishermen ($n = 6$) or civil servants working in sub-district government offices ($n = 7$), with the remainder being community leaders/activists ($n = 3$), cooperative organisation staff ($n = 2$),

Table 6.1 Demographic information of participants (Source: Modified from data participants)

Characteristics of Informants	Respondents (n = 19)
Gender	
Female	7
Male	12
Occupation	
Fisherman	6
Civil Servant	7
Community Activist/Leader	3
Cooperative Association	2
Housewife	1
Residence	
Muara Angke	8
Kalibaru	3
Cilincing	3
Marunda	1
Muara Gembong	2
Jakarta Province	2

and one housewife. Additionally, most participants had several side jobs, including clam peelers, small traders, and green mussel farmers. Most participants were residents of the four sub-districts in North Jakarta, with six from Muara Angke, three from Kalibaru, three from Cilincing, and two from Muara Gembong.

Climate Change Impact

The study participants reported similar impacts of climate change to those described in previous studies on the impact of climate change in the coastal areas of North Jakarta, noting that they have experienced both direct impacts of climate change—including tidal flooding and coastal erosion—and indirect impacts.

Tidal flooding in this area has been regular, as reported by one informant, and it disrupts the daily activities of coastal communities, damages infrastructure, and causes huge economic losses:

> [I]f the water rises, things will be damaged. I'm afraid our children's mobile phones … they get wet. Then, the impact is because rubbish is scattered around, it's flooded, the rubbish has been swept away, and the flood comes again the next day (AP1).

Coastal erosion was apparent in the Muara Gembong area, where we could see an electricity pole in the middle of the sea, a few hundred metres from the land, and where damaged houses were still standing (LA2). This was the impact of sea level rise and coastal erosion, reducing the area of residential land and threatening neighbouring coastal settlements.

As a result of climate change, informants also reported that fish catches have decreased and that they have to sail further to find fish. One informant stated, 'The impact of climate change is indeed devastating. Fishing catches are reduced because of big waves (NL5). Another stated, 'We have travelled longer distances. Since we left our anchorage, I said earlier, with the change of fish, with the bad weather, we have travelled further' (NL3). One fisherman said that it was difficult to get fish these days compared to previous years (NL1).

As reported, fishermen experienced difficulties due not only to large waves and monsoons but also to industrial and household waste impacting the ocean ecosystem. 'This condition is influenced by unfriendly sea conditions and pollution from industry' (NL5).

These climate change impacts—low fish catch, time spent at sea, and tidal flooding—had indirect impacts on local communities residing in the coastal areas of North Jakarta. Among the notable impacts reported by the informants were worsened economic conditions and lack of water, health and nutrition, and education for children.

Regarding the impact of climate change on communities' economic positions, an informant reported that they were forced to find second jobs to deal with financial problems: 'That's why many fishermen are looking for odd jobs. Doing odd jobs. To support income during climate change that does not get income' (NL 5). Since the communities in this area live very close to the shore, an informant reported a lack of clean water: '[the] water, [looks] like milk, then when we tasted it, it was salty, like salt' (AP1).

Poor economic conditions and lack of potable water, plus poor sanitation with rubbish scattered after flooding, also reportedly increased the risk of residents contracting water-related diseases like diarrhoea and itching. 'The impact is due to rubbish scattered by the flood, even though it has been cleaned up' (AP1).

Deteriorating economic conditions in this area also forced coastal communities to prioritise fulfilling basic needs over children's education. This has resulted in many children dropping out of school or experiencing barriers to accessing education. Children experienced delays in enrolling in public elementary schools because of the zoning system; parents were waiting until the children were aged eight or nine years to enrol in a public elementary school because at that age, they would be prioritised by the school district. 'Here, to get into a public school, one must be older. If he/she is younger, he/she cannot get into a public school. While private schools are paid, public schools are free. ... Some are up to 11 years old when they enter elementary school. If it's a private school, it is not free' (AP2). 'Because here most of them [children] are older [aged 8–9]. So [they] choose the free one [public school], it's okay to be late as long as it's free' (AP2). This could be problematic since it neglects children's rights to education.

Climate change has also impacted children's health and nutrition, with a high level of stunting reported in Muara Gembong. 'In August 2023, data was collected from the 6 villages, from a total of 687 babies checked. Of 687, it was found that [there were] 288 in the stunting category' (LA2). This was due to a number of problems reported in this area, including a lack of clean sanitation, since residents did not have toilets in their houses and used the nearby river to defaecate.

Another impact of climate change reported in Muara Gembong was soil salinisation, which means that farmers could not grow rice in their fields. As reported by one informant: '[It] has an impact on the pH of water that is high in acidity. As a result, many small fish in ponds die... In some locations, [there is] water shortage, and also water with high salt, so rice seeds die' (LA2). This exacerbated the decrease in household income from fishing, further impacting children's health and nutrition.

Another notable impact of climate change, as reported by the sub-district head, was a conflict between residents and newcomers from outside Jakarta who had relocated to work as fishermen. The reason they moved to the Jakarta area was that the local waters have more fish due to numerous islands (the area is known as the 'thousand islands'), and catches sell at higher prices in Jakarta than elsewhere. The informant noted this conflict and how they dealt with this issue: 'There was a small conflict, fishermen here prohibit fishermen from there [outside of Jakarta], [they] did not accept the ban, then the case was brought to the Provincial office, from the Marine service to reconcile this problem' (LA2). Due to this gradual urbanisation, residents in the coastal areas of North Jakarta and Bekasi were mostly fishermen from other districts, mainly from Indramayu and Cirebon.

Community Adaptation Strategies to Mitigate Climate Change Impact

Coastal communities have adapted to the current climate impacts. This can be seen in how they cope with tidal flooding that causes eroded alleys. The community used its resources to repair the alleys in their neighbourhood: 'The alley was repaired, sir, with the help from the community' (AP1).

Another effort to cope with the direct climate change impacts in this coastal area is the mangrove conservation initiative led by the Muara Angke Mangrove Community (KOMMA). One informant, a leader of the KOMMA, stated that they have planted mangroves gradually in this area since 2008:

> This is now approximately 2 hectares, which means that the planting in 2008 was gradual. Gradually, we planted 200 first, not all alive, about 70 per cent were alive … After that, we planted another 500, and Alhamdulillah, there were 50 per cent [of mangroves] alive. That were two plantings, that were from our own money, our own team, KOMMA. (AM1)

In addition to the above community initiatives, the Indonesian government has created a national project in this coastal area: the National Capital Integrated Coastal Development (NCICD) project. As part of cooperative organisation activity in Kalibaru, the cooperative leader noted that they provided input to the government on this project.

> [The] NCICD Development project … initially, we were not considered [by local government], … So at the time, I sent a letter, so we were allowed space here … because with the

> development of NCICD here, we can't do anchorage, it's still an open field. So we were given a space in the crematorium. Yes, that became our temporary anchoring. And there is access to the entrance. (NL3)

In addition to mitigation strategies to tackle direct climate change impacts, as described above, local communities must also adapt to indirect impacts. Most of the residents engaged in several types of income-generating activities, which can be sorted into three main categories. The first is informal jobs and activities, notably clam peeling, small staple food vending, or daily work at a factory. The second type of work is related to communities' involvement in cooperative organisations and small and medium enterprises, and the third type of adaptation strategy was obtaining government and non-government social assistance.

In the first category, adaptation strategies include informal jobs and other activities to increase communities' economic resources. Some people work in the informal sector, such as becoming green mussel peelers or green mussel farmers, as noted by one of the informants: 'Here, the mothers help with the [family] economy, so they help with more income, and the children also help their mothers if they have time, sometimes the father goes to the sea, the mother helps to peel the shells there' (LA1). Another informant stated that she and her husband farm green mussels to earn more income but will also do other odd jobs: '[W]e have green mussels farming too. However, we can't harvest it every day, ma'am. So, it's like 4 months before we harvest, but while waiting for green mussel farming to harvest, my husband does odd jobs' (AP1).

Other forms of informal work include running a small food stall and collecting plastic for recycling: 'Fishermen who were used to fishing, their [fishing] locations are now becoming local tourist spots, so they began to become small traders, so this helps with their livelihood' (LA3). One informant also reported that people in the Kalibaru area undertake informal work in local factories and other industries. 'Many work in Marunda, in Cakung, in garment factories, soap factories, and a [cooking] oil industry. But for the women, most of them go to garments, sewing in Marunda and Cakung' (AP3). Another informant stated that fishermen were willing to do everything to feed their families:'[H]e will do odd jobs, anything, if there is a demolition of land, he will do anything, … [including being] a construction worker' (NL5).

Secondly, to adapt to climate change impacts, people in the community create a cooperative organisation to improve the community's resilience and general well-being. As one of the informants in Kalibaru recounted, he formed a cooperative organisation and later formed a small enterprise called Cangkring. This initiative is a way to help fishermen's communities and their families. The Cangkring and cooperative organisations help their members by creating spaces for them to discuss community problems, earn more income through Cangkring, and help with access to government assistance. As one informant, the founder of the cooperative organisation stated:

> If [we] want [this Cooperative Organisation] to run, there is [an assumption] that it has a compulsory [contributions or savings], well that is precisely the mindset that [I want to change], we want to move forward so a cooperative can be formed at that time. But changing the mindsets of the members here is difficult … At that time, I had already formed a

cooperative organisation assisted by the sub-district and the Ministry, but to change the mindsets of the cooperative members, rules … I changed the cooperative organisation rules. So it doesn't have compulsory savings in this cooperative organisation, I eliminated it, but I changed it to groups' contributions (dues) for [showing] their cooperative memberships.

Finally, some people in this community overcame economic difficulties by utilising social assistance programmes provided by the government or other non-government institutions, as stated by one of the informants: 'In the past, the assistance could be twice a year, such as nets, crab nets, and there are shrimp nets. Now there is nothing' (NL5). This informant from the Marunda area noted that while they had received significant government assistance in the past, it was now lacking, possibly due to the cooperative organisation being inactive in that area.

In the Muara Angke area, fishermen relied on government assistance to improve their current housing conditions. It was reported by one informant that some residents received floating house assistance: 'There is a house program, houses on stilts and floating houses. Twenty units in total, I think it's a miracle, unexpectedly, we didn't have a proposal' (AP1). Others relied on government assistance for gasoline subsidies through the Nelayan Association (NL2). 'In Kalibaru, a female informant reported that they received assistance for needy families that includes the Family Hope Programme *(Program Keluarga Harapan)* and the Indonesia Smart Card Program *(Kartu Indonesia Pintar)*—a government scholarship programme for children from poor families' (AP3).

Climate Change Adaptation Among Women

As described above, in the discussion of how local communities adapt to climate change impacts, it is apparent that women are engaged in adaptation processes and activities. Women have utilised the available resources to navigate their life challenges, including through their active involvement in neighbourhood associations, doing odd jobs to generate income, and working in cooperative organisations that offer economic empowerment programmes. Additionally, as seen from these women's narratives below, they have shown their adaptive capacities to mitigate climate change impacts, both direct and indirect, by improving their capabilities, empowerment, and citizenship (collective action).

Capabilities

Women have shown their ability to adapt to the current climate change context by struggling to achieve their goal of helping their communities, regardless of some of the limitations they face. As one informant stated, 'I am an ordinary citizen, my schooling is also not high, my education is also low, but I want to fight' (AP1).

Women have been active in the community, seeking to mitigate community issues such as tidal floods and a series of resettlements as a result of government policies on poor urban areas. One informant noted that their capacity to deal with problems was evolving.

> As time goes by, there are many tragedies that people do not want, one of which is eviction. .... so that's why I want to learn a little, learn so that our neighbourhood, which at that time the neighbourhood was still ponds and we used a bridge, sir, using a bamboo bridge. Now as we are now working as green mussel peelers, we use these green mussel shells to [fill] the land, reclamation by the residents. (AP1).

One way to deal with climate change's impact on livelihood was through economic diversification, facilitated by small and medium enterprises' activities in which women were actively involved. Active involvement in these organisations opened up possibilities for women to gain skills and knowledge, achieve economic sustainability, and create opportunities to help their communities in general.

Women's capacity, skills, and knowledge were also improved through training, as narrated by a member of the Cangkring organisation mentioned earlier. Its members, represented by two females, reported that they gained knowledge and skills on how to reuse green mussel and clam shells for art and craft products, such as vases, paving blocks, cup holders, and toilet seats. Women also reported that they learnt how to access potential markets to sell their products and generate income. The head of the cooperative organisation recounted how women gain this knowledge and these skills through Cangkring:

> Cangkring grew one year here because I saw that it would be an added value, not only to the fishermen but also to the fisherwomen. So we raised the dry shell waste into a product. I started backing these mothers in the previous two years. So the mothers are from the fishermen's families, so [they] understand. For what purpose, to give an understanding that we are forming something or making this product so that the delivery will be widespread. Now with them, the benefit value is high. They meet in the social community, and can add [this activity] to their works, for example [by being active] in Cangkring. From waste to something, then create added value [income for the family]. That's why in the previous two years, we formed this [Cangkring] forum under the auspices of a cooperative organisation. To add [financial] value to the fisherwomen. Afraid that their husbands go to the sea [and come home] empty [no catch], we can use this for added value. That was the beginning, so we started to involve these women. (NL 3)

In this area, Cangkring activities are often supported by the private sector through their corporate and social responsibility (CSR) programmes, as narrated by the cooperative leader:

> Alhamdulillah, we have been able to do this for two years, initially supported by CSR Bahtera Adiguna, CSR Shipping. Initially, it was two years, but it is approaching the next three years. Now we also just entered again from [another] CSR SQ DKI [Jakarta]. The assistance is not only limited but intense, they are directly involved in the field. (NL 3)

Another example of improving women's capabilities through an organisation can be found in Muara Angke. Women there were involved in the activities of the previously mentioned KOMMA group and gained knowledge and skills, including processing mangrove fruits for syrup and jam and processing batik from mangrove leaves. One informant, the leader of this mangrove community, stated: 'For the empowerment itself from the community, we are indeed empowering more mothers … [and it has] begun around 2015. This one [we had now] was just only two or three years ago' (AM1).

He added that women learnt how to work in small business enterprises, and they gained access to potential markets: 'If batik [to produce batik cloth from mangrove leaves], is mostly [done by women] from RW 22 … because of the process, the time to produce the material, meaning the time we make it first market it first, then we get it [income]' (AM 1). The informant also added that it is important for women to learn and understand that the process will take time, from manufacturing to selling the product and finding the market; thus, not all women from the area can be involved in this activity (producing batik).

Empowerment

The empowerment of women living in coastal areas can be seen in their active involvement in the community. Women gained awareness of how to get power back from those traditionally more dominant groups (e.g. male community leaders, government officials). A female informant, who was known as a community leader, stated:

> [B]ecause I'm not from, I am not someone [with power]. I feel it because I've been around for a long time. So [people said] it's useless for you to form an RT [neighbourhood association], how many times [they] have [tried]. In the past, when the head of the neighbourhood used to ask for data again, RT was never formed. Now, since we have friends who supported [us], even though you don't understand, just participate, the important thing is that you represent the community. (AP1)

Women's involvement in the community resolved some of the community problems in Muara Angke. A female informant noted how she and her community worked together to deal with a new reclamation project located in a nearby area. This needed to be resolved, as it threatened the community's livelihoods. She stated, 'Imagine, sir, if the estuary [muara] is dredged, right? Yes, we fishermen, where will our sources of livelihood go?... Alhamdulillah, because we struggled with friends from NGOs and students... Alhamdulillah ma'am, we won' (AP1).

Another example of women's empowerment in this community is their involvement in a larger context. In Muara Angke, women were involved in 'a political contract' with the local government official. An informant noted:

> [T]o fight for the land when I was still the Alley Leader *(Ketua Gang/Kampung)*, I was not alone anymore, sir, I was guided. I also don't understand sir, the important thing is that I was there representing my neighbourhood, including representing the Empang Block RW No. 22, invited by the Urban Poor People's Network [*Jaringan Rakyat Miskin Kota*, or JRMK) … the important thing is that we struggled to fight for the place where we live too. Then there was someone who accompanied me, I couldn't understand anything, but I took part in representing [the community]. (AP1)

She continued: 'I have a building, have a place to live, have a family, have parents here sir. So, Alhamdulillah here, we were participating in a political contract [with the Governor of Jakarta]'. Through this political contract with the governor, as narrated by the informant, they were eventually officially recognised by the state as official Jakarta citizens and obtained Jakarta ID cards.

Another form of empowerment can be seen in one Cangkring member's description of their progress in this organisation. She noted how her involvement in Cangkring empowered her by getting her out of the house and doing something meaningful for the community.

> Because here we have a family system because this Cangkring/cooperative is newly established, we really want everyone to progress. So in any way, we have to try so that this Cangkring/cooperative can develop, for the future there can be other Cangkring to parse the garbage here as well as waste. If for the family, the important thing is that I come out and the children are ready for school, my job is to take them to school in the morning, if in the afternoon I pick up the children from school, when it's time to cook then I will cook, later I have free time, I will come here. (AP2)

Collective Action

Women's involvement in the community in tackling climate change at the structural level can be seen in the context of community struggles in Muara Angke and Kalibaru. A female informant narrated her struggles for the community's settlement and recognition from the Jakarta government, which involved establishing a new neighbourhood association (Rukun Tetangga, or RT) in the area. To form a new RT, one needs a Jakarta Governor's Decree that stipulates the regulation of the new RT structure. It was reported by the informant that her fight to get recognition was finally fruitful:

> Alhamdulillah, as 2019 progressed, I received a decree from Mr Anies Baswedan during his term of office. That was one of the proofs of our request for recognition and ratification of the formation of RT RW in Pluit Reservoir. (AP1)

The Muara Angke community not only engaged in collective action to form an RT but also participated in other collective actions to address community problems. This included repairing damaged roads and alleys, providing access to clean water, and advocating for floating housing in this area. An informant noted that to obtain clean water, they worked with the Government Water Company (PAM Jaya):

> To address the current clean water deficit, there is a collaboration with PAM JAYA to develop reservoir pools. Water kiosks are available in several locations, either through cooperatives or PAM itself. Later, the community will donate, or someone will assist with the operation. (LA5)

As part of women's citizenship or collective action, they assisted those in need by helping each other and caring for those who were sick or who had lost loved ones. As a result of such efforts, the community now has an ambulance to provide immediate assistance if anyone needs hospitalisation. 'We have provided it here, and the result of our jimpitan [self-help] is also to buy an ambulance. If a resident is sick, if we need a car, we have our own ambulance from the residents of RW 01' (AP2).

Discussion and Conclusion

This chapter provides insights into climate change adaptation in the coastal areas of North Jakarta and, in particular, how women adapt to climate change impacts. Utilising Kabeer's (2021) women's agency concept, the chapter highlights three elements of agency—capabilities, empowerment, and citizenship. Our interviews with local community leaders in four sub-districts of coastal North Jakarta suggest that local communities have shown their adaptive capacity to cope with climate change impacts.

This study finds that community adaptation strategies practised in this community show people's resilience, as defined in the Paris Agreement document (Article 2 and IPCC, 2014). The climate adaptation strategies identified include working with communities to repair eroded alleys, initiating mangrove conservation activities, advocating for clean water and floating houses, forming neighbourhood associations and cooperatives, and initiating small and medium enterprises. This also aligns with resilience in the climate context, defined as 'the capacity of social, economic, and environmental systems to cope with a hazardous event or trend or disturbance, responding or reorganising in ways that maintain their essential function, identity, and structure, while also maintaining the capacity for adaptation, learning, and transformation' (Ackerl et al., 2023).

Adaptation processes, as shown in our interviews with local communities, are not easy and take considerable effort and struggle to be successful. This study shows how women, as local actors, have worked with male actors to adapt to climate risks. Women have been active throughout the adaptation processes, from being involved in economic activities to being local leaders and initiating structural changes in the community by establishing neighbourhood associations. The results reveal how women have adopted different types of informal work as alternative sources of income, showing adaptive capacity by doing odd jobs to generate income and working in cooperative organisations for economic empowerment programmes.

In terms of women's agency, this study finds that women have shown agency by acquiring knowledge and skills and gaining the awareness needed to act through their involvement in communities and cooperative organisations. Making the choice to do something for their communities, to get out of their homes, and to learn something new are examples of women's agency, leading to their empowerment and active participation in the community—in Kabeer's (2021) terms, their personhood in society.

This study has found women's empowerment model emerges through their engagement in community life and their choice to make decisions regarding their lives (e.g. to get out of their houses, to help communities, to deal with community problems). This description fits with empowerment in the local context, as described in Widjajanti's (2011) study on empowerment in Indonesia—as 'a process and effort to acquire or provide power, strength or ability to vulnerable individuals and communities to identify, analyse, determine the needs and potentials as well as choose alternative solutions by optimising the resources and potentials possessed independently' (p. 16).

The results also show that climate adaptation strategies practised by local communities are influenced by communities' social identities, as found in previous studies on key social identities that affect different local adaptation strategies (Wannewitz & Garschagen, 2024). As described in this chapter, local communities' ability to form neighbourhood associations, undertake mangrove conservation, and build women's empowerment through Cangkring and KOMMA is not only a result of local community initiatives but is also influenced by cross-neighbourhood networks and NGO support. This chapter confirms Wannewitz and Garschagen's (2024) finding that family, close neighbourhoods, institutionalised local groups such as cooperative organisations, and cross-neighbourhood organisations such as CSOs/NGOs play important roles in communities' adaptive capacity and resilience.

Nevertheless, despite the progress that has been made, as mentioned above in the context of climate change adaptation, women still face several challenges in exercising agency, especially when dealing with gender structures that remain unequal, as narrated in the interviews. This includes the persistent view that women have the primary responsibility for child-rearing and domestic work, even while they are also active in communities and cooperative organisations. Similarly, women's agency is often highly dependent on external factors, as described in the interview data, including male community leaders who provide space for women to participate or external parties such as CSOs that support women's movements. Therefore, in order to truly achieve agency, especially in the sense of having the empowerment to bring about change in society, it is necessary to continually re-examine power relations both within the family and in communities where women are active.

Recommendation

The North Jakarta coastal community adaptation strategy described above aligns with the current development policies. As a result of the complex issues confronting the North Jakarta Coast, which is the focus of cross-ministerial and cross-regional government policies, the Ministry of Public Works and Public Housing (PUPR) has focused its strategy's development in these coastal areas, including an integrated coastal embankment as a means of preventing floods (Ministry of Public Works and Public Housing, 2023). In the interim, the Jakarta Province Provincial Government prioritises the vision of low-carbon development by enacting Jakarta Province Governor Regulation No. 90/202, which aims to achieve net zero emissions by 2050 through the Regional Low Carbon Development Plan. Our research finds that such policies and regulations could constrain the ability of local communities to adapt to climate risks. Thus, as described above, collective action, including leadership by female figures in the community, is a necessary response to the absence of public engagement by government officials when they design and implement national strategic plans such as the NCICD.

Additionally, regarding climate adaptation and communities' resilience, the Jakarta Province government and relevant ministries—such as the Ministry of Social Affairs, Ministry of Trade, and Ministry of Health—should prioritise these vulnerable communities and provide them with access to economic empowerment, potential markets to sell local products, access to education for school-aged children, and access to health checks and services for children and mothers.

In addition to the above policy recommendation, this study provides insight to civil society and non-government organisations working with vulnerable communities, including women in coastal areas, to understand how women's agency and empowerment models work in this community, despite women still facing unequal gender structures in society. This study has shown women's adaptive capacity and resilience in the face of climate risks.

Future studies could benefit from interviewing more women and specifically addressing how climate adaptation affects women's and their families' health and well-being. Additionally, future research could utilise both quantitative and qualitative approaches (mixed methods) to address power relations and climate change adaptation among women.

References

Ackerl, T., Weldemariam, L. F., Nyasimi, M., & Ayanlade, A. (2023). Climate change risk, resilience, and adaptation among rural farmers in East Africa: A literature review. *Regional Sustainability, 4*(2), 185–193. https://doi.org/10.1016/j.regsus.2023.05.004

Ahmed, S., & Eklund, E. (2021). Climate change impacts in coastal Bangladesh: Migration, gender and environmental injustice. *Asian Affairs, 52*(1), 155–174. https://doi.org/10.1080/03068374.2021.1880213

Alston, M. (2014). Gender mainstreaming and climate change. *Women's Studies International Forum, 47*(Part B), 287–294. https://doi.org/10.1016/j.wsif.2013.01.016

Carr, R., Kotz, M., Pichler, P. P., Weisz, H., Belmin, C., & Wenz, L. (2024). Climate change to exacerbate the burden of water collection on women's welfare globally. *Nature Climate Change, 14*, 700–706. https://doi.org/10.1038/s41558-024-02037-8

Chang, W., Diaz-Martin, L., Gopalan, A., Guarnieri, E., Jayachandran, S., & Walsh, C. (2020). What works to enhance women's agency: Cross-cutting lessons from experimental and quasi-experimental studies. *J-PAL Working Paper, 87*. Retrieved from https://www.povertyactionlab.org/page/what-works-enhance-womens-agency

Chindarkar, N. (2012). Gender and climate change-induced migration: Proposing a framework for analysis. *Environmental Research Letters, 7*(2), 025601. https://doi.org/10.1088/1748-9326/7/2/025601

Chinowsky, P., Hayles, C., Schweikert, A., Strzepek, N., Strzepek, K., & Schlosser, C. A. (2011). Climate change: Comparative impact on developing and developed countries. *The Engineering Project Organization Journal, 1*(1), 67–80. https://doi.org/10.1080/21573727.2010.549608

Creswell, J. W. (2014). *Research design: Qualitative, quantitative, and mixed methods approaches*. SAGE Publications.

Dewi, K. H., Sari, D. R., & Latifa, A. (2022, December). Sustainability of climate change issues in Indonesia: the voices of women's NGOs. In *IOP Conference Series: Earth and Environmental Science* (Vol. 1105, No. 1, p. 012005). IOP Publishing. https://doi.org/10.1088/1755-1315/1105/1/012005

Djoudi, H., Locatelli, B., Vaast, C., Asher, K., Brockhaus, M., & Basnett Sijapati, B. (2016). Beyond dichotomies: Gender and intersecting inequalities in climate change studies. *Ambio, 45*(Suppl 3), 248–262. https://doi.org/10.1007/s13280-016-0825-2

Donald, A., Koolwal, G., Annan, J., Falb, K., & Goldstein, M. (2017). Measuring women's agency. In *Measuring women's agency* (WPS 8148; Policy Research Working Paper). https://doi.org/10.1596/1813-9450-8148

Eastin, J. (2018). Climate change and gender equality in developing states. *World Development, 107*, 289–305. https://doi.org/10.1016/j.worlddev.2018.02.021

Ebi, K. L., Vanos, J., Baldwin, J. W., Bell, J. E., Hondula, D. M., Errett, N. A., et al. (2021). Extreme weather and climate change: population health and health system implications. *Annual Review of Public Health, 42*(1), 293–315. https://doi.org/10.1146/annurev-publhealth-012420-105026

IPCC. (2022). In H.-O. Pörtner, D. C. Roberts, M. Tignor, E. S. Poloczanska, K. Mintenbeck, A. Alegría, M. Craig, S. Langsdorf, S. Löschke, V. Möller, A. Okem, & B. Rama (Eds.), *Climate Change 2022: Impacts, Adaptation and Vulnerability. Contribution of Working Group II to the Sixth Assessment Report of the Intergovernmental Panel on Climate Change* (p. 3056. Retrieved from). Cambridge University Press. https://doi.org/10.1017/9781009325844

Kabeer, N. (1999). Resources, agency, achievements: Reflections on the measurement of women's empowerment. *Development and Change, 30*(3), 435–464. https://doi.org/10.1111/1467-7660.00125

Kabeer, N. (2021). Three faces of agency in feminist economics: Capabilities, empowerment and citizenship. In G. Berik & E. Kongar (Eds.), *The Routledge handbook of feminist economics* (pp. 99–108). Routledge. Retrieved from https://www.taylorfrancis.com/chapters/edit/10.4324/9780429020612-12/three-faces-agency-feminist-economics-naila-kabeer

Kikstra, J. S., Nicholls, Z. R., Smith, C. J., Lewis, J., Lamboll, R. D., Byers, E., et al. (2022). The IPCC Sixth Assessment Report WGIII climate assessment of mitigation pathways: from emissions to global temperatures. *Geoscientific Model Development, 15*(24), 9075–9109. https://doi.org/10.5194/gmd-15-9075-2022

Letta, M., Montalbano, P., & Paolantonio, A. (2022). Understanding the climate change-migration nexus through the lens of household surveys. *Journal of Economic Surveys, 38*, 1234–1275. https://doi.org/10.1111/joes.12584

MacGregor, S. (2009). A stranger silence still: The need for feminist social research on climate change. *The Sociological Review, 57*(2-suppl), 124–140. https://doi.org/10.1111/j.1467-954X.2010.01889.x

Mar'iyah, C. (2019). Redevelopment of Slum Area in Jakarta: Case study the impact of gender inequality on the sustainable livelihood assets post resettlement from Kampung to modern apartment style. In A. Munhanif, A. B. Ihsan, I. Halimatusa'diyah, B. Ruswandi, & K. Nisa (Eds.), *Proceedings of the 5th International Conference on Social and Political Sciences, ICSPS 2019, 12th November 2019, Jakarta, Indonesia*. EUDL. https://doi.org/10.4108/eai.12-11-2019.2293533

Ministry of Public Works and Public Housing. (2023). *Implementation and Concept of Integrated Coastal Flood Control in Jakarta Bay (Implementasi dan Konsep Pengendalian Banjir Terpadu Pesisir Teluk Jakarta)*.

Nhemachena, C., Nhamo, L., Matchaya, G., Nhemachena, C. R., Muchara, B., Karuaihe, S. T., & Mpandeli, S. (2020). Climate change impacts on water and agriculture sectors in Southern Africa: Threats and opportunities for sustainable development. *Water, 12*(10), 2673. https://doi.org/10.3390/w12102673

Palinkas, L. A., & Wong, M. (2020). Global climate change and mental health. *Current Opinion in Psychology, 32*, 12–16. https://doi.org/10.1016/j.copsyc.2019.06.023

Prana, A. M., Dionisio, R., Curl, A., Hart, D., Gomez, C., Apriyanto, H., & Prasetya, H. (2024). Informal adaptation to flooding in North Jakarta, Indonesia. *Progress in Planning, 186*, 100851. https://doi.org/10.1016/j.progress.2024.100851

Republic of Indonesia. (2022). Enhanced Nationally Determined Contribution Republic of Indonesia 2022. https://unfccc.int/documents/615082

Saputro, T., & Kurian, P. (2023). Gender and climate change vulnerability: A case study of a coastal community in Pramuka Island, the Seribu Islands. In A. Triyanti, M. Indrawan, L. Nurhidayah, & M. A. Marfai (Eds.), *Environmental governance in Indonesia* (Vol. 61, pp. 69–88). Springer. https://doi.org/10.1007/978-3-031-15904-6_5

Smith, E. S. (2020, June). Climate change in women, peace and security national action plans. *SIPRI Insights on Peace and Security*, 1–31. https://www.jstor.org/stable/resrep25311

Stringer, L. C., Mirzabaev, A., Benjaminsen, T. A., Harris, R. M., Jafari, M., Lissner, T. K., et al. (2021). Climate change impacts on water security in global DRYLANDS. *One Earth, 4*(6), 851–864. https://doi.org/10.1016/j.oneear.2021.05.010

Tickamyer, A. R., Kusujiarti, S., & Wornell, E. J. (2014). Gender justice, climate change, and sustainable development in Indonesia. *Globalization, Development and Security in Asia*. May. 67–91. https://doi.org/10.1142/9789814566582_0041

UNFCCC. (2015). *Paris Agreement*. Retrieved from https://unfccc.int/sites/default/files/english_paris_agreement.pdf

UNFCCC. (2023). *Sharm el-Sheikh work programme on the global goal on adaptation referred to in decision*. Retrieved from https://unfccc.int/sites/default/files/resource/cma5_auv_8a_gga.pdf

Van Daalen, K. R., Kallesøe, S. S., Davey, F., Dada, S., Jung, L., Singh, L., et al. (2022). Extreme events and gender-based violence: A mixed-methods systematic review. *The Lancet Planetary Health, 6*(6), e504–e523. Retrieved from https://www.thelancet.com/journals/lanplh/article/PIIS2542-5196(22)00088-2/fulltext

Wahyuni, R. (2022). The inevitability of double burden fisherwomen on the North Coast of Jakarta. In J. Nurkamto, K. Osiri, A. Q. Muslim, M. Alie, & I. Feofilova (Eds.), *Proceedings of the First Lekantara Annual Conference on Public Administration, Literature, Social Sciences, Humanities, and Education, LePALISSHE 2021, August 3, 2021, Malang, Indonesia*. EUDL. https://doi.org/10.4108/eai.3-8-2021.2315072

Wanjiru, L. (2012). Gender, climate change and sustainable development. *Fordham Environmental Law Review, 23*(2), 1–6. https://www.jstor.org/stable/44175841

Wannewitz, M., & Garschagen, M. (2024). The role of social identities for collective adaptation capacities–general considerations and lessons from Jakarta, Indonesia. *International Journal of Disaster Risk Reduction, 100*, 104194. https://doi.org/10.1016/j.ijdrr.2023.104194

Widjajanti, K. (2011) Model Pemberdayaan Masyarakat. *Jurnal Ekonomi Pembangunan Universitas Muhammadiyah Surakarta, 12*(1), 15–27. Retrieved from https://journals.ums.ac.id/JEP/article/view/202/189

Yoga Putra, G. A., Koestoer, R. H., & Lestari, I. (2019). Psycho-social performance towards understanding local adaptation of coastal flood in Cilincing Community, North Jakarta, Indonesia. *IOP Conference Series: Earth and Environmental Science, 243*(1). https://doi.org/10.1088/1755-1315/243/1/012005

Chapter 7
Evaluation of Infrastructure Development in Coastal Central Java: Reducing Climate Change, Environmental, and Socio-Economic Impacts

Agus Suntoro, Laely Nurhidayah, and Ade Angelia Yusniar Marbun

Introduction

Indonesia's Coordinating Ministry for Economic Affairs states that the north coast of Java contributes 20.7% to Indonesia's GDP and is home to more than 50 million Indonesians (NDPA, 2024). This region thus plays a strategic role in the national economy and people's livelihoods. As the economic and social backbone of the nation, the north coast of Java is highly vulnerable to the threat of sea level rise (SLR) and tidal waves that damage the environment and physical and economic facilities. This vulnerability is exacerbated by the fact that the region is densely populated (Rusmadi et al., 2017). These challenges require long-term solutions to maintain economic, social, and environmental stability in the region and nationally (Limanseto, 2024).

As a mitigation measure against the impact of climate change due to rising SLR, the government has planned the construction of a giant sea wall along the 600-km north coast of Java. The National Development Planning Agency (Bappenas) began the giant sea wall initiative in 1994, with the Deputy for Facilities and Infrastructure at Bappenas stating that the government has conducted eight studies on the giant sea wall since 2014. These studies cover technical aspects, ecology, environment, land subsidence, drinking water, groundwater, traffic, and so on. However, the estimated financing depends on the choice of technology, budget, and policy options (Agt, 2024).

The Ministry of Public Works estimates the project's cost to be IDR 600 trillion—or IDR 1 trillion per kilometre—with an additional IDR 200 trillion for sanitation, clean water supply, and other supporting facilities. Phase A of the 12.66 km

A. Suntoro (✉) · L. Nurhidayah · A. A. Y. Marbun
Research Center for Law, National Research and Innovation Agency (BRIN), Jakarta, Indonesia
e-mail: agus154@brin.go.id

L. Nurhidayah et al. (eds.), *Climate Change, Labour and Migration in Indonesia*, Springer Climate, https://doi.org/10.1007/978-3-032-15334-0_7

long western section of the sea wall was built in 2019 in the north of Jakarta, while Phase A of the 33.54 km long eastern section is still in the planning stage, from the initial target in 2020. (Bahfein, 2024). President-elect Prabowo Subianto also discussed the blue economy and the sea wall project with Chinese President Xi Jinping in early November 2024. It is estimated that if the sea wall project is delayed, climate change impacts could affect 70 industrial areas, 5 special economic zones, 28 industrial allotment areas, 5 new industrial growth areas, and millions of hectares of agricultural land (Sukamsari, 2024).

Infrastructure development is also carried out based on regional characteristics, such as sheet pile and flood canal options in Semarang City, tidal embankments and extended storage in Pekalongan, and tidal embankments with polder systems in the Demak Regency (Juana, 2024). These infrastructures address regional issues and will be integrated with the national sea wall project. Another infrastructure initiative by the Central Java Government and Semarang City is the construction of retention ponds in eastern Semarang, such as in the Terboyo area bordering the Demak Regency. Retention ponds store water securely using embankments (piles) as barriers (Rochim, 2024).

Various infrastructure projects to control tidal floods and SLR have also become targets of civil society groups and environmental activists, especially the construction of giant sea walls and toll roads. The Maleh Dadi Segoro Coalition in Central Java has criticised the immediate impact of an infrastructure development project that threatens 45 hectares of mangrove areas, leading to widespread flooding due to the closure of five rivers and creating changes in sea currents to the east, exacerbating erosion in the Demak coastal area. Toll roads and sea wall projects have also caused many residents in Sayung Sub-district to be evicted from villages such as Tambaksari, Morodadi, Timbulsloko, and others (WALHI, 2024).

Methodology

This chapter uses a qualitative approach, which includes a literature review, policy analysis, and empirical research on the study of tidal floods and SLR control infrastructure development, especially in the north coast area of Central Java. Field research was conducted in Central Java Province, which was critical for data collection, involving interviews with the central government (Ministry of Public Works), the Central Java River Basin Centre, provincial and municipal governments, local communities, non-governmental organisations (NGOs) and academics. Direct observations were made in Tambaklorok, Tanjung Emas Village, Semarang City.

Climate Resilience Infrastructure

The impacts of climate change, specifically in the form of SLR and tidal waves, have increased the burden on island nations, including Indonesia. Adaptation and mitigation, including climate-resilient infrastructure development, are key to resilience (Maskun et al., 2023). Adaptation refers to a range of measures to adjust to current climate conditions and anticipate future conditions by reducing adverse impacts and capitalising on potential. Mitigation comprises efforts to slow the rate of climate change, such as by reducing carbon emissions (Zen, 2023). Preventing global warming and climate change will require significant investment in low-carbon technologies and the redesign of existing or new infrastructure. The complexity of adaptation and mitigation at global and national scales makes it a breakthrough solution (Falkner, 2016).

According to Hirschman's concept of social overhead capital, the category of infrastructure is defined by several features: (a) the development or services provided from these activities become the basis of economic and social activities, (b) the provision of services or development is carried out by public institutions (the state) and partly by private institutions that are subject to public control, and (c) the services provided are free of charge or at a reasonable price (Srinivasu, 2013). Infrastructure can include dams, roads, sea walls, and electricity systems. Many developing countries have constraints on funding infrastructure development. Private sector involvement has become a rational option to bridge the gap in investment needs. Infrastructure investment will directly impact economic gains from taxation, sales, and property and strengthen regional connectivity and resilience (Yoshino et al., 2018).

The Organisation for Economic Co-operation and Development (OECD) states that climate resilience and infrastructure are closely linked. First, climate change impacts pose direct and indirect risks to infrastructure assets and the provision of services to society. Second, accelerating infrastructure development without controls can exacerbate climate-related risks, such as electricity transmission line failures and dam collapses. Third, climate change creates new needs for infrastructure, such as sea walls, flood defences, retention areas, and various other infrastructures (OECD, 2024).

Climate-resilient infrastructure is therefore designed, constructed, and operated in a way that anticipates, prepares for, and adapts to climate change. Despite the adverse weather impacts of climate change, such as SLR and tidal waves, if a climate resilience infrastructure is in place and functioning, it can reduce these impacts so that people and businesses are protected and operate normally (OECD, 2024). Climate-resilient infrastructure planning starts with setting goals or targets, using appropriate technologies, setting a budget, and operating an infrastructure project. Climate-resilient infrastructure generally involves more expensive techniques, technologies, and costs. However, the additional financing required to create more resilient infrastructure assets is only about 3% of the total investment. Moreover, these costs can be offset by reduced spending on maintenance and repair (Zen, 2023).

Fig. 7.1 Relationship between climate change and infrastructure development (*Source: Modified from Theoretical Framework by Author*)

However, the massive development of significant infrastructure, such as dams, ports, and roads, and the expansion of urban and industrial areas can, if not properly managed, exacerbate the crisis by increasing environmental degradation and leading to the forced eviction of people from their homes and habitats (HLRN 2007). Figure 7.1 outlines the relationship between climate change and infrastructure development.

Research Findings

Hard Infrastructure Development

Sheet Pile

On 17 June 2024, the then-President Joko Widodo visited Tambaklorok to check the construction of 3.6 kilometres of sheet piles[1] and the arrangement of the fishing village. The president emphasised that the sheet piles would be able to withstand tidal waves and tidal floods for more than 30 years, with this infrastructure project serving as a model for other regions to handle or control tidal floods and waves (Wicaksono, 2024). The Central Java Regional Secretariat's Assistant for Economics and Development, Peni Rahayu, stated that the sheet pile construction itself functions as the primary retainer of the sea wall.

The construction of sheet piles in the Tambaklorok area, Semarang City, a reservoir for tidal waves and tidal flooding, has been hoped for by residents for decades. Mr Subowo, one of the figures who initiated the project application, said that the proposal had been submitted three times to the Ministry of Public Works, with the process taking years. Political intervention involving members of Commission IV of the House of Representatives, as well as officials in the State Palace circle, became necessary to accelerate the approval process and construction of sheet piles. Finally, in 2021–2022, the Commission approved a budget of around IDR 300 billion proposed by the Ministry of Public Works for the construction of sheet piles and their supporting facilities (Suntoro, 2025a).

[1] A sheet pile is a vertical wall that holds the soil and prevents air from entering the excavation hole (Province, 2019).

Vivin Kurnia Setyawan, Secretary of Tanjungemas Village Head, Semarang City, stated that the sheet pile has covered the areas of Citizens' Association (RW) 12, 13, 14, 15, and 16, with the result that there has been no tidal flooding or high waves in the Tambaklorok area since 2024. Vivin confirmed that the proposal to build the sheet pile was an initiative by local residents that was submitted to the Mayor of Semarang and forwarded to the Ministry of Public Works. The entire budget for the sheet pile construction is entirely from the State Budget funds, including land acquisition and construction of other facilities, including two pumps on the west and east sides of the sheet pile (Suntoro, 2025a). Sheet pile infrastructure is important because it also protects the port area, which is the heart of the local economy. In May 2022, the embankment at PT Lamicitra Nusantara in the Tanjung Emas port area broke down, causing tidal flooding for three days. This affected operations in the port area, and many companies in the industrial area were also affected by damage to machinery and raw materials (Yusuf, 2022a). Figure 7.2 depicts a sheet pile.

Pump House

The Head of the Pemali Juana River Basin Centre (BBWS), Harya Muldianto, explained that the infrastructure project in the Tambaklorok area includes the construction of a sea wall and two retention ponds with a total area of around 20 hectares, as well as two pump houses to reduce the risk of flooding. The two pump houses also have a pump capacity of 3 × 500 litres per second each, accelerating the discharge of tidal floods into the sea (Setpres, 2024). Pump houses, although helpful in reducing flooding, incur high operational costs. Some pump houses cannot operate because they do not receive subsidies from the government. The pump house

Fig. 7.2 Sheet piles located in the Tambaklorok area, Tanjungemas Village, Semarang City (*Source: Agus Suntoro, BRIN Documentation*)

function will be disrupted if flooding and SLR occur at the same time, because the flood cannot be diverted to the sea areas (Sofaniadi et al., 2015).

Despite the government's view of the pump house as a success, its current condition is concerning. Presidential Instruction ('Inpres') Prabowo Subianto Number 1 of 2025 imposed 'efficiency' budget cuts of more than 306 trillion IDR, resulting in operational funds for pump house fuel ceasing, a number of pump house officers being laid off and arrears in employee wages. This has created considerable potential for unmanaged tidal flooding in Semarang (Nurhidayah, 2025).

Semarang–Demak Toll Road and Sea Wall

The Semarang–Demak Toll Road and Sea Wall Project (TTLSD) is one of the central government's designated national strategic projects (PSN). As a basis for implementation, an ecological study was prepared through an Environmental Impact Analysis (ANDAL) for the Business Plan and/or Integration of Semarang City Sea Wall Development with Semarang-Demak Toll Road Development, which was ratified in March 2018. Based on ANDAL, the Governor of Central Java issued Environmental Permit Number 660.1/32 of 2018 concerning the Activity Plan for Integrating Semarang City Sea Wall Development with Semarang-Demak Toll Road Development of Central Java Province.

In the opinion of environmental activists, however, this project will actually worsen flooding in coastal areas (due to the closure of five rivers and the destruction of 46 hectares of mangrove forest areas) and adversely affect the livelihoods of shrimp and crab farmers along the Semarang–Demak coastal area (WALHI, 2024). In response to the damage to the mangrove ecosystem area, the Central Java Provincial Environmental Service has announced 46 hectares of replacement land spread around the sea toll project, at Sidogemah (8 ha), Kalimati (7 ha), and Terboyo Kulon (20 ha), with locations for the remaining 11 hectares still being identified (Fuaziah, 2025). Figure 7.3 depicts the toll road and sea wall.

Sugito, Head of Trimulyo Village, Genuk District, Semarang City, stated that in his area, there is a mangrove forest that has been affected by the construction of the Semarang-Demak Toll Road and Sea Wall. However, the village government does not yet know the details of the affected area—whether it is to be used for a toll road or a water reservoir polder (Marbun, 2025). Based on the public information posted at the Trimulyo Village Office, there are only three people whose shrimp ponds are affected by land clearing for sea toll and sea wall projects—and they are not residents, according to their ID cards, but entrepreneurs from areas outside Semarang.

Development of Marine Village

Tambaklorok was inaugurated as a Marine Village (Kampung Bahari) by President Joko Widodo in 2015 and by the Semarang city government in 2016. Although it has experienced various obstacles in its implementation, the city government has

Fig. 7.3 Construction of the Semarang-Demak Toll Road and Sea Wall project located in Trimulyo Village, Genuk District, Semarang City (*Source: Agus Suntoro, BRIN Documentation*)

Fig. 7.4 The Kampung Bahari programme, in the form of raising roads in the RW 15 and 14 areas, supports access for Tambaklorok residents (*Source: Ade Angelia Marbun, BRIN Documentation*)

continued to strive to complete the programme. The Mayor of Semarang stated that the Kampung Bahari programme plan includes the construction of a pier, an entrance road, a market, a fish auction venue, and a green open space (Jateng, 2015). The legal basis or policy for the development of the Marine Village programme is stated in Regional Regulation Number 11 of 2017 concerning Amendments to Regional Regulation Number 6 of 2016 concerning the Semarang City Medium-Term Development Plan (RPJMD) for 2016–2021 and is based on Semarang City Regional Regulation Number 14 of 2011 concerning the Semarang City Regional Spatial Plan (RTRW) for 2011–2031, and explained in detail in the Tambaklorok 2016 Building and Environmental Plan (RTBL) master plan (Authori et al., 2019). Figure 7.4 depicts the road being raised.

The development of Kampung Bahari is an appropriate model considering the findings from research by Diponegoro University (UNDIP) that there is a slum settlement in Tambaklorok dominated by light (RW 12, 13, and 15) and medium (RW 14) slums. The parameters that most influence the level of slums are the drainage and garbage conditions. Residential areas are still experiencing inundation, and the garbage transportation system has not reached all households (Rais & Hariyanto, 2021). Referring to the RTBL, there are four stages to realise the Kampung Bahari Tambaklorok programme: (1) improving environmental quality; (2) utilising reclamation potential for economic activities; (3) community empowerment programmes; and (4) training and community empowerment programmes. The four stages were confirmed by Mr Subowo, one of the community leaders in Tambaklorok. The first and second stages are focused on improving the area through sheet pile infrastructure to protect against tidal flooding and waves, which now envelop the entire Tambaklorok area. The next stage is environmental management, sanitation, drainage, village roads, and community economic empowerment (Suntoro, 2025b).

Nature-Based Solutions

One effective solution to prevent coastal flooding is nature-based solutions, such as restoring mangrove forests in coastal areas. However, the government has to date overlooked this aspect, instead prioritising complex infrastructure. For example, the effort to restore mangroves in Semarang is being hampered because no more public land is available, with most of the land in coastal areas owned by companies. In Tambaklorok, mangroves have been planted by communities on land owned by PT Pelindo. The community manages this restoration and receives funding to maintain the mangroves through corporate social responsibility (CSR) funding from companies such as PERTAMINA or Pelindo. In other locations, mangrove restoration is initiated by local communities; for example, Pak Sururi has been restoring and planting mangroves in Mangkang Kulon, Semarang City. Figure 7.5 depicts mangrove regeneration.

With limited local government funding to restore mangroves, most projects are initiated by local communities with assistance from NGOs such as the Organization for Industrial, Spiritual Cultural Advancement (OISCA)–Bintari and universities such as Diponegoro University, after the communities recognise the importance of mangroves for reducing flood risk. In the past, most mangroves were cleared to open shrimp or aquaculture farms. Not all locations suit mangrove plantations, and most mangroves are located in non-forested land. For example, in Bedono Village, near Semarang City, all community land subject to SLR-related subsidence has been planted with mangroves. This has co-benefits for the community in terms of livelihood in the form of ecotourism. For example, in Mangkang Kulon Semarang, women's groups use mangroves to create batik, crackers, syrup, and coffee products.

Fig. 7.5 Mangrove restoration in Mangkang Kulon, Semarang City (*Source: Laely Nurhidayah, BRIN Documentation*)

Discussion

Socio-Economic Impacts

Reducing Tidal Flooding and Waves

The aim of creating climate-resilient infrastructure is to reduce vulnerability to climate change and uncertainty and limit its adverse impacts. One of the considerations in developing such an infrastructure is how to respond to and prevent significant losses. Based on data from Bappenas, economic losses due to disasters in the marine and coastal sectors from 2020 to 2024 averaged IDR 81 trillion. This amount continues to increase, from IDR 81.3 trillion in 2020 to IDR 81.8 trillion in 2024 (Zen, 2023). The infrastructure built, such as sea walls, breakwaters, water gates, pumps, and storm barriers, will coexist with coastal features, such as dunes, beaches, and ponds, as structures needed for flood management (Hill, 2015).

The sheet pile infrastructure constructed in the 3.6 km coastal area of Tambaklorok can withstand tidal waves and tidal flooding. Its effectiveness was confirmed by statements from local figures in the area, including the Tanjungemas village head and secretary, the heads of RW 14, 15, and 17, and community representatives. Dicky Andrian Rudiawan, Head of RW 14, stated in an interview that before the construction of sheet piles, the community was regularly flooded during heavy rain or high seas. Since 2024, however, residents have been protected from tidal flooding, with only a few locations experiencing seepage. Before the project, around 900 families in the Tambaklorok area had to provide tens of millions in funding for home repairs every five years to avoid being submerged in floodwaters.

The Urgency of Providing Supporting Utilities

The Semarang City Housing and Settlement Agency stated that the government has no specific local budget allocated for constructing a sea wall to address SLR in coastal areas. However, it does have a programme to improve environmental facilities and infrastructure, such as drainage construction, road repairs, tap water installation, electricity, and other supporting facilities, including in the Tambaklorok area. Several other interventions were also carried out as a result of regional planning, such as providing flats as alternative housing for evicted residents affected by regional planning. However, the Semarang City Government reported that it could not manage the tasks of controlling SLR and tidal waves on the north coast without budget contributions from the national government.

According to the RTBL for the Tambaklorok area, three area planning agendas will be carried out following the completion of the sheet pile project: (a) the Pemali Juana River Basin Centre at the Directorate General of Water Resources is preparing activities to normalise the Banger Lama River channel along with protective embankments and drainage in Tambaklorok village; (b) the Strategic Settlement Area Development Work Unit at the Directorate General of Human Settlements is conducting area planning design, especially for main road access and revitalisation of the Tambak Mulyo market; and (c) the Semarang City Government is carrying out land acquisition and resolution of social problems (Nugroho, 2020).

The head of RW 16, Slamet Riyadi, confirmed that the local situation improved following the second phase of sheet pile construction. There are still other development projects needed in the form of raising roads in the village, improving sanitation, and encouraging a change in the culture of littering. An additional problem is residents' heavy use of artesian wells in Tambaklorok, which can accelerate land subsidence. There are two reasons Tambaklorok residents choose to use artesian wells in preference to supplies from government water supply projects: (a) installation is easier and cheaper, and (b) water payments to artesian well owners can be paid in instalments (debited) according to the residents' ability to pay (Nurhidayah, 2025).

Company Relocation

The negative impacts of tidal flooding and tidal waves include submerging industrial areas in Semarang City, especially in the Tanjungemas port area and the Terboyo industrial area of 300 hectares, which accommodates more than 60 companies. Entrepreneurs in both industrial areas have spent billions on the same strategy employed by the community—namely, raising the land to prevent flooding. For example, PT Nayati Indonesia, a company that produces kitchen equipment and furniture, was established in the 1990s and has experienced flooding almost every year, especially since 2015. The company, which had 600 workers, was disrupted by tidal flooding that interrupted goods distribution, closed the factory area, and caused workers to take leave (Bambani, 2024).

The Lamicitra Industrial Area in Semarang has 90 industries or factories, each with an average land area of around 3000 m^2. The entire area has a workforce of 15,000 people. The Indonesian Employers Association (APINDO) confirmed that many electronics and manufacturing factories chose to relocate from the Tanjungemas Port industrial area managed by PT Lamicitra Nusantara due to the frequent impact of tidal floods and the collapse of the sea wall in 2022 (Fardianto, 2023). The Semarang Legal Aid Institute assesses that relocating companies from Semarang City is a rational choice amid the climate crisis and tight business competition. The argument for relocation is based on a cost-benefit analysis comparing the high cost of adapting to tidal flooding versus the benefits in the form of cheaper production materials and labour wages (Rahayu, 2023).

Loss of Livelihood

The Head of Trimulyo Village, Genuk District, Semarang City, stated that his area has various companies, hundreds of hectares of community ponds, and mangrove conservation areas. One issue that needs to be considered in constructing the Semarang-Demak Toll Road and Sea Wall is the community's livelihoods. In the context of ponds, it is necessary to ensure that residents can still carry out cultivation and that the government's plan to replace the mangrove areas affected by the project maintains the ecological system and fish life sources. The information provided to the village government indicates that the inside of the sea wall and sea toll road will also be used for mangrove rehabilitation (Marbun, 2025).

The impact of the construction of the toll road and sea wall, although not yet fully completed, has affected hundreds of fishermen in Trimulyo Village, who have had to give up fishing because they no longer have access to the sea. One such fisherman is Mr Nurseh, who must find alternative work to support his family. Luckily, he receives support from one of his children, who sends food and money. Based on the interviews, as a result of the ongoing construction of toll road infrastructure and sea walls, the natural flow of water was blocked. Consequently, the floodwater from rivers or land cannot flow into the sea, thus causing major flooding in residential areas and shrimp ponds, as occurred in 2023 and 2024. (Marbun, 2025).

The Maleh Dadi Segara Coalition reported that the government's environmental impact analysis for the Semarang-Demak Toll Road and Sea Wall lacked public participation and ignored the potential for residents to lose their jobs due to the destruction of mangrove forests and aquatic ecosystems.

Legal Certainty of Land Rights

Land plays a significant role in human life, as it is both a place to live and a source of livelihood. Land can also be assessed as a permanent property, meaning that it has a use value and an element of ownership rights. The agrarian sector's legal system guarantees every person's right to control or own their land, including Basic

Agrarian Law Number 5 of 1960 and Article 36 of Law Number 39 of 1999 concerning Human Rights. Maria SW Sumardjono, an agrarian expert from Gajah Mada University, emphasised that rights, including land rights, entail an obligation/responsibility for the state/government to protect, advance, uphold, and fulfil them (Suntoro, 2018).

From the collection of field data and analysis, there are three typologies of community land rights status in the Tambaklorok area. The first is the land owned by the community, as evidenced by the fundamental rights in the form of a land rights certificate from the Semarang City Land Office. The second is the river boundary area, claimed as government land—in this case, the Semarang Juana River Basin Center. Residents affected by the infrastructure development project were moved to rented flats or, in some cases, accommodated in temporary housing on the riverbank. Finally, there is disputed land, such as the problem of land ownership claims between the community and the Indonesian Shipping Company (Pelindo), which has resulted in residents being unable to prove their ownership status at the Semarang City Land Office.

Subowo, one of the community leaders, said that the Tambaklorok area is a colonial-era Dutch village built in 1942; the toponym 'Nglorok' became 'Tambaklorok'. Administratively, in 1982, it was changed to Tambakmulyo, which attempted to change the identity of Tambaklorok as a coastal area. The Tambaklorok area was then 'included' in the port development plan by Pelindo (Suntoro, 2025b). In terms of land rights, the Tambaklorok community is disadvantaged. Residents, even if they have a legal claim in the form of a land title certificate, can no longer transfer their title to their heirs or descendants. Similarly, if the land is sold, the new owner cannot change the status of the land title certificate. The Head of RW 14 has personal experience of this issue; he cannot transfer the title of the land for his residence, even though he has a land title certificate, because of 'blocking' by Pelindo at the Semarang City Land Office. The community is still allowed to live in the Tambaklorok area, but the uncertainty over land claims causes corresponding uncertainty in the lives of residents (Suntoro, 2025c).

The Role of Local Government

Coastal Area Infrastructure Budget

The Semarang City Government plays a central role in developing infrastructure, particularly in the coastal areas that are highly vulnerable to impacts such as sea level rise, tidal flooding, and ecosystem shifts. The coastal area of Semarang holds a strategic position, as it encompasses various vital functions, including ports, densely populated residential areas, industrial zones, and ecosystem conservation areas. Therefore, the approach to climate change adopted by the Semarang City government considers technical factors and comprehensively incorporates social, economic, and ecological aspects.

A tangible form of this adaptation effort can be seen in managing and allocating the infrastructure budget for coastal areas. The Coastal Area Infrastructure Budget is one of the government's strategies to strengthen infrastructure in response to the impacts of climate change. This effort is oriented towards physical development and considers sustainable natural resource management. Any analysis of the impacts on coastal areas must consider various environmental change scenarios that may result from climate change, including flood risks caused by rising sea levels and shifting weather patterns (Nicholls et al., 2013). The emphasis on managing these risks, in line with international guidelines on adaptation, demonstrates the local government's commitment to developing a disaster-resilient infrastructure.

The massive development of flood and tidal wave control infrastructure in the northern area of the city is still highly dependent on national government policies and budgets. The budget is allocated for projects such as the construction of sea walls, strengthening drainage systems and water pumps, river normalisation, road improvements, and retention pond construction (Redaction, 2023). Developing sheet piles in the northern Tambaklorok coastal area has become a strategic regional and national project. These sheet piles not only function to prevent seawater overflow due to high tides but are also equipped with pump systems and automatic sluice gates to regulate water flow from the land (Table 7.1).

The Semarang City Government has gradually allocated funds to the Regional Revenue and Expenditure Budget for infrastructure development and strengthening to increase coastal resilience to the impacts of climate change, as stated in the Semarang City Medium-Term Development Plan (RPJMD) 2021–2026. However, the Mayor of Semarang, Agustina Wilujeng, stated that this year, various budget efficiencies were carried out, which focused on scholarships, education, and health, and which reduced the funding for infrastructure (Sekar, 2025). Deputy Chairman of the Semarang City Regional People's Representative Council (DPRD), Muhammad Afif, emphasised that the city government would not be able to handle

Table 7.1 Several flood and tidal control projects in Semarang City with funding from the Central Government

No	Project	Budget (IDR)	Years
1	DED of the retention pond and pump house of Kampung Bahari Tambaklorok	850 million	2016
2	Flood and tidal control in the Tambaklorok area, Phase II	225.1 billion	2020–2024
3	Flood and tidal control in Semarang Package I on the Babon River and Sayung River, Semarang City-Demak Regency	131.2 billion	2021–2022
4	Semarang Flood and Tidal Control Package II to reduce the impact of tidal flooding in the Terboyo Industrial Area	207 billion	2021–2022
5	Procurement of mobile pumps	6 billion	2017

Source: Semarang City LPSE, Central Java Provincial Government LPSE, Pemali-Juana 2022 BBWS Report

tidal flooding in coastal areas independently. Therefore, it still relies on intervention and budget allocation from the central government (Yusuf, 2022b).

Clean Water Program

Clean water provision programmes are an important component in mitigating the impacts of climate change, especially in increasingly unpredictable hydrological cycles. As observed in Semarang, increased rainfall intensity and changing precipitation patterns require special attention from the local government to provide and maintain adequate access to clean water (Egaputra et al., 2022). Several studies have shown that poor water quality can exacerbate the impacts of climate change, such as worsening public health and reducing resilience to disasters (Tu et al., 2022). The government focuses on environmental sanitation in addition to clean water provision. This aligns with the mandate of Semarang City Regulation No. 4 of 2024 on the Environmental Protection and Management Plan for 2024–2025, which emphasises the importance of improving sanitation systems to maintain river water quality in urban areas. One concrete step taken is the addition of sludge truck fleets and the strengthening of the Tambaklorok Sludge Treatment Plant (IPLT) with the provision of adequate supporting facilities.

The clean water crisis in Semarang's coastal areas has been a long-standing structural issue. One of the leading causes of this crisis is land subsidence due to massive groundwater exploitation, accelerating seawater intrusion into residents' water sources, including domestic wells. As an intervention to address this situation, the Semarang City Government, through PDAM Tirta Moedal, has implemented a clean water network expansion programme in coastal areas. Previously unserved areas, such as Bandarharjo, Tanjung Mas, and parts of Terboyo Kulon, have gradually gained access to piped water services. This effort has been accompanied by policies to subsidise household connections in low-income communities as a strategy to promote equitable access to basic services. PDAM Tirta Moedal Semarang allocated 1000 clean water connections for Tambaklorok residents with a discount on installation costs of only IDR 625 thousand from the normal price of IDR 1.3 million. The President Director of PDAM, Tirta Moedal, Etty Laksmiwati, said the commitment aimed to ensure the provision of clean water, not mixed with seawater (Budiasto, 2014). Observations show that most residents in the coastal areas still rely on artesian wells for their clean water supply. Factors driving the use of this water source include flexibility in payment systems, where residents can delay payment, unlike the routine billing system in piped water services.

Marine Area Management

Marine area management is central to preserving coastal ecosystems and promoting sustainable economic activities. Garschagen et al. (2018) note that many coastal cities are experiencing the adverse effects of climate change, and effective adaptation will require careful spatial planning to minimise impacts. This includes the restoration of mangrove ecosystems, which serve as natural buffers against storms and flooding and improve the environmental quality of coastal communities (Martuti et al., 2019). Community involvement in this area management process is also critical, not only in terms of utilisation but also in maintaining its sustainability.

The Semarang City Government also strives for holistic marine area management in response to climate vulnerability and as part of sustainable regional development. The city government's marine area management is based on the principles of ecosystem-based adaptation, considering the environmental carrying capacity and social-cultural characteristics of coastal communities. One of the main initiatives in this regard is the revitalisation of Kampung Bahari Tambaklorok, a fishing village facing chronic tidal flooding, poor sanitation, and limited basic infrastructure. As part of this revitalisation, the government has built concrete-based environmental roads, pumping stations, floating docks, and public spaces that support economic activities such as markets, fish auction places, craft centres, and parks within the area. This revitalisation not only improves the physical environmental conditions but also drives local economic transformation through the development of culture- and marine-based tourism.

This transformation has fostered the development of community-based marine tourism, with local communities now involved in productive economic activities such as boat tour services, seafood-based culinary experiences, and marine-themed handicrafts. Community involvement in tourism management strengthens the participatory and sustainable development approaches. Furthermore, infrastructure improvements in coastal areas facilitate easier access to tourist destinations, such as traditional docks, mangrove conservation sites, and educational fishpond areas, which were previously difficult to reach due to poor roads and minimal supporting facilities.

Public Participation

Public and community participation is a crucial component in the policy formulation process at the local government level. Wamsler (2016) explain that a participatory approach based on local communities in implementing adaptation strategies can strengthen community adaptive capacity and ensure long-term sustainability. Involving the community in planning and monitoring policies not only fosters a sense of ownership but also enhances the effectiveness of the programmes being implemented, as the process is more responsive to the community's real needs, which often get overlooked in highly centralised technocratic approaches.

However, the effectiveness and inclusiveness of public participation in infrastructure development remain uneven. In several coastal communities, particularly in areas vulnerable to tidal flooding, community involvement in infrastructure planning has been largely reactive and self-initiated rather than facilitated by the government. For instance, Mr Subowo—one of the community leaders in the Tambaklorok area—explained that the construction of sheet piles in certain neighbourhoods was initiated by local residents themselves, following delays and inaction from the municipal government. In these cases, communities organise collectively and escalate their demands directly to national-level political channels, bypassing local institutions. This underscores both the potential community agency and the limitations of formal participatory mechanisms at the local level.

Public participation becomes even more constrained in the context of large-scale infrastructure under the National Strategic Projects (Proyek Strategis Nasional, PSN), such as the Sea Toll and Giant Sea Wall. In Semarang, for example, this project meant that fishermen could not work for a month, as access to the sea was blocked. These projects are typically initiated by central government agencies with limited local consultations. Although the legal framework for Environmental and Social Impact Assessments (AMDAL) aims to promote stakeholder engagement and environmental responsibility, the actual implementation often falls short. AMDAL documents are usually prepared by third-party consultants hired by project proponents, raising concerns about independence, transparency, and accountability.

Public consultations tend to be symbolic—poorly publicised, overly technical, and often exclude critical stakeholders such as fisherfolk, the urban poor, and grassroots environmental groups. In practice, most feedback gathered during these consultations is not thoroughly integrated into the final reports, indicating a form of lip service rather than genuine engagement (Utama et al., 2023). The influence of political and economic interests further complicates the process, as expediency is frequently prioritised over inclusive participation and environmental sustainability (Zuhri, 2022). Reports from the Maleh Dadi Segara community highlight that affected communities are routinely excluded, and their livelihoods are overlooked. Consequently, the efficacy of AMDAL as a tool for sustainable development in Semarang is increasingly being questioned, underscoring the need for reforms that ensure more equitable participation and stronger oversight in environmental governance (Utama et al., 2023; Susanto, 2013).

These observations suggest an urgent need to strengthen participatory infrastructure governance in Semarang and beyond. Community involvement should not be limited to consultative forums but must be institutionalised throughout the entire infrastructure development, from planning and budgeting to implementation, monitoring, and evaluation. Inclusive participation not only improves policy legitimacy and programme effectiveness but also ensures that infrastructure development supports, rather than undermines, the resilience and rights of coastal communities. Addressing these issues aligns with the legal and ethical imperatives of sustainable development and enhances communities' capacity to effectively advocate for their interests and environmental rights.

Potential of Hybrid Engineering

Many institutions, such as NGOs and universities, have attempted to restore mangroves using hybrid engineering methods. Hybrid engineering is a technology adopted from the Netherlands that combines engineering and nature-based solutions in mangrove restoration. For example, sediment traps are permeable structures that allow mangroves to grow within trapped sediments (Nurhidayah et al., 2022). Moh. Nurhadi from the NGO Bintari attempted to apply this hybrid engineering technology to Pekalongan. The Ministry of Environment and Fishery, UNDIP, and Wetland Indonesia have also attempted to build sediment traps for mangroves using hybrid engineering technologies. Some have been successful, while others have failed due to the strong winds and waves in coastal Java.

Long-term and integrated approaches to using complex infrastructure- and nature-based solutions offer win–win solutions for flood risk prevention and biodiversity conservation, as noted by Nurhadi, from Bintari, during an interview. Collaboration between institutions and the community, government, NGO, and academic, and private sectors is essential for mangrove restoration and conservation. Collaboration between the companies that own the lands and the communities that plant mangroves is likewise essential. Conversely, uncertainty regarding law and policy in promoting mangrove conservation creates concern among local communities about the mangrove restoration areas being converted into other uses. The relevant legislation on mangrove protection needs to provide protection for the existing mangrove forest and extend this into, for example, ecotourism or empowering local communities to benefit from mangroves for their livelihood and utilising this sustainably.

Conclusion

Complex infrastructures, such as sheet piles and sea walls, have proven effective in protecting communities from coastal flooding. However, the problem is not only caused by the rise in sea level; Semarang and Demak, in particular, are also vulnerable to land subsidence problems. Despite efforts by the government to prevent coastal flooding, land subsidence remains a central issue in Tambaklorok, Semarang. Every 5 years, local communities need to raise the level of their house floors to address flood risks. In addition, roads also need to be raised even higher, as subsidence means that they will continue to sink. Waste, clean water, sanitation, and pollution remain significant community problems. Local communities must use groundwater daily, as surface water is expensive and unreliable, yet it is this need for water extraction by the communities and industries that has triggered the land subsidence problem in Semarang.

Despite the effectiveness of infrastructure in reducing flood risks, there are also negative impacts on fishermen. Some fishermen in the Pond, Trimulyo Village,

Genuk, and Semarang City areas have been blocked from accessing the sea due to the development of the sea wall and toll roads. The government does not provide alternative routes or compensation for those thus affected. In addition, some of the mangroves have been cut down or damaged due to this development. Prior informed consent and public participation are needed to reduce the negative implications of infrastructure development on communities affected by mangroves.

Involving communities in decision-making and improving community-based initiatives, such as mangrove restoration or urban farming, is one way to increase the community's ownership of this development or project. The combination of infrastructure- and nature-based solutions can achieve an outstanding balance between biodiversity conservation and the United Nations' Sustainable Development Goals. There should be collaboration between local governments, communities, the private sector, NGOs, and academics to achieve synergies and establish infrastructure- and nature-based solutions. Key among the recommendations is the need for greater collaboration between policymakers, practitioners, and local communities to integrate nature into 'mainstream' infrastructure planning and design, provide win–win solutions for communities, and reduce the negative implications for their livelihoods.

References

Agt. (2024). *Bappenas Has 8 Studies on the Giant Sea Wall since 2014*. CNN Indonesia. 2024. https://www.cnnindonesia.com/ekonomi/20241111205240-92-1165500/bappenas-punya-8-kajian-soal-giant-sea-wall-sejak-2014

Authori, N., Kismartini, d. R., & Slamet, S. (2019). *Development of Marine Villages from the Perspective of SDGs in the Coastal Area of Tambak Lorok, Semarang City. Departemen Administrasi Publik Fakultas Ilmu Sosial Dan Ilmu Politik, Universitas Diponegoro*. https://ejournal3.undip.ac.id/index.php/jppmr/article/view/33559%0A; https://ejournal3.undip.ac.id/index.php/jppmr/article/viewFile/33559/26717

Bahfein, S. (2024). 'Giant Sea Wall' Needs Rp 800 Trillion, Save Pantura Java.' *Kompas.Com*, 2024. https://www.kompas.com/properti/read/2024/11/14/160000821/giant-sea-wall-butuh-rp-800-triliun-selamatkan-pantura-jawa

Bambani, A. (2024). *Semarang Flood Causes Losses of Rp 1.5 Trillion, Billions of Rupiah Drained to Save People*. Kontan. 2024. https://pulitzercenter.org/id/stories/banjir-semarang-rugikan-rp-15-triliun-milyaran-rupiah-pun-terkuras-agar-selamat

Budiasto, B. B. (2014). *PDAM Gives Discount to Tambaklorok Residents of IDR 625*. Tribunnews.Com. 2014. https://jateng.tribunnews.com/2014/09/12/pdam-beri-diskon-warga-tambaklorok-rp-625-ribu

Egaputra, A. A., Ismunarti, D. H., & Pranowo, W. S. (2022). Inventarisasi Kejadian Banjir Rob Kota Semarang Periode 2012–2020. *Indonesian Journal of Oceanography, 4*(2), 29–40. https://doi.org/10.14710/jekk.v%vi%i.13240

Falkner, R. (2016). The Paris Agreement and the new logic of international climate politics. *International Affairs Journal, 5*(December), 1107–1125. https://doi.org/10.1111/1468-2346.12708

Fardianto, F. (2023). *Emergency! Frequently Flooded, Several Factories at Tanjung Emas Port Move to Demak*. IDN Times Jateng. 2023. https://jateng.idntimes.com/business/economy/fariz-fardianto/gawat-sering-kebanjiran-sejumlah-pabrik-pelabuhan-tanjung-emas-pindah-demak

Fuaziah, T. A. (2025). *Tens of hectares of mangroves displaced by Semarang-Demak Toll Road Project, marine ecosystems and fishermen threatened*. Kompas.Com. 2025. https://regional.kompas.com/read/2025/01/14/204500378/puluhan-hektare-mangrove-tergusur-proyek-tol-semarang-demak-ekosistem-laut?page=all

Garschagen, M., Surtiari, G. A. K., & Harb, M. (2018). Is Jakarta's new flood risk reduction strategy transformational?. *Sustainability, 10*(8), 2934. https://doi.org/10.3390/su10082934

Hill, K. (2015). Coastal infrastructure: A typology for the next century of adaptation to sea-level rise. *Frontiers in Ecology and the Environment, 13*(9), 468–476. https://doi.org/10.1890/150088

HLRN. (2007). *Handbook on United Nations basic principles and guidelines on evictions and displacement*. Housing and Land Rights Network.

Jateng, H. P. (2015). *Tambaklorok Towards Marine Village*. Jatengprov.Go.Id. 2015. https://jatengprov.go.id/beritadaerah/tambak-lorok-menuju-kampung-bahari/

Juana, B. B. W. S. P. (2024). *Information material for flood and flood control and security on the North Coast of Java*. Semarang.

Limanseto, H. (2024). *In collaboration with South Korea and the Netherlands, the Government Develops a Strategy for the Protection and Sustainable Development of Java's Coastal Zone*. Ekon.Go.Id. 2024. https://www.ekon.go.id/publikasi/detail/5985/bekerja-sama-dengan-korea-selatan-dan-belanda-pemerintah-susun-strategi-perlindungan-dan-pembangunan-berkelanjutan-pantura-jawa

Marbun, A. A. (2025). *Interview with the Head of Trimulyo Village*. Semarang.

Martuti, N. K. T., Setyowati, D. L., & Nugraha, S. B. (2019). *Mangrove Ecosystem (Diversity, Phytoremediation, Carbon Stock, Role and Management)*. Institute for Research and Community Service, Semarang State University. SEMARANG, Indonesia, https://lib.unnes.ac.id/33810/1/PDF_EKOSISTEM_MANGROVE.pdf

Maskun, M., Habaib, N., Mukarramah, A., Charmario, J., & Angi, W.. (2023). Policy strategies for infrastructure in climate change adaptation: Jurisdictional comparison between Chile and Indonesia. In *9th ICCC 2023*, 467:1–7. E3S Web of Conferences. https://doi.org/10.1051/e3sconf/202346705001

NDPA. (2024). *Law of the Republic of Indonesia Number 59 of 2024 Concerning The National Long-Term Development Plan 2024-2045*. Indonesia.

Nugroho, P. C. (2020). Study of Semarang City Government Policy in the Implementation of Tambak Lorok Marine Village. In *Green development and licensing: Diplomacy, equipment readiness and standardization patterns* (pp. 29–38). Sekolah Pascasarjana Universitas Diponegoro. https://mil.pasca.undip.ac.id/wp-content/uploads/2021/11/5.-Putut-Cahyo-Nugroho-29-38.pdf

Nurhidayah, L. (2025). Interview with the Head of Neighborhood Association 14, Tambaklorok.'

Nurhidayah, L., Davies, P., Alam, S. et al. (2022). Responding to sea level rise: challenges and opportunities to govern coastal adaptation strategies in Indonesia. *Maritime Studies, 21*, 339–352. https://doi.org/10.1007/s40152-022-00274-1

OECD. (2024). *Infrastructure for a Climate-Resilient Future*. OECD. www.oecd.org/about/publishing/corrigenda.htm. © OECD 2024 The.

Rahayu, Y. S.. (2023). *Industrial Relocation 'Questioning Industrial Relocation to the Job Creation Law Polemic*.' LBH Semarang. 2023. https://lbhsemarang.id/relokasi-industri-menyoal-relokasi-industri-hingga-polemik-uu-cipta-kerja/

Rais, M., & Hariyanto. (2021). Tourism development strategy with urban community based tourism (UCBT) concept approach in identified slum area of Bahari Tambak Lorok Village, Semarang City. *Geo Image, 10*(2), 95–106. http://journal.unnes.ac.id/sju/index.php/geoimage

Redaction. (2023). *Semarang's large flood control budget does not address the root of the problem*. Serat.Id. 2023. https://serat.id/2023/11/03/anggaran-besar-pengendalian-banjir-semarang-tak-sentuh-akar-masalah/

Rochim, A. (2024). Study on the depth of cantilever sheet pile wall based on the type of soil where sheet pile embedded. *IOP Conference Series: Earth and Environmental Science, 1321*(1). https://doi.org/10.1088/1755-1315/1321/1/012020

Rusmadi, R., et al. (2017). Gendering the climate change policy: A study of gender analysis on Semarang's integrated city climate strategy. *Advanced Science Letters, 23*(3), 2556–2558. https://doi.org/10.1166/asl.2017.8697

Sekar, N. (2025). *Budget efficiency, Semarang mayor prioritizes education and infrastructure*. Detik.Com. 2025. https://news.detik.com/berita/d-7830045/efisiensi-anggaran-walkot-semarang-prioritaskan-pendidikan-dan-infrastruktur

Setpres, BPMI. (2024). *President Jokowi Reviews Rob Flood Control Project in Tambak Lorok Semarang*. Presidenri.Go.Id. 2024. https://www.presidenri.go.id/siaran-pers/presiden-jokowi-tinjau-proyek-pengendalian-banjir-rob-di-tambak-lorok-semarang/

Sofaniadi, S., Rusmadi, & Sari, A. D.. (2015). *Protect, Adapt or Relocate? Responding to Climate Change in Coastal Indonesia*. Asian Cities Climate Resilience 14:2015. http://pubs.iied.org/10723IIED.html?f=BCAJISEDM&w=HS&b=d

Srinivasu, B. (2013). Infrastructure development and economic growth: Prospects and perspective. *Journal of Business Management and Social Sciences Research, 2*(1), 81–91. https://ianskerrett.wordpress.com/2016/04/14/profile-of-an-iot-developer-results-of-the-iot-developer-survey/

Sukamsari, N. M.. (2024). *Getting to know the Giant Sea Wall Project, one of the things Prabowo discussed with Xi Jinping*. Tempo.Co. 2024. https://www.tempo.co/ekonomi/mengenal-proyek-giant-sea-wall-salah-satu-yang-dibahas-prabowo-dengan-xi-jinping-1167008

Suntoro, A. (2018). *Study of Law Number 2 of 2012 Concerning Land Acquisition for Development in the Public Interest*. Komnas HAM RI. Vol. Pertama. Jakarta. https://www.komnasham.go.id/files/20200116-kajian-terhadap-uu-nomor-2-tahun-$QXA5.pdf

Suntoro, A. (2025a). *Interview with Secretary of Tanjungemas Village*.

Suntoro, A. (2025b). *Interview with Subowo, Sheet Pile Application Initiator*.

Suntoro, A. (2025c). *Interview with the Head of Neighborhood Association 14, Tambaklorok, Semarang City*.

Susanto, E. (2013). The effectiveness of environmental impact assessment for nuclear power plant. *Applied Ecology and Environmental Sciences, 1*(4), 61–66. https://doi.org/10.12691/aees-1-4-5

Tu, Q., Schneider, M., Hase, F., Khosrawi, F., Ertl, B., Necki, J., Dubravica, D., Diekmann, C. J., Blumenstock, T., & Fang, D. (2022). Quantifying CH4 emissions in hard coal mines from TROPOMI and IASI observations using the wind-assigned anomaly method. *Atmospheric Chemistry and Physics, 22*, 9747–9765. https://doi.org/10.5194/acp-22-9747-2022.

Utama, K., Saraswati, R., Putrijanti, A., & Sukmadewi, Y. (2023). Environmental protection barriers in Indonesia in the form of administrative effort. *Iop Conference Series Earth and Environmental Science, 1270*(1), 012019. https://doi.org/10.1088/1755-1315/1270/1/012019

WALHI, Central Java. (2024). *Ecological study of the Semarang-Demak Toll Road and Sea Wall Project?* Semarang. https://www.walhijateng.org/

Wamsler, C. (2016). From risk governance to city-citizen collaboration: Capitalizing on individual adaptation to climate change. *Environmental Policy and Governance, 26*(3), 184–204. https://doi.org/10.1002/eet.1707

Wicaksono, A. (2024). *President Jokowi Visits Sheet Pile Construction in Tambaklorok, Semarang*. Radar Semarang. 2024. https://radarsemarang.jawapos.com/semarang/724768027/presiden-jokowi-cek-pembangunan-sheet-pile-tambaklorok-semarang

Yoshino, N. et al. (2018). Land acquisition and infrastructure development through land trust laws: A policy framework for Asia. www.adbi.org

Yusuf, M. D. (2022a). *Semarang city government claims Embankment at Tanjung Emas Port has been repaired for 3 days*. Kompas.Com. 2022. https://regional.kompas.com/read/2022/11/05/183306878/sempat-buat-lumpuh-3-hari-pemkot-semarang-klaim-tanggul-di-pelabuhan

Yusuf, M. D. (2022b). *DPRD Asks Central Government to Help with Rob Flood Problems in Semarang City*. Kompas.Com. 2022. https://regional.kompas.com/read/2022/12/13/133733978/dprd-minta-pemerintah-pusat-bantu-permasalahan-banjir-rob-di-kota-semarang

Zen, F. (2023). Enhancing climate-resilient infrastructure development in Indonesia. *Infrastructure for Inclusive Economic Development, Lessons Learnt from Indonesia, 1*, 193–2015.

Zuhri, S. (2022). Green accounting in Indonesia pathways to sustainable economic development. *Journal of Economics and Business Letters, 2*(1), 10–12. https://doi.org/10.55942/jebl.v2i1.282

Chapter 8
The Critical Role of Local Leaders in Managing the Impacts of Climate Change in Demak, Indonesia

Inayah Hidayati, Bulan Prabawani, Aditya Rahmadhony, Adis Nur Hayati, and Jennifer Boddy

A Crisis-Ridden Coast

Rising sea levels are one of the most concerning impacts of global climate change. As polar ice sheets melt and glaciers on land shrink, ocean levels are rapidly rising (Dutton et al., 2015; Jevrejeva et al., 2016). The impact of climate change also reduces the amount of water that is stored on land, contributing to water shortages and a lack of access to clean drinking water (Reager et al., 2016). The effects of these changes are already being seen: coastlines are changing, flooding is getting worse, and ecosystems, infrastructure, and human settlements are at risk. Low-lying coastal regions are especially vulnerable. Storm surges and tidal dynamics make flooding more likely, and erosion and habitat loss will follow (Werner & Simmons, 2009). Studies have shown that the rises in the global mean sea level are associated with global carbon emissions (Clark et al., 2018). In the absence of efforts to reduce carbon emissions and implement effective climate adaptation measures, climate change may displace millions of people from their homes by the end of this century (Hummel & Stacey, 2021).

I. Hidayati (✉)
Directorate of Human Development, Demography, and Culture Policy, National Research and Innovation Agency, Jakarta, Indonesia
e-mail: inay001@brin.go.id

B. Prabawani
Diponegoro University, Semarang, Indonesia

A. Rahmadhony · A. N. Hayati
Research Center for Law, National Research and Innovation Agency (BRIN), Jakarta, Indonesia

J. Boddy
Griffith University, Brisbane, QLD, Australia

L. Nurhidayah et al. (eds.), *Climate Change, Labour and Migration in Indonesia*, Springer Climate, https://doi.org/10.1007/978-3-032-15334-0_8

Groundwater extraction has had significant impacts on coastal communities in Indonesia, particularly in low-lying urban areas. The geological and hydrological characteristics of Demak, which is primarily situated on alluvial plains, have led to a heavy reliance on groundwater resources (Bemmelen, 1970; Lo et al., 2021). Excessive pumping of groundwater to meet domestic, agricultural, and industrial demands has led to land subsidence, in which the ground gradually sinks (Nurhidayah et al., 2022). This subsidence, which can reach significant rates, increases the vulnerability of these communities to flooding, especially during high tides and heavy rainfall, while also posing serious threats to the surrounding ecosystems (Susilo et al., 2023). Although this issue is more documented in metropolitan areas like Jakarta and Semarang, some parts of Demak are also beginning to experience the compound effects of subsidence and rising sea levels. Over-extraction can also lead to saltwater intrusion, where seawater contaminates freshwater aquifers, jeopardising access to clean drinking water and affecting agriculture. These combined pressures, both climate-induced and human-driven, pose a major threat to infrastructure, public health, and long-term community sustainability in vulnerable coastal regions. Nor are the impacts solely physical; cultural heritage sites, especially in coastal areas, are at increasing risk from environmental damage and extreme weather events (Marković et al., 2018; Kantamaneni et al., 2022), as are local biodiversity and fishing stocks.

To deal with this, it is important to come up with ways to adapt that are both sustainable and based on local needs. Such adaptation measures could include protecting shorelines, planning how land is used, and restoring the mangroves and other ecosystems (Ghani & Hidayati, 2024). Local Indonesian governments and communities have taken various steps to deal with these problems, including building and maintaining coastal structures, raising roads to stop them from being washed away, and planning where people live and work to reduce exposure. Community-led efforts to plant more mangroves have also been successful, combining environmental protection with the restoration of people's livelihoods (Septiarani & Yesiana, 2020).

The impact of rising sea levels is not unique to Indonesia. The Intergovernmental Panel on Climate Change (IPCC) reports that rising sea levels will lead to problems such as erosion, land subsidence, and changes to the environment (Kantamaneni et al., 2022). Mangroves are extremely important for protecting coastlines, but they are also at a high risk (Koh et al., 2018). Warmer temperatures and rising sea levels also impact the spread of waterborne diseases and the extent to which people can grow crops (Walker et al., 2018).

To deal with these problems, plans that look at how changes in the ocean, air, ice, and water affect each other are needed (Suursaar, 2011). Melting polar ice and changes in the sea level highlight the need to act in different ways in different places (Stammer et al., 2013). It is important to understand how sea levels are changing so that good decisions can be made to protect communities and ecosystems that may be affected. In Central Java, these changes are already happening, with local community members taking the lead in addressing the impacts. The next few sections will look at how they are dealing with climate change in a brave, new, and community-focused way.

The Power and Role of Local Leadership

Local leaders play a pivotal role in guiding communities by escalating the challenges of climate change. Their leadership determines how effectively communities can respond, adapt, and, ultimately, thrive in an increasingly uncertain environment. One of their central functions is mobilising resources, both internally—such as community labour and funding—and externally, including government assistance and NGO support. Equally important is their ability to facilitate communication across all levels of society, ensuring that information about climate risks, policies, and adaptive strategies reaches even the most marginalised groups (Ntontis et al., 2018). Effective leadership not only enhances community participation in decision-making but also ensures that adaptation strategies are locally relevant and sustainable. Leaders act as essential bridges between local realities and institutional structures, helping communities access technical and financial resources while translating national agendas into culturally appropriate initiatives. Through strong social capital, such as networks of trust and shared local knowledge, leaders can identify context-sensitive solutions tailored to specific vulnerabilities. These leaders typically work closely with local experts to ensure that solutions are grounded in people's lived experiences, blending traditional practices with contemporary innovations (Cox & Hamlen, 2014).

Ensuring that every member of the community—including marginalised and vulnerable groups—is informed about climate risks and involved in decision-making is a hallmark of effective local leadership. Inclusive leadership encourages a sense of shared responsibility, builds collective commitment, and ensures that climate adaptation measures reflect the values and needs of the whole community, particularly women, the elderly, and low-income households. Culturally sensitive approaches that draw on local knowledge and customs enhance both the acceptance and sustainability of climate solutions (Barrett & Armstrong, 2023; Bryan et al., 2023; Grey et al., 2020).

Research strongly supports the value of local leadership in climate adaptation. Gupta et al. (2010) developed the concept of Adaptive Capacity Wheel to evaluate institutional attributes that enhance community resilience. Measham et al. (2011) and Agrawal (2008) emphasise that local institutions must strengthen their capacity to cope with climate impacts. Baker et al. (2012) argue that effective adaptation depends on the active involvement of stakeholders and community members. Makate (2020) highlights the role of indigenous knowledge in promoting the adoption of climate-smart practices. Meanwhile, Benulic et al. (2022) examine how leaders within polycentric climate governance structures contribute to adaptation at various levels.

These insights are echoed in practice. In Ras Sudr, Egypt, local leaders facilitated collaboration between tourism operators, communities, and the government to ensure sustainable climate responses in the tourism sector (Abdelwahab, 2023). In the UK, local governments have developed innovative climate strategies, even during periods of financial austerity, demonstrating the flexibility of local governance (Shaw & Theobald, 2011). Canada's climate resilience maturity matrix has been used in the health sector to integrate environmental and social vulnerability assessments into planning, emphasising community partnership (Thomson & Roach, 2023).

Community opinion leaders also play an instrumental role in communicating climate concerns related to land use, water management, and public services (Keys et al., 2014). In informal settlements, where municipal governance is often limited, local leaders help bridge the gap between residents and city authorities to ensure more inclusive adaptation strategies (Williams et al., 2020).

In Indonesia—and particularly along the vulnerable northern coast of Central Java—village heads and informal leaders have emerged as vital actors in mitigating the impacts of sea level rise, tidal surges, and flooding. Their work includes infrastructure improvements, relocation programmes, and mangrove restoration initiatives. These actions are not only practical responses to environmental challenges but also key to sustaining local economies, especially in coastal communities reliant on fisheries and marine ecosystems.

Local leadership is also essential for organising adaptation strategies, promoting participatory planning, and advocating for community needs in policy discussions. Local leaders' engagement ensures that adaptation is not imposed from outside but arises organically from community knowledge and agency. In Demak, this leadership continues to guide the region through an evolving crisis, fostering resilience through collaboration, innovation, and care.

Agents of Adaptation in Action

Local leaders are essential actors in shaping how communities respond to the escalating impacts of climate change, particularly in vulnerable coastal regions such as Demak. Their leadership is instrumental in facilitating communication, mobilising both internal resources (such as community labour and mutual funds) and external support (such as from governments and NGOs), and ensuring that community needs are prioritised. This leadership can build a shared vision of adaptation strategies that are inclusive, context-sensitive, and sustainable (Sienkiewicz-Małyjurek, 2022; Shaw & Theobald, 2011). Transformational leadership, in particular, fosters collaboration and collective action, both of which are vital for long-term resilience (Wang et al., 2021).

In Demak, several village heads have emerged as leaders in promoting adaptation to tidal floods and land subsidence due to the impact of climate change. The role of these leaders is crucial and reflects transformational leadership, in which the decisions made are oriented towards long-term socio-economic changes. Their leadership includes elevating frequently flooded village roads, building embankments to redirect tidal water away from residential areas, and relocating families from high-risk zones. These actions not only restore access and protect infrastructure but also safeguard livelihoods. In one community, roads that were frequently submerged during high tide were raised significantly, improving mobility and enabling access to emergency services. Elsewhere, embankments were constructed using sandbags and stones, supported by collective labour and local fundraising. These embankments play a critical role in diverting water and protecting vulnerable infrastructure from major floods. The head of Purworejo village in Bonang district explained:

> This road elevation physical programme is part of my first year in office; the issue of road repairs had not been addressed by the previous village head. Therefore, I am making road repairs because this is the most urgent issue and will affect the economic activities of the residents. After the road to the coast near the TPI is completed, residents selling fresh and salted fish will no longer be disturbed. (Mr Pw, village head)

Viewed through the lens of transformational leadership (Burns, 1978), Mr Pw, as the village head, exemplifies idealised influence by commencing road development and fostering trust and credibility as a leader attuned to the needs of his constituents. This leadership style can foster acceptance of change, with a village head who is attuned to local needs and adept at enhancing his community's resilience against the effects of climate change.

The relocation of households living in areas most affected by erosion and tidal flooding has also been a key component of local climate action. Guided by thoughtful planning, village heads selected safe relocation sites and helped residents rebuild their lives with the support of local authorities. The relocation of households living in areas most affected by erosion and tidal flooding is also carried out by the residents. Residents whose homes were routinely submerged by tidal floods, with flood levels increasing year on year, generally made efforts to adapt by raising their houses independently and avoiding moving to other places (Ghani & Hidayati, 2024). The decision to move can be difficult for households. Residents with better access to resources and financial capabilities are more able to relocate, while those with limited means often remain and adapt in place. However, when the government intervenes through the village head with a well-planned relocation programme and provides subsidy assistance, it can reduce psychological and financial barriers to moving. In this case, the presence of a credible local leader and policy support can boost the community's confidence, making the decision to relocate more acceptable and no longer perceived as a heavy burden. An interview with the head of Bedono Village in the Sayung District showed the importance of their role, as in the following quote:

> The process of relocating residents to a location outside the village is indeed quite lengthy. The assistance from the district government is only in the form of money for the purchase of building materials. However, the condition is that residents must already own land outside the area affected by tidal flooding. This certainly poses a challenge, because not all residents have access to land in safe locations. Finally, I took the initiative to buy rice field land outside the village with my personal money, then I 'lent' that land to the residents.

> They are free to decide when they can pay for the land to me. The important thing for me is that the residents can immediately move to a location safe from tidal flooding and access government assistance to build new houses. (Mr Bd, head of the village)

This quote illustrates how transformational leadership is applied in practice. Burns (1978) states that transformational leaders do not just manage but also inspire change by prioritising collective interests, promoting moral values and creating deep trust between leaders and their followers. The initiative described above not only provides technical solutions but also serves as a social intervention that fosters the self-efficacy of the residents. By creating a bridge between public policy (government subsidies) and residents' limitations (land access), the village head acts as a broker of adaptation—a link between formal structures and local realities.

However, informal community leaders—mangrove conservation activists—also address environmental degradation in their work. For instance, Mr Ns, a local leader in Timbulsloko village of Sayung district, initiated the planting of mangrove seedlings to strengthen natural embankments and support biodiversity. As he shared, 'Mangroves are more than just trees; they serve as natural barriers and also offer livelihoods and marine protection'. By linking environmental protection with economic benefits, Mr Ns not only reframes conservation as a shared responsibility but also nurtures collective ownership of adaptation strategies. The leadership fosters collective efficacy, reinforcing the idea that climate resilience is not just a technical task but also a deeply social and moral endeavour (Maak, 2007).

The initiatives of local leaders engaged in environmental conservation via mangrove restoration are also evident in Purworejo village in the Bonang district. On that site, Mr Mk, a local environmental activist and fisherman, has spearheaded mangrove planting for several decades. Besides being the primary driver, Mr Mk also emerged as a transformative figure inside the community. During the interview, he described his actions: 'I possess limited financial resources, yet I can procure mangrove seeds directly from the environment'. Mr Mk also invited residents who could assist with planting to join. The action taken by Mr Mk is an example of ideal influence and inspirational motivation (Burns, 1978), which occurs when local leaders become role models. The result of such leadership is that the community is willing to get involved without coercion.

In addition to local leadership from males, women, as informal community leaders in Demak, also contribute to climate change responses, particularly the increasingly frequent tidal floods that affect coastal areas. For example, a group of mothers in a coastal village in the Sriwulan–Sayung district is forced to allocate household income for home repairs each year due to the effects of tidal floods. Additional sources of income are necessary to meet this requirement. Mrs Ry, a housewife whose husband is a fisherman, emerged as a local leader amid these constraints. Mrs Ry established a milkfish processing enterprise to serve as a ready-to-eat food that could be distributed beyond the city, recognising the economic potential of the sea catch. In addition to starting production, she invited her neighbours to participate, establishing a micro-business ecosystem that empowers women in her community. Additionally, she is pursuing buyers and establishing marketing networks, ranging from local souvenir stores to orders from neighbouring cities. Mrs Ry stated in an interview:

> Initially, I was the sole individual who attempted to prepare presto milkfish and boneless milkfish; however, as time progressed, numerous other women expressed interest in participating. As a result of the tidal floods, household expenses increased. Consequently, we sought methods to continue assisting my spouse. Additionally, I assisted in identifying consumers by utilising social media and acquaintances. Thank God, there are now consistent mandates from government employees in Demak agencies. Additionally, some are dispatched to Kudus and Semarang. (Mrs Ry, local leader in micro-business)

Mrs Nn, from Purworejo Village, Bonang District, is also actively involved in the initiation and development of the salted fish processing business. Following the completion of her higher education, Mrs Nn resolved to return to her birthplace and implement a more contemporary managerial approach to the salted fish business management system, with a particular emphasis on the marketing aspect. Her perspective on entrepreneurship was also enhanced by the academic and social experiences acquired during her education outside the region. This not only enhances the competitiveness of local products but also creates an opportunity for women in their community to be empowered by actively participating in production and distribution activities.

The efficacy of women leaders' activities demonstrates that adaptation to climate change transcends mere technical or infrastructural concerns; it also requires the augmentation of the social and economic capacities of communities, especially marginalised groups such as women. Such examples reflect key components of community resilience: strong social capital (Meerow & Stults, 2016), access to financial and employment resources (Skidmore & Cohon, 2023), and resilient infrastructure (Dedekorkut-Howes et al., 2020). Governance that enables participation, supports knowledge sharing, and ensures inclusivity further enhances adaptive capacity (Bronen & Chapin, 2013). Effective communication ensures that even marginalised groups are informed and involved in climate-related decisions.

In Demak, community workshops, supported by local leaders, promote awareness and inclusive planning for ecosystem restoration and sustainable infrastructure. These efforts help reduce vulnerability while reinforcing livelihoods. Nonetheless, challenges persist. Cultural displacement, emotional distress, and shifts in economic activities can accompany adaptation, particularly relocation. Leaders must support community dialogue and ensure that no group is left behind (Hassan & Kanasan, 2019). Local leadership in Demak exemplifies how adaptation is not only a technical process but also a social one rooted in culture, collaboration, and the capacity to mobilise the community towards shared resilience.

Inclusive Community Participation

To encourage communities to participate in adaptation efforts, effective communication is essential. Local leaders must ensure inclusivity—that all community members can access and understand information about climate risks and adaptation strategies. This requires using multiple communication channels and engaging different groups to ensure that all voices are heard (Beauchamp et al., 2021). Inclusive participation increases the legitimacy of decision-making processes and community support for adaptation efforts. One of our research participants in Dombo Village, Sayung District, explained:

> But we need everyone's participation to make this happen. This is work that everyone can and should participate in, not just a few. Next week, we will start a training session that will teach how to plant mangroves. I ask everyone to participate, especially our younger generation. In addition, it is important for us to consider your ideas and perspectives. What problems might we encounter, do you think? How can we ensure that everyone enjoys this effort? (Mr Mar, village elder)

These grassroots actions reflect the strength of community-based adaptation. The villagers came together to build simple but effective embankments using shared funds and manual labour. One female champion villager in Bedono–Sayung district, Mrs D, recounted how at first we just accepted the flooding, but then we began talking and eventually built our own protection; everyone contributed, from money to labour and even food for the workers. This collective effort strengthens resilience from within, leveraging social ties and mutual trust.

The proactive engagement of informal community leaders in mangrove conservation highlights the transformative power of grassroots leadership in climate adaptation. In an interview with Mr Ns, a respected local leader and conservation activist in Purworejo village, Bonang district, he shared how his initiative began:

> At that time, we saw the coastline receding; land was starting to disappear. So I invited the residents to start planting mangroves. I said, this is not just about the environment—it is also about the future of our children. If mangroves grow, the fish will return, there will be lots of shrimp, and the coast can be protected.

This action reflects transformational leadership's idealised influence and inspirational motivation aspects (Burns, 1978). By appealing to shared environmental and economic concerns, Mr Ns cultivated a collective vision of sustainability beyond immediate concerns. Recognising that sustainability must be grounded in community capability, he took deliberate steps to build local capacity. He organised training sessions on mangrove planting and monitoring techniques, focusing especially on youth and women's groups—two demographics whose engagement is often limited in coastal resource management. As he explained, 'Young people and women have energy and attention. I invited them to join the training and go directly to the field. Now they are the most active in maintaining the mangroves that have been planted'. This reflects a form of inclusive leadership, maintaining trust and shared goals. To ensure that the needs and perspectives of marginalised groups are considered in climate adaptation planning, an equitable decision-making process is essential. According to research, inclusive leadership can increase the psychological security of community members and encourage them to actively participate in discussions and decision-making (Javed et al., 2017). Equitable decision-making processes often result in more sustainable and efficient adaptation strategies.

Effective communication, community engagement, and integration of local knowledge are some of the critical components required for a successful climate adaptation strategy. Community engagement is essential to building trust and ensuring that adaptation strategies are in line with people's specific needs and priorities. Effective communication is also essential to ensure that community members are well informed about climate risks and the adaptation measures being undertaken. Involving communities in decision-making processes increases the legitimacy of these efforts and enhances a sense of ownership, which results in more cooperation and participation in adaptation implementation.

For successful adaptation strategies, the recognition and inclusion of local knowledge and traditional practices are also necessary. Communities often have important insights into their local environment gained from historical experiences with climate variation. These insights can be useful for current adaptation efforts (Huang et al., 2019). Local knowledge increases the effectiveness and sustainability of strategies, making them more context-appropriate and culturally sensitive. In addition, these strategies should be flexible, adaptable to changing situations, and resilient to uncertainty.

Local leaders in the coastal district of Demak have played an important role in managing climate adaptation initiatives. As they protect community needs, facilitate communication, and mobilise resources, they contribute significantly to resilience building. Demak's local leaders have emerged as agents of change, leading their communities through transformative adaptation, as their communities are increasingly threatened by climate change, particularly sea level rise. They have a collective vision to address these challenges, and their actions are key to long-term sustainability. For example, Demak village heads have improved infrastructure, such as providing emergency response services, upgrading roads that are often flooded, and improving accessibility. These leaders have also led efforts to build embankments to divert water flow away from vulnerable areas, thereby increasing the village's resilience to flooding.

Local leaders have recognised the importance of moving families from high-risk areas to safer places. One village head specifically started a relocation programme, which helped affected families rebuild their homes with construction assistance. By involving communities in selecting the location, local leaders ensured that residents could relocate with dignity; they were not regarded as passive subjects but as active agents in determining their future. Residents are encouraged to participate in selecting the relocation site, design, and livelihood support mechanisms. Relocation can be transformed from a disempowering process to a form of empowerment when residents are involved. These efforts demonstrate the proactive role that local leadership plays in helping communities adapt to climate change.

Mangrove ecosystem restoration is one of the biggest successes of Demak. Local leaders have played an important role in initiating the mangrove conservation project. This project will support coastal biodiversity and health and provide economic opportunities for the local community. The concerted efforts of Demak residents to build embankments and raise roads are an additional success for local leaders. This self-driven initiative shows how local leaders can come together to tackle flooding. These initiatives not only reduce coastal erosion and flooding but also improve livelihoods, demonstrating the dual benefits of environmental restoration and economic resilience.

Navigating Barriers, Harnessing Local Strength

Building climate resilience at the community level requires more than just technical solutions—it demands a nuanced understanding of cultural values, financial realities, and community dynamics. Cultural factors such as traditional beliefs, norms, and practices shape how people perceive climate risks and how willing they are to accept adaptation measures. In many cases, these cultural contexts can either support or hinder the adoption of resilience strategies. Therefore, local leaders must possess cultural literacy—an ability to engage respectfully with community traditions while guiding transformation (Nalau & Handmer, 2018). In areas such as Demak, where customary norms remain strong, leadership that respects local values is essential to ensure that adaptation is not seen as disruptive but empowering.

Economic challenges further complicate resilience efforts. Vulnerable communities often lack access to the financial resources and markets needed to implement large-scale or technology-driven adaptation strategies (Liu et al., 2022). Local leaders in such contexts face the uphill task of mobilising support, identifying funding streams, and advocating for programmes that prioritise resilience. This might involve partnering with NGOs, lobbying for governmental support, or working to align local initiatives with broader policy frameworks. Proactivity in policy engagement is critical to unlocking financial assistance for adaptation investments.

Despite these difficulties, there are promising opportunities, especially through community-led approaches. Studies show that when local communities are given ownership of adaptation planning, their sense of agency and accountability increases significantly, enhancing project success (Liu & Mostafavi, 2023). In Demak, for instance, local leaders harnessed this spirit by involving residents in mangrove restoration, road elevation, and embankment construction. These bottom-up strategies not only enhance resilience but also foster community cohesion and local creativity. The use of digital tools, such as social media and mobile messaging, has also expanded communication and participation, particularly among youth (Buzzanell, 2010).

Transformational leadership is central to this process. Leaders who foster collaboration, shared vision, and mutual trust are better positioned to unite communities around adaptation goals (Wang et al., 2021). In Demak, village heads have risen to this challenge through championing initiatives that address sea level rise, tidal surges, and flooding. One such leader spearheaded the elevation of a frequently submerged village road, enhancing mobility and ensuring emergency access. Another led to the construction of embankments made from sandbags and stone, successfully redirecting floodwaters and protecting vulnerable infrastructure.

These efforts reflect the understanding that adaptation must be inclusive and participatory. Effective communication is essential, ensuring that all community members—regardless of age, gender, or socio-economic status—can understand risks and contribute to solutions (Beauchamp et al., 2021). Demak leaders have also conducted training and workshops on mangrove planting, environmental management, and flood mitigation. These events, which target youth and women, fostered a strong sense of ownership and increased the long-term sustainability of the resilience measures.

One example comes from a resident who shared her experience in organising community action. 'At first, we just resigned ourselves to the daily tidal flooding', she explained. 'But over time, we began discussing with neighbours and decided to build our own embankments. Everyone contributed—some gave money, others gave time and labour. Mothers cooked for workers. It became a shared mission'. This spirit of gotong royong—collective action—is a defining feature of Demak's approach to resilience.

Integrating local knowledge with adaptation strategies is another key pillar. Leaders combine traditional wisdom with scientific approaches to design solutions that are both culturally appropriate and technically effective. By blending modern adaptation methods with community practices, they improved not only the physical outcomes of projects but also social trust and engagement.

Knowledge dissemination has also been prioritised. Seminars and public education campaigns on sustainable coastal restoration and seawall construction helped residents understand both why and how to adapt. These efforts reduce vulnerability and create a culture of preparedness. The case of Demak demonstrates that successful adaptation hinges on multi-dimensional leadership—empathetic, informed, and willing to lead from the ground up.

As noted, however, challenges persist. Cultural displacement, economic disruptions, and psychological stress—especially related to relocation—must be managed sensitively (Hassan & Kanasan, 2019). Leaders in Demak have addressed these issues by encouraging open dialogue and ensuring that displaced families receive support throughout the process.

Ultimately, the success of Demak's climate resilience efforts is founded on the leaders' holistic approach. Effective communication, inclusive participation, integration of local knowledge, and visionary leadership have all contributed to building adaptive capacity. Demak offers a compelling example of how community-driven strategies, guided by local leaders, can effectively address the profound challenges posed by climate change, particularly in coastal areas that are most at risk (Sienkiewicz-Małyjurek, 2022; Shaw & Theobald, 2011; Javed et al., 2017).

Conclusion

Sea level rises has significant impacts on coastal communities, infrastructure, and ecosystems. As global temperatures rise and polar ice caps melt, the need for effective adaptation measures is becoming more apparent. Coastal areas are particularly vulnerable to flooding and erosion, especially those at low elevations, such as along the north coast of Java. Addressing these issues requires strong coastal management, community participation, and sustainable practices.

Local leaders play a critical role in organising adaptation efforts, mobilising resources, and integrating local knowledge into policies and practices. Their involvement ensures that adaptation strategies are not only context-specific but also culturally sensitive, thereby increasing the overall effectiveness of adaptation measures.

Incorporating local knowledge and traditional practices into policy frameworks is important to successfully combat sea-level rise and coastal sea inundation. To meet local needs and adapt to change, policies should consider community engagement, capacity building, and continuous monitoring. Policymakers can create more resilient and efficient adaptation strategies by incorporating local wisdom and encouraging collaborative approaches. This method not only strengthens community relations but also ensures that adaptation measures address the specific problems of vulnerable areas.

Future research should focus on climate adaptation and community resilience, particularly in the context of sea level rise. There is a need to examine the long-term impacts of climate change on coastal communities and assess the effectiveness of different adaptation approaches. Studies should also look at how local leadership drives new solutions and encourages community participation in climate resilience efforts. Understanding how local adaptation practices interact with global climate change impacts will be critical to developing comprehensive strategies to reduce the impacts of sea level rise and increase community resilience.

We recommend that national and local governments prioritise flexible funding mechanisms for local adaptation initiatives, institutionalise knowledge exchange platforms between districts, and develop leadership training for grassroots actors.

References

Abdelwahab, M. (2023). Negative consequences of climate change on marine and coastal tourism activities in Ras Sudr: Evidence from stakeholders. *Journal of Association of Arab Universities for Tourism and Hospitality, 24*(2), 709–731. https://doi.org/10.21608/jaauth.2023.224420.1487

Agrawal, A. (2008). The role of local institutions in climate change adaptation. *Social Dimensions of Climate Change*. The World Bank, Washington DC.

Baker, I., Peterson, A., Brown, G., & McAlpine, C. (2012). Local government response to the impacts of climate change: An evaluation of local climate adaptation plans. *Landscape and Urban Planning, 107*(2), 127–136. https://doi.org/10.1016/j.landurbplan.2012.05.009

Barrett, H., & Armstrong, J. (2023). Climate change adaptation and the back of the invisible hand. *Philosophical Transactions of the Royal Society B: Biological Sciences, 378*(1889). https://doi.org/10.1098/rstb.2022.0406

Beauchamp, E., Sainsbury, N., Greene, S., & Chaigneau, T. (2021). Aligning resilience and well-being outcomes for locally-led adaptation in Tanzania. *Sustainability, 13*(16), 8976. https://doi.org/10.3390/su13168976

Bemmelen, R. W. (1970). *The geology of Indonesia: Economic geology* (Vol. 2). Martinus Nijhoff.

Benulic, K. S., Kropf, M., Linnér, B. O., & Wibeck, V. (2022). The meaning of leadership in polycentric climate action. *Environmental Politics, 31*(6), 1016–1036. https://doi.org/10.1080/09644016.2021.1970087

Bronen, R., & Chapin, F. (2013). Adaptive governance and institutional strategies for climate-induced community relocations in Alaska. *Proceedings of the National Academy of Sciences, 110*(23), 9320–9325. https://doi.org/10.1073/pnas.1210508110

Bryan, E., Elias, M., Kosec, K., Kyle, J., Morgan, M., & Najjar, D. (2023). *Women's leadership and implications for climate resilience: A conceptual framework*. https://hdl.handle.net/10568/138888

Burns, J. M. (1978). *Leadership*. Harper & Row.

Buzzanell, P. (2010). Resilience: Talking, resisting, and imagining new normalcies into being. *Journal of Communication, 60*(1), 1–14. https://doi.org/10.1111/j.1460-2466.2009.01469.x

Clark, P. U., Mix, A. C., Eby, M., Levermann, A., Rogelj, J., Nauels, A., & Wrathall, D. J. (2018). Sea-level commitment as a gauge for climate policy. *Nature Climate Change, 8*(8), 653–655. https://www.nature.com/articles/s41558-018-0226-6

Cox, R., & Hamlen, M. (2014). Community disaster resilience and the rural resilience index. *American Behavioral Scientist, 59*(2), 220–237. https://doi.org/10.1177/0002764214550297

Dedekorkut-Howes, A., Torabi, E., & Howes, M. (2020). When the tide gets high: A review of adaptive responses to sea level rise and coastal flooding. *Journal of Environmental Planning and Management, 63*(12), 2102–2143. https://doi.org/10.1080/09640568.2019.1708709

Dutton, A., Carlson, A. E., Long, A. J., Milne, G. A., Clark, P. U., DeConto, R., et al. (2015). Sea-level rise due to polar ice-sheet mass loss during past warm periods. *Science, 349*(6244), aaa4019. https://doi.org/10.1126/science.aaa4019

Ghani, M. W., & Hidayati, I. (2024). Adapting to sea level rise: Migration decisions and disaster mitigation strategies in small-scale fishing communities. In *IOP Conference Series: Earth and Environmental Science* (Vol. 1314(1), p. 012013). IOP Publishing. https://doi.org/10.1088/1755-1315/1314/1/012013.

Grey, M., Masunungure, C., & Manyani, A. (2020). Integrating local indigenous knowledge to enhance risk reduction and adaptation strategies to drought and climate variability: The plight of smallholder farmers in Chirumhanzu District, Zimbabwe. *Jàmbá: Journal of Disaster Risk Studies, 12*(1). https://doi.org/10.4102/jamba.v12i1.924

Gupta, J., Termeer, C., Klostermann, J., Meijerink, S., Van den Brink, M., Jong, P., et al. (2010). The adaptive capacity wheel: A method to assess the inherent characteristics of institutions to enable the adaptive capacity of society. *Environmental Science & Policy, 13*(6), 459–471. https://doi.org/10.1016/j.envsci.2010.05.006

Hassan, S., & Kanasan, V. (2019). *Expecting the unexpected: Climate resilience in Malaysian local government land use planning*. https://doi.org/10.15405/epms.2019.12.19

Huang, Y., Wong, H., & Fu, Y. (2019). Resilience and depression among the survivors of the 2013 Yaan earthquake. *Journal of Social Work, 20*(6), 817–833. https://doi.org/10.1177/1468017319852593

Hummel, M. A., & Stacey, M. T. (2021). Assessing the influence of shoreline adaptation on tidal hydrodynamics: The role of shoreline typologies. *Journal of Geophysical Research: Oceans, 126*(2), e2020JC016705. https://doi.org/10.1029/2020JC016705

Javed, B., Naqvi, S., Khan, A., Arjoon, S., & Tayyeb, H. (2017). Impact of inclusive leadership on innovative work behavior: The role of psychological safety. *Journal of Management & Organization, 25*(1), 117–136. https://doi.org/10.1017/jmo.2017.3

Jevrejeva, S., Jackson, L. P., Riva, R. E., Grinsted, A., & Moore, J. C. (2016). Coastal sea level rise with warming above 2 °C. *Proceedings of the National Academy of Sciences, 113*(47), 13342–13347. https://doi.org/10.1073/pnas.1605312113

Kantamaneni, K., Christie, D., Lyddon, C. E., Huang, P., Nizar, M., Balasubramani, K., et al. (2022). A comprehensive assessment of climate change and coastal inundation through satellite-derived datasets: A case study of Sabang Island. *Indonesia. Remote Sensing, 14*(12), 2857. https://doi.org/10.3390/rs14122857

Keys, N., Thomsen, D., & Smith, T. (2014). Adaptive capacity and climate change: The role of community opinion leaders. *Local Environment, 21*(4), 432–450. https://doi.org/10.1080/13549839.2014.967758

Koh, H. L., Teh, S. Y., Kh'Ng, X. Y., & Raja Barizan, R. S. (2018). Mangrove forests: Protection against and resilience to coastal disturbances. *Journal of Tropical Forest Science, 30*(5), 446–460. https://www.jstor.org/stable/26514436

Liu, C. F., & Mostafavi, A. (2023). Revealing hazard-exposure heterophily as a latent characteristic of community resilience in social-spatial networks. *Scientific Reports, 13*(1), 4817. https://www.nature.com/articles/s41598-023-31702-9

Liu, W., Zhang, J., & Qian, L. (2022). Measuring community resilience and its determinants: Relocated vulnerable community in western China. *International Journal of Environmental Research and Public Health, 20*(1), 694. https://doi.org/10.3390/ijerph20010694

Lo, W., Purnomo, S. N., Sarah, D., Aghnia, S., & Hardini, P. (2021). Groundwater modelling in urban development to achieve sustainability of groundwater resources: A case study of Semarang City. *Indonesia. Water, 13*(10), 1395. https://doi.org/10.3390/w13101395

Maak, T. (2007). Responsible leadership, stakeholder engagement, and the emergence of social capital. *Journal of Business Ethics, 74*, 329–343. https://doi.org/10.1007/s10551-007-9510-5

Makate, C. (2020). Local institutions and indigenous knowledge in adoption and scaling of climate-smart agricultural innovations among Sub-Saharan smallholder farmers. *International Journal of Climate Change Strategies and Management, 12*(2), 270–287. https://doi.org/10.1108/IJCCSM-07-2018-0055

Markovic, M., Komljenovic, I., Kovacevic, V., Radic, V., Jovic, J., Trbić, G., Pesevic, D., & Markovic, M. (2018). Alleviation of Negative Climate Change Effects on Maize Yields in Northern Bosnia by Liming and Phosphorus Fertilization. In *Climate Change Adaptation in Eastern Europe: Managing Risks and Building Resilience to Climate Change* (pp. 169–183). Cham: Springer International Publishing. https://doi.org/10.1007/978-3-030-03383-5_12

Measham, T. G., Preston, B. L., Smith, T. F., Brooke, C., Gorddard, R., Withycombe, G., & Morrison, C. (2011). Adapting to climate change through local municipal planning: Barriers and challenges. *Mitigation and Adaptation Strategies for Global Change, 16*, 889–909. https://doi.org/10.1007/s11027-011-9301-2

Meerow, S., & Stults, M. (2016). Comparing conceptualizations of urban climate resilience in theory and practice. *Sustainability, 8*(7), 701. https://doi.org/10.3390/su8070701

Nalau, J., & Handmer, J. (2018). Improving development outcomes and reducing disaster risk through planned community relocation. *Sustainability, 10*(10), 3545. https://doi.org/10.3390/su10103545

Ntontis, E., Drury, J., Amlôt, R., Rubin, G. J., & Williams, R. (2018). Emergent social identities in a flood: Implications for community psychosocial resilience. *Journal of Community & Applied Social Psychology, 28*(1), 3–14. https://doi.org/10.1002/casp.2329

Nurhidayah, L., Davies, P., Alam, S., Saintilan, N., & Triyanti, A. (2022). Responding to sea level rise: Challenges and opportunities to govern coastal adaptation strategies in Indonesia. *Maritime Studies, 21*(3), 339–352. https://doi.org/10.1007/s40152-022-00274-1

Reager, J. T., Gardner, A. S., Famiglietti, J. S., Wiese, D. N., Eicker, A., & Lo, M. H. (2016). A decade of sea level rise slowed by climate-driven hydrology. *Science, 351*(6274), 699–703. https://doi.org/10.1126/science.aad8386

Septiarani, B., & Yesiana, R. (2020). Affecting factors on community based mangrove replantation programs in Semarang coastal area. *Ind J Plann Develop, 5*(2), 87–94. http://dx.doi.org/10.14710/ijpd.5.2.87-94

Shaw, K., & Theobald, K. (2011). Resilient local government and climate change interventions in the UK. *Local Environment, 16*(1), 1–15. https://doi.org/10.1080/13549839.2010.544296

Sienkiewicz-Małyjurek, K. (2022). Social capital and transformational leadership in building the resilience of local governance networks. *Sustainability, 14*(8), 4720. https://doi.org/10.3390/su14084720

Skidmore, T. A., & Cohon, J. L. (2023). A multicriteria decision analysis framework for developing and evaluating coastal retreat policy. *Integrated Environmental Assessment and Management, 19*(1), 83–98. https://doi.org/10.1002/ieam.4662

Stammer, D., Cazenave, A., Ponte, R. M., & Tamisiea, M. E. (2013). Causes for contemporary regional sea level changes. *Annual Review of Marine Science, 5*(1), 21–46. https://doi.org/10.1146/annurev-marine-121211-172406

Susilo, S., Salman, R., Hermawan, W., Widyaningrum, R., Wibowo, S., Lumban-Gaol, Y., Meilano, I., & Yun, S. (2023). GNSS land subsidence observations along the northern coastline of Java, Indonesia. *Scientific Data, 10*(1). https://doi.org/10.1038/s41597-023-02274-0

Suursaar, Ü. (2011). Sea level variations along the Estonian coast of the Baltic Sea. *Sea Level Rise, Coastal Engineering, Shorelines and Tides*, 105–122. Nova Science Publisher Inc, New York.

Wang, Y., Sun, L., & Zhang, L. (2021). Evaluation of community adaptation measures to climate change: A case study in Chinese coastal cities. *Cities, 112*, 103177. https://doi.org/10.1016/j.cities.2021.103177

Werner, A. D., & Simmons, C. T. (2009). Impact of sea-level rise on sea water intrusion in coastal aquifers. *Groundwater, 47*(2), 197–204. https://doi.org/10.1111/j.1745-6584.2008.00535.x

Williams, D. S., Celliers, L., Unverzagt, K., Videira, N., Máñez Costa, M., & Giordano, R. (2020). A method for enhancing capacity of local governance for climate change adaptation. *Earth's Future, 8*(7), e2020EF001506. https://doi.org/10.1029/2020EF001506

Chapter 9
From Buyers to Planners: Understanding Market Behaviour in Climate Adaptation for Urban Indonesia

Sariffuddin, Ira Irawati, and Datuk Ary A. Samsura

Introduction

The government of Indonesia needs to expand the role of spatial planning—which is currently regulated by both Law No. 26/2007 on Spatial Planning and Law No. 24/2007 on Disaster Management—as an adaptation measure in addition to mitigation to address increasingly complex climate change issues, particularly in coastal cities. This extended function will be quite efficient if the property market plays a role in realising autonomous adaptation to support efficient land use. In following the market, property buyers' decisions shape land use and determine the spatial risk of climate change. Thus, navigating their decision by signalling the risk through price will enhance spatial adaptation and complement the planned adaptation from state governments. Given that the property market's response to climate change is lower property prices and displacement, property development could be navigated through market-based instruments (Filatova, 2014). As a result, urban planners can also steer investments.

Investment control, especially by buyers of property in flood-prone areas, can help prevent neighbourhood change and decline. This argument is based on global and local experiences regarding the response of the property market to flood-prone

Sariffuddin (✉)
Department of Urban and Regional Planning, Faculty of Engineering, Diponegoro University, Semarang, Indonesia
e-mail: sariffuddin@live.undip.ac.id

I. Irawati
Department of Urban and Regional Planning, National Institute of Technology, Bandung, Indonesia

D. A. A. Samsura
Department of Geography, Planning, and Environment, Radboud University, Nijmegen, The Netherlands

L. Nurhidayah et al. (eds.), *Climate Change, Labour and Migration in Indonesia*, Springer Climate, https://doi.org/10.1007/978-3-032-15334-0_9

areas. Most international studies have revealed the shaping of housing submarkets in flood-prone areas that were investigated, for example, in Australia (Rajapaksa et al., 2017) and in Indonesia (Sariffuddin et al., 2024b). Those scholars argue that the gathering of property buyers in particular areas does not happen by chance; rather, these actors cluster in specific areas to share locational benefits under their budget constraints. The decreasing property value in response to flooding has arguably emboldened lower-income buyers to purchase and reside in flood-prone areas, while leaving adaptation costs to the government (Filatova, 2014). Our recent local studies in Indonesia have also shown that the *flood-risk housing* submarket continues to decline and spread to nearby areas (Sariffuddin et al., 2024a; Sariffuddin et al., 2024b), due to disinvestment in flood-prone locations. These findings help explain why the pace of increase in coastal vulnerability in the Global South often outstrips the rate of change in hazard intensity (IPCC, 2012), as well as the argument that uncontrolled land-use change contributes to increasing exposure over time, thereby intensifying vulnerability in coastal areas. These global and local findings enhance the current understanding of the property market's critical role in shaping spatial variability risk, complementing the established knowledge that flood events have a significant correlation with uncontrolled land use change (Wolff et al., 2020).

Moreover, global literature indicates that coastal cities in Indonesia face at least three significant challenges related to complex climate change issues. First, uncontrolled land use changes have significantly contributed to higher rates of coastal flooding. Sejati et al. (2024) examined the spatial correlation between the dynamics of land use change and flood events in Indonesia. They concluded that urban growth often expands into low-lying areas, which can sometimes experience flooding. A second significant challenge is the rapid and uncontrolled urbanisation in coastal cities. Most scholars argue that this urbanisation is significantly related to rapid growth and industrialisation in Global South countries. The United Nations (2018) predicts that the urban population will reach 68% of the total population by 2050. Third, a lack of climate-related adaptation has been identified in most coastal cities in developing countries, including Indonesia. Hallegatte et al. (2013) argue that this lack of adaptation relates to the government budget. At the same time, land subsidence also significantly affects coastal areas of Indonesia, leading to flooding across more than 4000 km^2 (Murray et al., 2019). Although anomalous weather has been identified in the form of monthly seasonal changes and geographical variation in rain intensity, the rainfall over Java Island during 1985–2014 was relatively constant at an average of 60–80 mm/day (Yanto et al., 2017). Yet the number of flood events in Java continues to increase (Sariffuddin et al., 2019), with the resulting economic loss predicted to exceed 54% of Indonesia's GDP.

While the discussion on spatial adaptation to climate change is well established, the crucial role of the property market, especially in Indonesia, is rarely recognised. This gap is understandable because many scholars see climate change as a global issue requiring local action (Filatova, 2014). However, it is becoming increasingly urgent to adopt a more flexible approach to climate change adaptation, particularly by harnessing the role of property markets. Most countries, including the Indonesian government, are more concerned with engineered protection and spatial adaptation from the perspective of planning by law or land use planning—a command-and-control instrument that is constantly revised every five years. Considering that urban development is the result of the interplay between planning by law and property rights over land (Hartmann & Needham, 2012; Li et al., 2019), this chapter aims to share the critical role of the property market driven by private law to support spatial adaptation. This active role of property buyers through market forces is often understood as an autonomous adaptation (Filatova, 2014). Voluntary adaptation based on buyers' decisions before purchase will likely complement the spatial-based adaptation, which is regulated by planning law.

This chapter, which is based on a literature review, is divided into three sections. The first section explains the current issues related to climate change in Indonesia. The second section explains the research method, followed by the conceptual framework. The final section explains the position of the property market in relation to spatial adaptation.

Research Method

We employed a structured narrative review to explore the critical role of autonomous adaptation in supporting land use planning for adaptation instruments in Indonesia from a property market perspective. The review was designed to conceptualise theoretical frameworks, synthesise empirical findings, and review regulations relevant to spatial planning and disaster management. We focused our literature review on the interconnection between market-based instruments for climate adaptation, as conceptualised by Filatova (2014), and individual equilibrium, which explains autonomous adaptation from rational economic behaviour, as conceptualised by Mendelsohn (2000). Combining these two theoretical frameworks enabled us to conceptualise spatial adaptation from the property market perspective and the role of autonomous adaptation in shaping land use, which, in turn, influences the spatial climate risk. For this purpose, we also reviewed Schanze's (2006) flood risk management.

In the first step of the review, we searched peer-reviewed articles on the Web of Science and Semantic Scholar. We searched the Web of Science for scientific articles related to current studies on Indonesian coastal cities, with a focus on the property market, climate change, and coastal flooding. We employed specific keywords, such as the following combinations: (1) 'Coastal flooding' and 'Indonesia'; (2) 'Climate change' and 'Indonesia'; and (3) 'Climate adaptation' and 'Indonesia'. This targeted approach enabled us to identify and analyse current studies related to climate-related coastal disasters. After searching the articles, we connected the dots between spatial adaptation and the property market by tracking their link using semanticscholar.com based on their citations, references, and related papers. This tracing began with the idea of market-based instruments, as proposed by Filatova (2014), and the concept of efficient adaptation to climate change from an economic perspective, as outlined by Mendelsohn (2000). To conduct an evidence-based investigation, we employed a multilayered comparative approach by examining recent empirical studies on coastal disasters in Indonesia. This observation involved reviewing global and local research through the selection process, utilising a database search in the Web of Science. We prioritised peer-reviewed journal articles published in the last 5–10 years.

In the final step, we discussed the conceptual framework and recent studies on Indonesian coastal cities, with a focus on Indonesian policy—specifically, Law No. 26 of 2007 on Spatial Planning and Law No. 24 of 2007 on Disaster Management. Based on this discussion, we suggest recommendations and policy implications. Thus, our study incorporates a policy evaluation of spatial planning by synthesising insights from conceptual frameworks, empirical evidence on coastal disasters, and national regulatory directives. We analyse the policy implications to form recommendations to determine the essential role of the property market in land use control. We recommend the indispensable role of the property market in shaping land use control, as outlined in Law No. 26 and Law No. 24.

Recent Studies on Indonesian Coastal Flooding

The complex issues of coastal cities in Indonesia are at least partly due to (1) climate-related hazards and spatial drivers; (2) policy responses and their limitations; and (3) the property market's response to coastal risk. Many scholars have associated coastal flooding in Indonesia, such as in Semarang, Jakarta, and Demak, with rapid urbanisation and land use change (Marfai, 2014a; Sejati et al., 2024). Moreover, Abidin et al. (2013) and Andreas et al. (2017) have also revealed that land subsidence significantly influences the growing rise in sea level and coastline changes. These results are in line with earlier research by Murray et al. (2019), which identified land subsidence as a major contributor to coastal flooding across roughly 4,000 square kilometres in Indonesia. In addition, the expansion of built-up areas and rapid population growth along the coast have further exacerbated subsidence, accelerating the decline of many coastal neighbourhoods. Additionally, uncontrolled land use changes and population growth in coastal areas contribute to land subsidence, which in turn causes a decline in coastal neighbourhoods. Consequently, there is no room for water to absorb seawater runoff. These empirical findings highlight that climate change is not only a hydrometeorological problem but also a spatial issue (Wolff et al., 2020). The following section elaborates on these three issues in more detail.

First, coastal flooding in Indonesia is not only a hydrometeorological issue but also a spatial driver. Yanto et al. (2017) revealed that although weather anomalies were identified in the form of monthly seasonal changes and geographical variations in rain intensity, the rainfall over Java Island from 1985 to 2014 was relatively constant, averaging 60–80 mm/day. Yet the number of flood events in Java continues to increase (Sariffuddin et al., 2024a). In other words, although the weather has arguably remained constant, the frequency of flood events has been increasing.

Recent studies by Sejati et al. (2024) and Handayani et al. (2020) have shown that flood events are spatially correlated with land use changes. Their findings strengthen the idea that spatial drivers also influence the increasing frequency of flooding. In global discussions, Wolff et al. (2020) have even highlighted that uncontrolled land use change will lead to subsequent exposure, particularly in developing countries.

Second, the Indonesian government currently focuses on engineered flood defences, with spatial adaptation arguably being overlooked. This response is understandable because Indonesian regulations (Law No. 26 of 2007 and Law No. 24 of 2007) prioritise land use as a mitigation instrument rather than as an adaptation measure. Hence, the government allocated IDR 17 trillion under Presidential Regulation No. 18/2020 to finance coastal dykes in Jakarta, Cirebon, Pekalongan, Semarang, and Demak. While this step represents a substantial commitment, it still reflects a reliance on engineered, top-down solutions. In this measure, the government is more active than the private sector. Governments often employ command-and-control measures to mitigate the risk of climate-related hazards, such as constructing embankments and floodgates in Jakarta (Purnomo et al., 2024). While this approach offers strategic and political advantages, its implementation is often delayed by administrative complexities and budgetary constraints. For example, revising an urban master plan in Indonesia can take at least five years, thus limiting timely adaptation. In addition, rapid urbanisation requires anticipating global losses estimated to rise from USD 6 billion in 2005 to USD 60–63 billion in 2050. If only socioeconomic changes are considered, Indonesia is expected to lose USD 1.139 billion (0.14% of GDP) by 2050. However, if land subsidence and sea-level rise are also taken into account, then the loss is projected to be USD 1.750 billion (0.54% of GDP) by 2050 (Hallegatte et al., 2013).

Third, a recent study on the property market also revealed the adverse effects of coastal flooding. Using hedonic property modelling, Cobián Álvarez and Resosudarmo (2019) found that property prices decreased in response to coastal flooding in Jakarta. They calculated a decrease of at least 0.0952% in property value in response to a 1% increase in coastal flooding. Similar findings have also been found in the small coastal cities of Pekalongan. Property prices decreased by 4% due to coastal flooding and its spillover effects on nearby areas. In an earlier work, using a spatial data panel spanning eight years—from 2013 to 2020—we found that property prices continued to decrease, with no indication of a return to regular prices (Sariffuddin et al., 2024a). Demand shifts occur from higher-income residents to lower-income buyers in coastal neighbourhoods, affecting their limited ability to conduct maintenance. As a result, changes and declines in coastal neighbourhoods have been identified among coastal residents of Pekalongan. In this regard, Li et al. (2020) argue that urban development is the result of an interplay between government planning and property market dynamics over land. The property market responses described above provide complementary evidence that flood risk from climate change is not merely a matter of hydrometeorology, engineered flood defences, or rapid urbanisation, but also a property market issue. In other words, behind uncontrolled land use changes lies the growth of the property market in flood-prone areas, which has even become segmented into new submarkets.

Conceptual Framework

Spatial Adaptation and the Property Market

Spatial adaptation has emerged as a prominent mechanism for addressing climate-related risks, particularly in coastal flood-prone areas. Filatova (2014) even argues that spatial planning is at the heart of flood risk management, because the potential loss is also significantly determined by the spatial pattern. Drawing on the Source-Pathway-Receptor-Consequence (SPRC) framework, this chapter highlights the importance of managing receptors in mitigating the adverse risks associated with coastal flooding, particularly among property buyers, who play a central role in reducing exposure and vulnerability.

The SPRC model describes the process of flood risk management. This model was first developed in environmental studies to explain the movement of pollutants from a source transmitted through a pathway and received by a receptor. This model was later adapted for flood risk management (Schanze, 2006). The basic idea is that flood risk management can be addressed at the source, including pathways such as rivers and coastal receptors, to minimise adverse consequences (van der Plank et al., 2021).

In the SPRC model, land use planning contributes to receptor management in environmental risk assessment. Its role differs from that of source and pathway management, which involves reducing risk by controlling hazards. In receptor management, risk-takers—including property owners, businesspeople, and governments—are managed through land use planning and other relevant strategies. For property purchasers, avoiding risky areas is an alternative to risk management. As a result, efficient spatial planning can be accomplished to prevent spatial externalities. Different types of receptors also face varying levels of potential losses. For instance, industrial areas are likely to suffer greater economic damage than agricultural zones and often have a lower capacity to recover quickly after a disaster. We agree that the risk perception of coastal communities regarding dissonance can be addressed through property prices, which signal risk through price (Filatova, 2014; Miranda Sara et al., 2016; Moser et al., 2022).

Climate-related risks, however, are unevenly distributed across space. Spatial adaptation is increasingly being framed as part of the broader debate on land for climate adaptation, particularly in light of the recent evidence that climate change amplifies spatial variability in risk. For example, Suri et al. (2023) report that mountainous areas currently receive more rainfall than low-lying locations due to rising temperatures. These geographical risks encourage some urban scholars and geographers to adopt spatial-based adaptation strategies to cope with spatial risk variation. The basic idea is to manage risk receptors, including avoiding hazardous areas, to reduce the potential for adverse effects. Thus, some scholars utilise the SPRC model to explain the position of land use planning in adaptation measures (Carter & Sherriff, 2016; Hurlimann & March, 2012).

In response to spatial variability in climate risk, most governments have adopted planned adaptation approaches based on command-and-control mechanisms. These approaches allow or restrict specific land use at a particular location through spatial planning regulations. This instrument is often referred to as planning by law or planned adaptation (Filatova, 2014). Some governments complement this instrument with building codes, zoning regulations, and land use planning for disaster risk reduction. These instruments have been identified in several countries in the Global South, such as Malaysia, with its National Climate Change Policy (Chitsa et al., 2022); Indonesia, with spatial planning for disaster risk reduction (Prana et al., 2024); and the Philippines, which integrates climate change adaptation into local planning (Cuevas, 2016). They argue that involving disaster risk reduction in spatial planning can support government budgeting for disaster management. However, some studies on disaster management have also revealed issues with lock-in situations in the implementation of planned adaptation due to normative or strategic constraints (Matthews et al., 2015; Thoidou, 2021). The rigidity of these regulatory instruments often leads to a situation in which spatial plans cannot be revised swiftly enough to respond to evolving risk dynamics, especially in rapidly urbanising contexts like Indonesia.

These limitations underscore the need for more flexibility, particularly through the autonomous decisions of property buyers. Therefore, spatial adaptation requires the active participation of property owners as key players in urban development (Li et al., 2019). Behavioural change is essential to anticipate future uncertainties. Market forces, which serve as the basis for selecting property locations, can be considered as part of market planning (Filatova, 2014) and are also subject to complementary planning by law (Hartmann & Needham, 2012). Property owners can contribute to spatial adaptation before flooding occurs. Before a disaster, for instance, they could have chosen safer areas instead of risky sites. Command-and-control measures through planned adaptation are ineffective in guiding property markets in flood-prone areas (Filatova, 2014). Instead, they base their rights on owner initiatives, which often consider the market conditions.

The spatial structure of cities is also shaped by economic behaviour and property rights. Hartmann and Needham (2012) demonstrated that regulations and property rights shape spatial patterns through landowners' initiatives to utilise their land. Economic factors sometimes become the primary consideration when choosing a home location to maximise utility or profits within budget limits (Filatova, 2014), rather than regulatory directives that shape the land. The monocentric urban model, proposed by Alonso (1964), explains how the property market creates spatial organisation (Filatova, 2014), as indicated by house price variations that decrease with distance from the city centre. Although Ahlfeldt (2011) concluded that Alonso's urban model still performs well, the land value gradient changes dynamically in flood-prone areas, as indicated by price discounts in particular submarkets (Rajapaksa et al., 2017). Floods, as environmental disamenities, may have affected these results as part of a bundle of property prices according to the hedonic price model. The above description explains how the property market responds to flood-prone areas, which can be examined through a discussion of spatial adaptation.

These spatial dynamics suggest that flood risk is increasingly internalised into property values, manifested through price discounts in vulnerable submarkets, indicating a market-led adaptation process that both complements and challenges planning by law.

Autonomous Adaptation

Spatial planning has long been recognised as a crucial tool for managing flood risk, particularly in coastal areas vulnerable to the impacts of climate change. However, planning by law—typically guided by government policies—often faces lock-in institutional effects, which limit its ability to respond to localised and dynamic flood risks. In this context, autonomous adaptation refers to voluntary, self-directed actions taken by individuals or households to reduce their exposure to climate hazards, such as relocating, retrofitting homes, or avoiding high-risk areas, typically without direct government intervention (Filatova, 2014). Understanding how such autonomous responses emerge, particularly through the behaviour of property buyers and price signals, is crucial for complementing formal planning mechanisms. This section explores how household behaviour and price dynamics reflect autonomous adaptation processes and how these market-based behaviours interact with spatial risk patterns in flood-prone areas.

Filatova (2014) argues that both planned adaptation by the government and autonomous adaptation by the property market should take place for effective climate-related disaster management. Behavioural changes are essential for autonomous adaptation to voluntarily avoid flood-prone areas to decrease property loss and damage. In the absence of autonomous adaptation, people collectively tend to choose similar sites, getting a favourable location quality, while leaving the adaptation cost to the government. Moreover, the government sometimes experiences a 'lock-in situation' due to normative or strategic constraints (Thoidou, 2021). Instead of being based on regulation, property markets that form spatial patterns have 'private' or 'civil' laws enforced by citizens' initiative to use their land. Therefore, understanding how autonomous adaptation works is essential to complement planned adaptation and to support spatial adaptation.

Active adaptation of buyers is the goal of behaviour change through house location decisions that avoid flood-prone areas to decrease the damage (Filatova, 2014). This action is part of policy-based adaptation using land use planning, often referred to as planned adaptation (Filatova, 2014). The residential buyout program in the United States is an excellent example of how spatial adaptation interventions through housing relocation can save citizens from flood risks (Loughran & Elliott, 2019). Autonomous adaptation has also been investigated through relocation (Buchori et al., 2018; Call & Gray, 2020; Gray & Mueller, 2012) and house renovation (Marfai, 2014b). This active role of property purchasers in climate risk adaptation can theoretically be explained by their efficient adaptation to climate change.

To implement this framework, we integrate economic theories of behaviour and housing behaviour, resulting in land use dynamics. Mendelsohn (2000) explains that efficient adaptation can be achieved when the marginal benefit (MB) exceeds or at least equals the sum of the marginal cost (MC) and marginal externalities (ME), or $MB \geq MC + ME$. Rosen (1974) developed the hedonic property model, later expanded by Freeman (1979) to consider environmental quality, drawing on Lancaster's earlier work (1966) on consumer demand. At its core, the model suggests that people assign value to a product based on the specific characteristics they care about rather than the product as a whole. Applied to housing markets, this framework helps explain how buyers respond to flood risk by adjusting their willingness to pay based on how they perceive associated environmental threats. According to Tobin and Newton (1986), buyers internalise flood risk by evaluating the severity and frequency of past events, as well as the land use characteristics of the area. Both flood severity and flood frequency influence the quality of building structures and environmental quality, which is part of a hedonic regression model.

Moreover, scholars in hedonic property modelling have revealed that risk perception influences purchasers' behaviour. The individual household chooses a set of different houses with varying prices and utilities by selecting a specific location to maximise the utility under their budget constraint (Filatova, 2014; Fujita & Thisse, 2013). Household utilities are highly influenced by the location and environmental qualities of urban facilities (Helbich et al., 2013). Freeman (1979) accounted for house characteristics, location, neighbourhood, and environmental quality as vectors of property prices in the hedonic price equation. Therefore, property prices can be considered the basis for rational decisions by consumers and suppliers to ensure individual equilibrium.

Building on this theoretical foundation, empirical studies have examined how variations in risk information influence actual market behaviour, particularly in flood-prone regions. Samarasinghe and Sharp (2010) found that flood risk information highly influences property market behaviour through price discounts within flood-prone areas in New Zealand. The value of the expected utility of a property depends on the risk information, which then triggers the behaviour of the property market. Stern (2008) also underlines that imperfect information in the property market is the main constraint to running effective autonomous adaptation. If imperfect information is disseminated, inequalities in the market chain between buyers and sellers proliferate. When there is no information about risk, buyers and sellers will make normal transactions while ignoring the negative consequences of flooding. Samarasinghe and Sharp (2010) conclude that the subjective risk assessment by buyers is internalised into their willingness to pay. Consequently, imperfect information about flood risk triggers imperfect property market chains in the flood-prone areas. The provision of risk information can potentially help mitigate the risk (de Koning et al., 2017). Given the above explanation, it is hardly surprising that households, as price takers, are likely to decide on reasonable adaptations based on economic rationality.

Conclusion and Policy Implications

Spatial adaptation calls for active adaptation by the property owner as a player in urban development (Li et al., 2019). A behavioural change from passivity to proactiveness is required in players for them to anticipate future uncertainty during adaptation. Market forces, which form the basis for the selection of property locations, can be considered a part of planning by the market (Filatova, 2014) and law (Hartmann & Needham, 2012). Property owners can contribute before and after flood events to facilitate spatial adaptation. Before the disaster, they could choose safer areas instead of risky sites. For properties already located in flood-prone areas, residential mobility is an alternative that should be considered. Command-and-control measures through planned adaptation are not effective in guiding property markets in flood-prone areas (Filatova, 2014). Instead, they have their right based on owner initiatives that often consider market conditions. Although voluntary mobility caused by natural disasters has been investigated (Call & Gray, 2020; Loughran & Elliott, 2019), the behavioural pathways that shape such decisions in flood-prone areas are still insufficiently explored.

This chapter highlights the complementary role of property buyers in spatial adaptation, particularly in Indonesia, where rapid urbanisation and uncontrolled land use change heighten vulnerability to climate-related risk. We introduce the conceptual framework of voluntary adaptation based on rational economic choice, mainly drawing from the hedonic pricing model by Rosen (1974) and efficient adaptation to climate change by Mendelsohn (2000). This chapter argues that property purchasers' decision to adopt spatial adaptation is a rational economic behaviour characterised by their consideration of private equilibrium to balance marginal costs, marginal benefits, and marginal externalities. Their decision of location selection shapes the spatial pattern and is consequent to the spatial variability of the spatial risk. The dynamism and proactivity of property buyers in terms of spatial adaptation may complement planned adaptation by the government, which often experiences lock-in situations.

In Indonesia, urban development involves three key steps: urban planning, implementation, and land use control. There are four instruments for land use control: permissions, sanctions, zoning regulations, and incentives and disincentives. To support effective spatial adaptation, land-use control mechanisms must incorporate the role of the property market. We argue that the navigation of the property market in the context of climate change adaptation can be facilitated through the use of incentives and disincentives. Moreover, we advocate for a new research agenda on land management for climate adaptation, with particular emphasis on the crucial role of the property market in addressing climate change. Policymakers and urban planners must manage and control property markets to effectively navigate urban growth. As a result, the spatial variability of climate-related risks can be controlled, thereby enhancing urban climate resilience.FundingThis study was funded outside the State Budget (non-APBN) of Diponegoro University for the fiscal year 2025, based on the Decree of the Head of the Institute for Research and Community

Service of Diponegoro University (Number: 610/UN7.D2/HK/III/2025 dated 18 March 2025) and the Contract Agreement (Number: 222-776/UN7.D3/PP/IV/2025 dated 8 April 2025).

References

Abidin, H., Andreas, H., Gumilar, I., Sidiq, T. P., & Fukuda, Y. (2013). Land subsidence in the coastal city of Semarang (Indonesia): Characteristics, impacts and causes. *Geomatics, Natural Hazards and Risk, 4*, 226–240.

Ahlfeldt, G. (2011). If Alonso was right: Accessibility and explaining the residential land gradient. *Journal of Regional Science, 51*, 318–338.

Alonso, W. (1964). *Location and land use: Toward a general theory of land rent*. Massachusetts, Harvard University Press.

Andreas, H., Abidin, H. Z., Gumilar, I., Sidiq, T. P., & Yuwono, B. (2017). Adaptation and mitigation of land subsidence in Semarang. *Aip Conference Proceedings,* 1857.

Buchori, I., Pramitasari, A., Sugiri, A., Maryono, M., Basuki, Y., & Sejati, A. W. (2018). Adaptation to coastal flooding and inundation: Mitigations and migration pattern in Semarang City, Indonesia. *Ocean & Coastal Management, 163*, 445–455.

Call, M., & Gray, C. (2020). Climate anomalies, land degradation, and rural out-migration in Uganda. *Population And Environment, 41*, 507–528.

Carter, J., & Sherriff, G. (2016). Adapting to climate change: Getting more from spatial planning. In W. Leal Filho (Ed.), *Innovation in climate change adaptation*. Springer International Publishing.

Chitsa, M., Sivapalan, S., Singh, B. S. M., & Lee, K. E. (2022). Citizen participation and climate change within an Urban community context: Insights for policy development for bottom-up climate action engagement. *Sustainability, 14*, 3701.

Cobián Álvarez, J. A., & Resosudarmo, B. P. (2019). The cost of floods in developing countries' megacities: A hedonic price analysis of the Jakarta Housing Market, Indonesia. *Environmental Economics And Policy Studies, 21*, 555–577.

Cuevas, S. C. (2016). The interconnected nature of the challenges in mainstreaming climate change adaptation: Evidence from local land use planning. *Climatic Change, 136*, 661–676.

De Koning, K., Filatova, T., & Bin, O. (2017). Bridging the gap between revealed and stated preferences in flood-prone housing markets. *Ecological Economics, 136*, 1–13.

Filatova, T. (2014). Market-based instruments for flood risk management: A review of theory, practice and perspectives for climate adaptation policy. *Environmental Science & Policy, 37*, 227–242.

Freeman, A. M. (1979). Hedonic prices, property values and measuring environmental benefits: A survey of the issues. *The Scandinavian Journal of Economics, 81*, 154–173.

Fujita, M., & Thisse, J.-F. (2013). *Economics of agglomeration cities, industrial location, and globalization*. Cambridge University Press.

Gray, C. L., & Mueller, V. (2012). Natural disasters and population mobility in Bangladesh. *Proceedings of the National Academy of Sciences, 109*, 6000–6005.

Hallegatte, S., Green, C., Nicholls, R. J., & Corfee-Morlot, J. (2013). Future flood losses in major coastal cities. *Nature Climate Change, 3*, 802–806.

Handayani, W., Chigbu, U. E., Rudiarto, I., & Putri, I. H. S. (2020). Urbanization and increasing flood risk in the Northern Coast of Central Java—Indonesia: An assessment towards better land use policy and flood management. *Land, 9*, 343.

Hartmann, T., & Needham, P. D. B. (2012). *Planning by law and property rights reconsidered*. Ashgate Publishing Limited.

Helbich, M., Brunauer, W., Vaz, E., & Nijkamp, P. (2013). Spatial heterogeneity in hedonic house price models: The case of Austria. *Urban Studies, 51*, 390–411.

Hurlimann, A. C., & March, A. P. (2012). The role of spatial planning in adapting to climate change. *Wires Climate Change, 3*, 477–488.

IPCC. (2012). In C. B. Field, V. Barros, & T. F. Stocker (Eds.), *Managing the risks of extreme events and disasters to advance climate change adaptation*. Cambridge University Press.

Lancaster, K. J. (1966). A new approach to consumer theory. *Journal of Political Economy, 74*, 132–157.

Li, K., Dethier, P., Eika, A., Samsura, D. A. A., Van Der Krabben, E., Nordahl, B., & Halleux, J.-M. (2019). Measuring and comparing planning cultures: Risk, trust and co-operative attitudes in experimental games. *European Planning Studies, 28*, 1118–1138.

Li, K., Perrine, D., Anders, E. A. S. D. A., Erwin, V. D. K., Berit, N., & Halleux, J.-M. (2020). Measuring and comparing planning cultures: Risk, trust and co-operative attitudes in experimental games. *European Planning Studies, 28*, 1118–1138.

Loughran, K., & Elliott, J. R. (2019). Residential buyouts as environmental mobility: Examining where homeowners move to illuminate social inequities in climate adaptation. *Population and Environment, 41*, 52–70.

Marfai, M. A. (2014a). Impact of sea level rise to coastal ecology: A case study on the northern part of Java Island, Indonesia. *Quaestiones Geographicae, 33*, 107–114.

Marfai, M. A. (2014b). Impact of sea level rise to coastal ecology: A case study on the Northern part of Java Island, Indonesia. *Quageo, 33*, 107–114.

Matthews, T., Lo, A. Y., & Byrne, J. A. (2015). Reconceptualizing green infrastructure for climate change adaptation: Barriers to adoption and drivers for uptake by spatial planners. *Landscape And Urban Planning, 138*, 155–163.

Mendelsohn, R. (2000). Efficient adaptation to climate change. *Climatic Change, 45*, 583–600.

Miranda Sara, L., Jameson, S., Pfeffer, K., & Baud, I. (2016). Risk perception: The social construction of spatial knowledge around climate change-related scenarios in Lima. *Habitat International, 54*, 136–149.

Moser, D., Steiglechner, P., & Schlueter, A. (2022). Facing global environmental change: The role of culturally embedded cognitive biases. *Environmental Development, 44*, 100735.

Murray, N. J., Phinn, S. R., Dewitt, M., Ferrari, R., Johnston, R., Lyons, M. B., Clinton, N., Thau, D., & Fuller, R. A. (2019). The global distribution and trajectory of tidal flats. *Nature, 565*, 222–225.

Prana, A. M., Dionisio, R., Curl, A., Hart, D., Gomez, C., Apriyanto, H., & Prasetya, H. (2024). Informal adaptation to flooding in North Jakarta, Indonesia. *Progress in Planning, 186*, 100851.

Purnomo, A. H., Kurniawan, T., Farandy, A. R., Apriliani, T., Nurlaili, Imron, M., & Sajise, A. J. (2024). Revisiting the climate change adaptation strategy of Jakarta's coastal communities. *Ocean & Coastal Management, 253*, 107112.

Rajapaksa, D., Zhu, M., Lee, B., Hoang, V.-N., Wilson, C., & Managi, S. (2017). The impact of flood dynamics on property values. *Land Use Policy, 69*, 317–325.

Rosen, S. (1974). Hedonic prices and implicit markets: Product differentiation in pure competition. *Journal of Political Economy, 82*, 34–55.

Samarasinghe, O., & Sharp, B. (2010). Flood prone risk and amenity values: A spatial hedonic analysis. *Australian Journal Of Agricultural And Resource Economics, 54*, 457–475.

Sariffuddin, S., Astuti, K. D., & Pangi, P. (2019). Blue space: Bridging urban spatial planning as a document of flood disaster mitigation. *IOP Conference Series: Earth and Environmental Science, 248*, 012022.

Sariffuddin, S., Samsura, D. A. A., Van Der Krabben, E., Setiyono, B., & Pradoto, W. (2024a). Distressed property and spillover effect: A study of property price response to coastal flood risk. *Land Use Policy, 147*, 107379.

Sariffuddin, S., Samsura, D. A. A., Van Der Krabben, E., Setiyono, B., & Pradoto, W. (2024b). Toward obsolete housing: A complementary explanation of increasing coastal vulnerability. *International Journal of Disaster Risk Reduction, 111*, 104709.

Schanze, J. (Ed.). (2006). *Flood risk management - A basic framework*. Springer Netherlands.
Sejati, A. W., Buchori, I., Lakshita, N. M., Wiratmaja, I. G. A., & Ulfiana, D. (2024). The spatial analysis of urbanization dynamic impacts in a 50-year flood frequency In Java, Indonesia. *Natural Hazards, 120*, 2639–2659.
Stern, N. (2008). The economics of climate change. *American Economic Review, 98*, 1–37.
Suri, A., Jena, P., & Azad, S. (2023). Rainfall wet spells variability across temperature mean-change years in the Northwestern Himalayan region. *Earth and Space Science, 10*, E2022ea002568.
Thoidou, E. (2021). Spatial planning and climate adaptation: Challenges of land protection in a Peri-Urban area of the Mediterranean City of Thessaloniki. *Sustainability, 13*.
Tobin, G. A., & Newton, T. G. (1986). A theoretical framework of flood induced changes in Urban land values. *Journal of the American Water Resources Association, 22*, 67–71.
United Nations. (2018). *68% of the World Population Projected To Live In Urban Areas By 2050, Says Un* [Online]. Available: Https://Www.Un.Org/Development/Desa/En/News/Population/2018-Revision-Of-World-Urbanization-Prospects.Html [Accessed]
Van Der Plank, S., Brown, S., & Nicholls, R. J. (2021). Managing coastal flood risk to residential properties in England: Integrating spatial planning, engineering and insurance. *International Journal of Disaster Risk Reduction, 52*, 101961.
Wolff, C., Nikoletopoulos, T., Hinkel, J., & Vafeidis, A. T. (2020). Future Urban development exacerbates coastal exposure in the mediterranean. *Scientific Reports, 10*, 14420.
Yanto, Livneh, B., & Rajagopalan, B. (2017). Development of a gridded meteorological dataset over Java Island, Indonesia 1985–2014. *Scientific Data, 4*, 170072.

Chapter 10
Disaster Management Strategies in Response to Climate Change: A Focus on Climate-Vulnerable Coastal Areas in Indonesia

Febi Dwirahmadi

Introduction

Climate change affects all sectors of life in Indonesia, with significant implications for agriculture and food security, water resources, public health, urban development and infrastructure, forestry and land use, as well as fisheries and coastal ecosystems (Aldyan, 2023; Djalante, 2016). These impacts manifest in various forms, including declining agricultural productivity, increased incidence of waterborne and vector-borne diseases (Fatmawati & Sulistyawati, 2019), stress on water availability (Pawitan, 2018), damage to infrastructure owing to extreme weather events, degradation of forest resources, and disruption of marine and coastal livelihoods (Gan et al., 2021; Heryani et al., 2022).

With disaster management, climate change has intensified climate-related disasters in Indonesia, including floods, landslides, droughts, tropical cyclones, extreme weather events, and forest fires (Djalante, 2016; Yamamoto et al., 2021). It has also contributed to the progression of slow-onset disasters, such as sea-level rise, coastal erosion, saltwater intrusion, and coral bleaching, across various regions of the country (Karlina & Johan, 2020; Zikra, 2015).

In response to these climate change-related hazards, numerous efforts have been undertaken in Indonesia, encompassing both policy-level interventions and community-based practices (Handayani et al., 2019). These efforts include the development of national and local adaptation plans, disaster risk reduction strategies, and community resilience programmes (Djalante et al., 2017). Addressing such complex and multifaceted challenges requires the active involvement of multiple stakeholders, including government agencies, non-governmental

F. Dwirahmadi (✉)
School of Medicine and Dentistry, Griffith University, Brisbane, QLD, Australia
e-mail: f.dwirahmadi@griffith.edu.au

L. Nurhidayah et al. (eds.), *Climate Change, Labour and Migration in Indonesia*, Springer Climate, https://doi.org/10.1007/978-3-032-15334-0_10

organisations, academic institutions, the private sector, and local communities (Dwirahmadi et al., 2023). It is widely recognised that the responsibility for responding to climate-related hazards cannot rest solely with the government; instead, it requires a collaborative, multisectoral approach to ensure effective and sustainable outcomes (Dwirahmadi et al., 2019; Yoseph-Paulus & Hindmarsh, 2018).

This chapter examines the challenges and opportunities in disaster management within the context of climate change in Indonesia, with a particular emphasis on coastal regions. Specifically, the study aims to address the following research question: **"How do policy frameworks and local stakeholder perspectives intersect in shaping disaster resilience in climate-vulnerable coastal regions of Indonesia?"**

This study explores both policy-level and community-based dimensions, highlighting the intersection between national frameworks and local realities. Drawing on a combination of a comprehensive desktop review and a series of roundtable discussions with key stakeholders, this chapter aims to provide an in-depth analysis of the current practices, gaps, and potential strategies for enhancing resilience. By integrating insights from policy documents and field experiences, we aim to contribute to a more comprehensive understanding of disaster risk management in climate-vulnerable coastal areas.

Study Design

Method

This study employed a secondary qualitative analysis approach, combining a comprehensive policy document review with secondary data derived from eight stakeholder roundtables conducted in September 2024. These roundtables included the participation of government representatives and community members and were held across four coastal locations in Indonesia: North Jakarta, Demak, Pekalongan, and Semarang. The roundtables were part of a collaborative research initiative involving the National Research and Innovation Agency of the Republic of Indonesia (BRIN), Universitas Diponegoro, and Griffith University, with funding support from the Knowledge and Innovation Exchange (KONEKSI), an initiative supported by the Australian Department of Foreign Affairs and Trade (DFAT).

The policy analysis component focused on identifying and examining key legislative and regulatory instruments relevant to disaster management and climate change adaptation in Indonesia. Policy documents were primarily sourced from official and authoritative databases, including the Indonesian Regulation Database (https://www.regulasip.id/), the Regulation Database (https://peraturan.bpk.go.id/), and the website of the Indonesian National Disaster Management Agency (BNPB) (https://www.bnpb.go.id/). The analysis encompassed national-level policies as well as subnational regulations, with particular attention paid to the provinces of

Jakarta Province and Central Java. The types of policy documents reviewed included national laws (*Undang-Undang*), government regulations (*Peraturan Pemerintah),* presidential decrees (*Keputusan Presiden*), ministerial regulations (*Peraturan Menteri),* and governor regulations (*Peraturan Gubernur*).

In analysing the results, this study employed both content analysis for policy mapping and thematic analysis for examining secondary data from the roundtable discussions. According to Krippendorff (2018, p. 18), "content analysis is a research technique for making replicable and valid inferences from texts (or other meaningful matter) to the contexts of their use." Thematic analysis, on the other hand, considers the recorded messages, in this case, the secondary data from the roundtables, as the primary data. Codes are developed by the researcher through a detailed and iterative review of the texts, allowing key themes to emerge inductively from the data (Neuendorf, 2018).

Study Settings: North Jakarta, Semarang, Demak, and Pekalongan

Indonesia's coastal regions rank among the most at-risk areas globally in terms of climate-related hazards. Projections indicate that by 2050, up to 12 million Indonesians could be exposed to devastating coastal floods, and as many as 2000 small islands may be submerged (Takagi et al., 2016). Climate change has intensified the frequency and severity of flooding and landslides, with over 90% of Indonesia's natural disasters over the past decade being linked to climate-related factors (Djalante & Garschagen, 2017). These hazards threaten not only lives and livelihoods but also critical infrastructure, public health, and long-term development, particularly in densely populated coastal settlements (Purnomo et al., 2024).

North Jakarta, Semarang, Demak, and Pekalongan are among the most densely populated coastal regions and are increasingly subject to frequent and severe climate-related disasters (Asrofi et al., 2024; Bott et al., 2021). The management of disaster risks in these areas is particularly complex due to the interplay of environmental and social factors. Environmental challenges include land subsidence, issues related to access to potable water, coastal inundation caused by tidal surges, and rising sea levels. In addition, social challenges such as poverty, livelihood vulnerability, marginalisation, and low levels of education exacerbate these risks and further complicate effective disaster risk management (Handayani et al., 2020).

Another complexity faced by North Jakarta, Semarang, Demak, and Pekalongan is the presence of slums and informal settlements, commonly referred to as *kampungs*. These areas often experience substandard housing conditions and a lack of adequate basic infrastructure, including limited access to clean water, sanitation, electricity, and waste management services (Roitman & Rukmana, 2022; Suhartini & Jones, 2023). Previous literature also highlights a range of public health issues associated with *kampung* environments, such as a high risk of vector-borne, waterborne, and sanitation-related diseases, including dengue, diarrhoea, and respiratory infections, as well as close proximity to contaminated rivers or waste sites (Irda Sari et al., 2018; Otsuka et al., 2019).

Despite the various challenges they face, urban *kampungs* in Indonesia are widely recognised for their inherent strengths. One of the most prominent is the enduring value of *gotong royong* (mutual cooperation), which remains central to daily life (Hawken & Sunindijo, 2018; Mardiasmo & Barnes, 2015). These communities are also characterised by high levels of trust and familiarity among neighbours, fostering strong social cohesion and a deep sense of solidarity. Such social dynamics enable kampung residents to collectively manage local issues, support one another in times of need, and maintain resilient and adaptive community structures (Mardiasmo & Barnes, 2015).

Findings and Discussions

Mapping of Government Policies on Disaster Management in the Context of Climate and Coastal Vulnerabilities

This study identified 20 key government policies related to disaster management in the context of climate and coastal regions. Following the hierarchy of laws and regulations in Indonesia, as outlined by the Government of Indonesia (2011), the policies included in this study range from the highest level, national laws (*Undang-undang)*, to the lowest level in this context, namely, the governor's instruction (Table 10.1).

As can be seen in Table 10.1, government policies were grouped into three dominant policy framings that shape Indonesia's approach to climate-related coastal vulnerabilities: (1) disaster management and disaster risk reduction, (2) risk-informed development, and (3) climate change adaptation. Each of these framings is embedded in distinct policy documents and led by different government institutions.

The disaster risk reduction framing is led by the National Disaster Management Authority (BNPB), the risk-informed development approach is promoted by the Ministry of National Development Planning (Bappenas), and the Ministry of Environment and Forestry advances the climate change adaptation framing. These institutional silos reflect fragmented yet complementary perspectives, highlighting the need for more comprehensive policy integration to address the complex and interlinked nature of climate and coastal risks.

In conducting the content analysis, this study employed a set of selected keywords to identify and interpret the presence of key thematic dimensions within the policy texts. These dimensions included comprehensive disaster management, climate-related hazards, the mainstreaming of climate change adaptation, the inclusion of vulnerable populations, and the empowerment of local communities. Given that all the policy documents analysed were written in Indonesian, the keywords

Table 10.1 Key policies in Indonesia, Jakarta, and Central Java Provinces related to disaster management, climate change, and coastal areas

No	Policy document	Title of document	Sectors	Type of document	Level
1	UU No. 24/2007	Disaster Management Act	DM/DRR	Law	National
2	PP No. 21/2008	Disaster Management Governance	DM/DRR	Government Regulation	National
3	Perka BNPB No. 1/2012	General guidelines of disaster-resilient village/ sub-district (destana)	DM/DRR	Head of BNPB Regulation	National
4	Perban BNPB No. 1/2018	Displaced people management in the aftermath of a disaster	DM/DRR	Ministerial Regulation	National
5	Perpres No. 87/2020	National disaster management master plan (RIPB) 2020–2044	DM/DRR	Ministerial Regulation	National
6	Perka BNPB No. 1/2025	National Disaster Management Plan 2025–2029	DM/DRR	Head of BNPB Regulation	National
7	Perda Jateng No. 11/2009	Disaster management governance for the Central Java province	DM/DRR	Government regulation	Provincial
8	Kepgub Jakarta Province No. 121/2022	Procedure of disaster management for Jakarta Province	DM/DRR	Governor decree	Provincial
9	Pergub Jateng No. 59/2023	Disaster emergency contingency plan for Central Java province	DM/DRR	Government regulation	Provincial
10	Pergub Jateng No. 28/2023	Disaster contingency plan for Central Java province	DM/DRR	Government regulation	Provincial
11	Pergub Jakarta Province No 1/2024	Jakarta province disaster management plan 2023–2027	DM/DRR	Government regulation	Provincial
12	UU No. 1/2014	Coastal and small island management	Development	Law	National
13	Bappenas: Climate Resilience Disaster Policy (2021)	National guidelines on climate-resilient development	Development	Policy Guidelines	National
14	UU No. 59/2024	National long-term development planning 2025–2045	Development	Law	National
15	Perda Jakarta Province No. 8/2024	Jakarta provincial long-term development planning 2025–2045	Development	Government regulation	Provincial
16	Perda Jatang No. 6/2024	Central Java provincial long-term development planning 2025–2045	Development	Government regulation	Provincial

(continued)

Table 10.1 (continued)

No	Policy document	Title of document	Sectors	Type of document	Level
17	UU No. 32/2009	Environmental management and protection	Env/CCA	Law	National
18	Permen LHKP.84/ MENLHK-SETJEN/ KUM.1/11/2016	Climate kampung (proklim) programme	Env/CCA	Ministerial Regulation	National
19	Pergub Jatang No. 51/2019	Implementation of the climate kampung (proklim) program in Central Java province	Env/CCA	Government regulation	Provincial
20	Governor of Jakarta Province Instruction No. 127/2017	Climate kampung (proklim) programme in Jakarta	Env/CCA	Governor Instruction	Provincial

used in the analysis were exclusively in Indonesian. This approach ensured linguistic consistency and contextual relevance, allowing for a more accurate reflection of how these themes are articulated within national and subnational policy frameworks.

Table 10.2 presents the results of the content analysis conducted on the 20 key government policy documents. The table displays the frequency with which the selected keywords appear in each document, serving as indicators of the extent to which various thematic dimensions, such as disaster management, climate change adaptation, climate-related hazards, and community inclusion, are addressed. A value of zero indicates that relevant keywords were absent from the document, suggesting that the theme was not discussed or considered within the policy text. While the frequency of selected keywords is recorded, this study also examines their context in depth, as recommended by Zhang and Wildemuth (2009).

The primary legislative framework for disaster risk reduction in Indonesia is UU (*Undang-undang* - Law) No. 24/2007 on Disaster Management. As suggested by Shaw and Cui (2025), disaster law encompasses a broad spectrum of statutes, policies, and regulations designed to mitigate the impacts of disasters and ensure effective management when they occur, which include activities such as disaster preparedness, response, recovery, and mitigation. Indonesia's Disaster Management Law (UU No. 24/2007) was enacted in the aftermath of the 2004 Indian Ocean Earthquake and Tsunami, one of the most devastating natural disasters in history, which claimed the lives of more than 150,000 Indonesians (Ariyaningsih et al., 2025). The introduction of this law signified the government's commitment to shifting the disaster management paradigm from a reactive, response-focused approach to one that emphasises preparedness and mitigation (Butt, 2013). However, the content analysis conducted in this study indicates that disaster emergency response remains the dominant focus across various stages of disaster management within policy documents. Specifically, keywords related to disaster response appeared 908 times across multiple policy texts, whereas keywords related to disaster preparedness and mitigation appeared only 183 and 180 times, respectively.

Table 10.2 Content analysis of key government policy documents

No	List of keywords (English and Indonesian)	Disaster management (Disaster risk reduction)											Risk-informed development					Climate change adaptation				Total
		National						Province					National			Province		National		Province		
		UU No. 24/2007	PP No. 21/2008	Perka BNPB No 1/2012	Perban BNPB No 3 /2018	Perpres No. 87/2020	Perban BNPB No 1/2025	Perda Jateng No. 11/2009	Kepgub Jakarta Province No. 121/2022	Pergub Jateng No. 59/2023	Pergub Jateng No 28/2023	Pergub Jakarta Province o 2024	UU No 1/2014	Bappenas: Climate Resilience Development Policy (2021)	UU No. 59/2024	Perda Jakarta Province No 8/2024	Perda Jateng No 6/2024	UU No 32/2009	Permen LHK P.84/ MENLHK-SETJEN/ KUM. 1/11/2016	Pergub Jateng No. 51/2019	Governor of Jakarta Province Instruction No 127/2017	
1	**Comprehensive Disaster Management**																					
1.1	Disaster preparedness (*kesiapsiagaan*)	4	7	18	0	6	36	5	21	10	8	54	1	0	9	3	1	0	0	0	0	183
1.2	Disaster mitigation (*mitigasi/ pencegahan bencana*)	4	5	9	0	8	40	5	9	0	9	23	0	0	30	10	26	2	0	0	0	180
1.3	Disaster response (*respon/tanggap darurat*)	31	75	24	26	14	37	37	9	46	544	54	0	0	6	2	3	0	0	0	0	908
1.4	Disaster recovery (*pemulihan*)	13	46	9	0	11	44	24	31	16	5	20	0	0	2	1	6	0	0	0	0	228
1.5	Disaster reconstruction (*rekonstruksi*)	6	14	3	0	4	43	9	12	3	2	14	0	0	0	0	0	0	0	0	0	110
1.6	Early warning system (*peringatan dini*)	7	7	10	0	6	60	6	9	3	17	12	0	0	15	6	4	0	0	0	0	162

(continued)

Table 10.2 (continued)

No	List of keywords (English and Indonesian)	Disaster management (Disaster risk reduction)											Risk-informed development					Climate change adaptation				Total
		National						Province					National			Province		National		Province		
		UU No. 24/2007	PP No. 21/2008	Perka BNPB No 1/2012	Perban BNPB No 3 /2018	Perpres No. 87/2020	Perban BNPB No 1/2025	Perda Jateng No. 11/2009	Kepgub Jakarta Province No. 121/2022	Pergub Jateng No. 59/2023	Pergub Jateng No 28/2023	Pergub Jakarta Province o 2024	UU No 1/2014	Bappenas: Climate Resilience Development Policy (2021)	UU No. 59/2024	Perda Jakarta Province No 8/2024	Perda Jateng No 6/2024	UU No 32/2009	Permen LHK P.84/ MENLHK-SETJEN/ KUM. 1/11/2016	Pergub Jateng No. 51/2019	Governor of Jakarta Province Instruction No 127/2017	
1.7	Disaster risk reduction (*Pengurangan risiko bencana*)	8	18	65	0	15	60	19	3	1	41	54	1	1	5	1	6	0	1	0	0	299
1.8	Resilience (*Ketangguhan/ ketahanan*)	0	5	80	0	12	93	2	0	0	73	38	0	3	60	44	45	1	0	2	0	458
2	**Mainstreaming of climate change adaptation and climate-related hazards in the coastal region**																					
2.1	Climate change (*Perubahan iklim*)	0	0	0	0	16	76	0	1	0	0	20	1	62	75	63	47	19	21	34	6	441
2.2	Climate change adaptation (*Adaptasi perubahan iklim*)	0	0	0	0	7	24	0	0	0	0	3	0	10	5	9	11	4	30	23	4	130
2.3	Flood (*Banjir*)	1	0	0	0	1	44	4	5	18	147	62	0	9	42	65	9	1	1	1	1	411
2.4	Extreme events (*Cuaca ekstrim*)	0	0	0	0	1	11	0	0	11	0	26	0	3	2	7	9	0	1	1	0	72

2.5	Storm surge *(Gelombang tinggi/ekstrim, badai)*	0	0	0	0	2	5	0	0	10	0	26	0	9	3	1	3	1	1	1	0	62
2.6	Heatwave *(Gelombang panas, panas ekstrim)*	0	0	0	0	0	0	0	0	1	0	0	0	1	1	1	0	0	0	0	0	4
2.7	Coastal Flooding (*Banjir rob*)	0	0	0	0	0	0	0	0	0	1	0	0	1	4	11	8	0	1	1	0	27
2.8	Diseases sensitive to climate (*Penyakit terkait iklim, mosquito-borne diseases*)	0	0	0	0	0	2	0	0	1	1	0	0	14	10	1	4	0	1	1	2	37
2.9	Abrasion (*abrasi, erosi pantai*)	0	0	0	0	1	8	0	0	9	0	28	0	4	17	3	2	0	1	1	0	74
2.10	Sea-level rise (*Kenaikan muka air laut*)	0	0	0	0	0	2	0	0	0	1	0	0	5	7	2	4	2	1	1	0	25
2.11	Land subsidence (*Penurunan muka tanah*)	0	0	0	0	0	7	0	0	0	0	1	0	0	5	10	6	0	0	0	0	29
2.13	Climate crisis (*krisis iklim*)	0	0	0	0	0	0	0	0	0	0	0	0	0	3	0	0	0	0	0	0	3

(continued)

Table 10.2 (continued)

No	List of keywords (English and Indonesian)	Disaster management (Disaster risk reduction)											Risk-informed development					Climate change adaptation				Total
		National						Province					National			Province		National		Province		
		UU No. 24/2007	PP No. 21/2008	Perka BNPB No 1/2012	Perban BNPB No 3 /2018	Perpres No. 87/2020	Perban BNPB No 1/2025	Perda Jateng No. 11/2009	Kepgub Jakarta Province No. 121/2022	Pergub Jateng No. 59/2023	Pergub Jateng No 28/2023	Pergub Jakarta Province o 2024	UU No 1/2014	Bappenas: Climate Resilience Development Policy (2021)	UU No. 59/2024	Perda Jakarta Province No 8/2024	Perda Jateng No 6/2024	UU No 32/2009	Permen LHK P.84/ MENLHK-SETJEN/ KUM. 1/11/2016	Pergub Jateng No. 51/2019	Governor of Jakarta Province Instruction No 127/2017	
3	**Inclusion of vulnerable populations and empowerment of local communities**																					
3.1	Vulnerable groups: disabled, elderly, ethnic minority (*Kelompok rentan, penyandang disabilitas, lansia, etnik minoritas*)	6	7	15	3	0	11	7	6	8	23	13	0	0	16	14	12	0	0	0	0	141
3.2	Community empowerment/ participation (*pemberdayaan/ pelibatan/ partisipasi masyarakat*)	4	10	5	10	3	16	5	13	2	5	25	3	0	27	14	39	6	22	4	11	224

In addition, while Law No. 24/2007 establishes the foundation for disaster management in Indonesia, this study found that it does not explicitly recognise climate change as a contributing factor to disaster risk. No keywords related to climate change were identified in this policy document. Previous studies have suggested that this omission reflects the limited awareness among policymakers at the time regarding the need to integrate climate change considerations into disaster legislation (Djalante & Thomalla, 2012; Sufri & Lassa, 2024).

In recent years, however, policy developments have begun to integrate climate-related considerations into disaster management frameworks. Notably, Presidential Regulation No. 87/2020 on the National Disaster Management Master Plan (*Rencana Induk Penanggulangan Bencana*, RIPB) 2020–2024, which serves as the overarching guideline for disaster management implementation in Indonesia, explicitly acknowledges climate change as a critical factor that exacerbates disaster risk and complexity (Government of Indonesia, 2020). In the analysis, the essential keywords reflecting the presence of the climate change adaptation dimension were found multiple times within the RIPB, with "climate change" appearing 16 times and "climate change adaptation" appearing seven times, indicating a clear policy commitment to addressing these challenges. RIPB highlights the increasing frequency and intensity of climate-related hazards, underscoring the importance of adopting integrated, adaptive, and forward-looking strategies to strengthen national resilience.

As outlined in the RIPB, the Indonesian government has articulated a clear vision: "To build a disaster-resilient Indonesia for sustainable development." In this context, disaster resilience refers to the nation's capacity to withstand, absorb, adapt to, and recover from the impacts of disasters and climate change in a timely, effective, and efficient manner (Government of Indonesia, 2020). Furthermore, the RIPB mainstreams Disaster Risk Reduction (DRR) and Climate Change Adaptation (CCA) by aligning its principles and actions with key international frameworks, including the Paris Agreement, the Sendai Framework for Disaster Risk Reduction, and the Sustainable Development Goals (SDGs). This alignment reflects Indonesia's commitment to adopting a comprehensive, multi-level approach to risk governance that bridges global priorities with national and local implementation.

The keywords 'resilience (*ketangguhan*)' and 'disaster risk reduction (*pengurangan risiko bencana*)' appeared frequently in disaster management and development-related policies at both national and provincial levels. For example, Law No. 59/2024 concerning the National Long-Term Development Planning 2025–2045 highlights the need to strengthen the resilience of the northern coastal areas of Java, including local communities, against climate change threats, such as tidal flooding and coastal erosion (Government of Indonesia, 2024). This emphasis reflects a growing political commitment to mainstream disaster risk reduction and climate change adaptation within regional development planning, particularly in areas highly vulnerable to climate-related hazards. As suggested by Hadi (2020) and Dwirahmadi et al. (2023), the inclusion of these priorities within a long-term national planning document demonstrates recognition of the interconnectedness between development goals and the need for proactive measures to address climate risks, which is essential for safeguarding both communities and economic development in the face of escalating climate impacts.

However, this study found that climate change adaptation policies in Indonesia generally did not employ the keyword *resilience* within their texts. An exception to this is found in Law No. 32/2009 on Environmental Protection and Management and the Central Java Governor Regulation on the Climate Village Program, both of which explicitly reference *resilience*. Notably, it is not surprising that climate change adaptation policies tend to use the term *'adaptive'* rather than *'resilient'* to describe the intended outcomes of intervention programs. This reflects a framing that emphasises the capacity to adjust and respond to climate-related impacts, rather than the broader concept of resilience, which encompasses not only adaptation but also the ability to absorb, recover, and transform in the face of climate-related hazards.

Furthermore, this study found that discussions on 'coastal abrasion, sea-level rise, and land subsidence' are mainly absent from key disaster management policy documents. These critical climate-related hazards are explicitly addressed only in a limited number of policies, such as the Jakarta Disaster Management Plan 2023–2027 and development-related policies, including the National Long-Term Development Planning (Law No. 59/2024), the Central Java Long-Term Development Planning (Perda Jateng No. 6/2024), and the Jakarta Long-Term Development Planning (Perda Jakarta Province No. 8/2024). The limited inclusion of these hazards in disaster management policies is concerning given the significant risks they pose to the coastal regions of Indonesia. Incorporating these issues into disaster management frameworks is crucial to ensure that risk reduction and adaptation strategies effectively address the complex realities of climate change's impacts on coastal communities. The presence of these topics in development planning documents indicates emerging awareness at the planning level. However, this awareness needs to be consistently integrated into disaster management policies to enhance community resilience and inform targeted mitigation and adaptation measures.

Another interesting finding is that the *Kampung Iklim* (Climate Village) Program (CVP), the flagship initiative of the Ministry of Environment (previously the Ministry of Environment and Forestry) launched in 2016, does not comprehensively address keywords related to climate-related hazards in coastal regions. Key policy documents underpinning the implementation of the CVP, including Permen LHK P.84/MENLHK-SETJEN/KUM.1/11/2016 at the national level, Pergub Jateng No. 51/2019 for Central Java, and Governor of Jakarta Province Instruction No. 127/2017 for Jakarta Province, do not discuss land subsidence, despite its clear and ongoing occurrence in many coastal areas. This omission indicates a gap in the policy framework, as land subsidence is a critical issue that exacerbates the impacts of tidal flooding and coastal erosion, particularly in urban coastal areas. Addressing land subsidence is essential for the effective implementation of climate adaptation strategies (Esteban et al., 2020; Siriwardane-de Zoysa et al., 2021). The absence of this consideration from key policy documents suggests the need for a more comprehensive approach to climate risk management within the CVP, ensuring that all relevant hazards are considered when building community resilience in coastal regions.

Table 10.3 compares **Kampung Iklim (KI)** and **Desa Tangguh Bencana (DTB)**, two community-based programs in Indonesia that enhance local resilience but with distinct focuses. KI, under the Ministry of Environment and Forestry, aims to strengthen the community's capacity for climate adaptation and mitigation through participatory activities and sustainable environmental management, primarily by addressing climate-related hazards. In contrast, Destana, governed by the BNPB, focuses on building village-level disaster preparedness and risk reduction capacity, addressing all types of hazards with formal regulatory support and integration into village development planning. While both programs emphasise community engagement and capacity building, KI centres on climate literacy and sustainable livelihoods, whereas DTB prioritises preparedness, emergency response, and post-disaster recovery. Recognition and institutional frameworks in each program help sustain community initiatives, ensuring that local resilience efforts align with broader national strategies on climate change adaptation and disaster risk reduction. These two programs can complement one another in building holistic community

Table 10.3 Comparison between Kampung Iklim (Climate Village) and Desa Tangguh Bencana (Disaster Resilient Village) Program

Aspect	Kampung Iklim (KI)	Desa Tangguh Bencana (Destana)
Governance	Ministry of Environment and Forestry; technical guidance, climate adaptation focus.	BNPB; formal regulation (Perka BNPB No. 1/2012); multi-level government roles.
Objective	Build community capacity for climate adaptation and mitigation.	Build village capacity for disaster risk reduction (DRR) and preparedness.
Key Activities	Community-based adaptation, climate risk assessment, climate literacy, and documentation for recognition.	Risk assessment, DRR and contingency planning, forming DRR forums, simulations, and integrating DRR into village plans.
Community Engagement	Participatory planning, behaviour change, local wisdom, and NGO/ government support.	Community empowerment, DRR forums, risk mapping, and inclusive decision-making (musyawarah).
Recognition	"Kampung Iklim" status after the ministry assessment.	"Desa Tangguh Bencana" status is based on BNPB criteria, and regularly evaluated.
Integration with Development	Encourages climate action in development, focusing on sustainability.	Legally requires DRR integration into village planning and budgeting.
Scope of Hazards	Climate-related hazards (flood, drought, sea-level rise, and abrasion).	All hazards (natural, non-natural, and social).
Capacity Building	Climate literacy, adaptation, and sustainable livelihoods.	DRR knowledge, preparedness, emergency response, and recovery.
Sustainability	Relies on community motivation, local champions, and ministry evaluation.	Institutionalised in local regulations and national disaster management systems.

resilience; however, it is important to ensure effective coordination and collaboration between the stakeholders involved to maximise their impact and avoid overlap.

Insights from Government and Community Roundtables

Regarding roundtables, this study conducted two rounds in each research setting: one with government stakeholders and one with community stakeholders, resulting in a total of eight roundtables. In total, there were 69 participants from the government (37 males and 32 females), and 59 participants were from the community (32 males and 27 females). Details of the number of participants are presented in Table 10.4.

The government stakeholder roundtables included participants from local government offices in Jakarta, Pekalongan, Demak, and Semarang, as well as from ministries and agencies involved in disaster management, water management, and community development. Specifically, the participating institutions included the Ministry of Maritime Affairs and Fisheries (KKP); the Ministry of Public Works and Public Housing (PUPR); the Ministry of Villages, Development of Disadvantaged Regions, and Transmigration (MoV); the Ministry of Environment and Forestry; and the National Disaster Management Authority (BNPB).

For the community stakeholder roundtables, participants were drawn from grassroots community organisations that play key roles in local governance and community mobilisation, such as *Rukun Tetangga* (RT or Neighbourhood Associations) and *Rukun Warga* (RW or Citizen Associations). Other participants included representatives from the Family Welfare Empowerment Program (*Pemberdayaan Kesejahteraan Keluarga* or PKK), Fishermen's Associations, Youth Associations (*Karang Taruna*), Disaster Preparedness Cadets (*Tagana*), and faith-based community organisations such as Muhammadiyah and Aisyiyah.

Through roundtable discussions, this study identified five key themes central to the implementation of coastal disaster management: (1) relocation as a last resort; (2) livelihood vulnerability and adaptation; (3) gender and social equity; (4) health and social protection; and (5) inclusive decision-making. These themes reflect the

Table 10.4 Roundtable participants

Research setting	Roundtable group	Number of participants	Code
North Jakarta	Community stakeholders	15	Roundtable #1
	Government stakeholders	16	Roundtable #2
Semarang	Community stakeholders	16	Roundtable #3
	Government stakeholders	24	Roundtable #4
Demak	Community stakeholders	14	Roundtable #5
	Government stakeholders	10	Roundtable #6
Pekalongan	Community stakeholders	14	Roundtable #7
	Government stakeholders	19	Roundtable #8

complex and interconnected challenges faced by coastal communities, highlighting the need for holistic, people-centred approaches in disaster risk reduction and climate adaptation planning.

Both stakeholders and community roundtables viewed migration and relocation as a last resort for coping with the impacts of climate change and disasters in their areas. "*People may voluntarily relocate to a new location when their current area becomes uninhabitable due to flooding, as seen in Pekalongan Utara*", Roundtable #7.

"*[We] did not want to relocate, but due to natural conditions, we were forced to do so. Three RWs [sub-villages] were relocated: Dombo, Tambaksari Baru, and Rejosari. These areas have now been moved to Sidogemah Village and Gemulak Village as part of a village programme*", Roundtable #5.

Livelihood is one of the primary concerns expressed by affected communities, particularly in the context of relocation due to coastal hazards. For many, the ability to maintain or restore a sustainable livelihood is regarded as the key indicator of successful adaptation. Communities consistently emphasise that relocation efforts must not compromise their access to essential livelihood resources. As such, they call on the government to ensure that any resettlement or adaptation strategy includes clear guarantees that their economic activities and means of subsistence will be preserved, supported, or appropriately replaced in the new location. "*Relocation can be considered if the new location offers the same livelihood [opportunities as the previous one]. For example, fishermen should be able to continue working as fishermen [in the new location*", Roundtable #8.

Issues related to gender and social equity emerged consistently across all roundtable discussions, highlighting the disproportionate impacts of coastal hazards on women and children. Participants emphasised that women are not only more vulnerable to the immediate effects of disasters—such as displacement, loss of livelihood, and disruption of caregiving responsibilities—but are also more likely to experience heightened risks of domestic violence in the aftermath of such events. This study found that domestic violence is linked to economic stress/unemployment due to disaster impacts. "*The majority of wives [continue] to work [while their] husbands have lost their jobs due to climate change and remain at home, [which can] trigger [a risk] of domestic violence*", Roundtable #6. In addition to this, "*women work in unsafe conditions, especially during disaster situations. [There is also] an increase in early marriages*", Roundtable #2.

Concerning health and social protection, participants expressed concern over the growing risks of climate-induced health impacts, including illness, injury, and long-term mental health consequences. "*[Climate change] increases [the risk of] disease spread*", Roundtable #1. Also, "*[due to the increase in disaster risk], people often feel surrendered, depressed, struggle, and doubtful*", Roundtable #7. They then emphasised the urgent need to strengthen social security systems, such as health insurance and labour protections, particularly for vulnerable and informal workers who are often excluded from formal support mechanisms. *[There is a need] to establish an insurance programme for fishermen*, Roundtable #2.

Governance-related concerns have focused on the inclusiveness of decision-making processes in coastal adaptation strategies and relocation planning, particularly regarding the meaningful participation of women and children. Participants emphasised that these groups are often excluded from consultations and planning discussions despite being among the most affected by climate-related risks and displacement. [When making decisions], it is important to involve women and children to ensure that policies are well-targeted [and inclusive], Roundtable #2. The community roundtable highlighted the importance of *gotong royong* (community self-help) as a vital component of local disaster preparedness efforts. They encourage *"[promote] community self-help both within the area and in collaboration with neighbouring areas"*, Roundtable #7.

Discussion and Recommendations: Potential Gaps Analysis and Key Recommendations

Indonesia has produced a wide range of policy documents related to coastal disaster management developed by various governmental stakeholders. Three key institutions play central roles in this domain: disaster management agencies (e.g. BNPB and BPBD), planning and development agencies (e.g. Bappenas and Bappeda), and environmental management agencies (e.g. the Ministry of Environment and Forestry). While these agencies share a common policy agenda aimed at addressing the challenges facing climate-vulnerable coastal areas, they approach this issue from distinct institutional perspectives. As can be seen in Fig. 10.1, disaster management agencies primarily frame the problem through the lens of emergency response and disaster risk reduction. In contrast, planning and development agencies emphasise risk-informed development as a strategic response to coastal vulnerability. Meanwhile, environmental agencies focus on climate change adaptation as a key mechanism to address climate-induced risks in coastal regions. This divergence in policy framing was also reported by Mardiah et al. (2017), who illustrate the need for greater policy coherence and integration across sectors to ensure effective and comprehensive coastal resilience strategies (Wijaya et al., 2017).

This study suggests that disaster management policy in Indonesia tends to focus predominantly on rapid-onset, climate-related disasters such as flooding, storm surges, and extreme winds. In contrast, slow-onset anthropogenic climate hazards, such as sea level rise and land subsidence, receive far less attention and are not yet formally recognised as disasters within the current disaster management framework. On the ground, however, the local community perceives these slow-onset hazards as serious and escalating threats with significant impacts on their lives and livelihoods (Pereira et al., 2019; Saputra et al., 2025). This lack of recognition in policy not only limits the development of appropriate planning and response measures but also undermines long-term resilience efforts in vulnerable coastal communities (Nurhidayah et al., 2022).

Another potential gap lies in the disconnection between policy frameworks and community experiences. Coastal communities in areas such as North Jakarta,

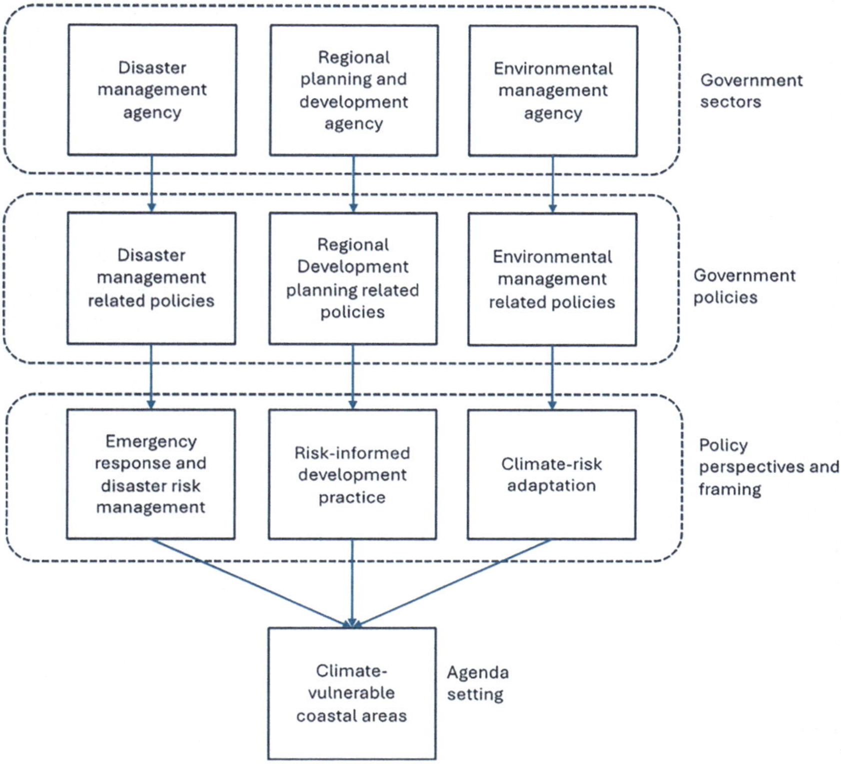

Fig. 10.1 Policy perspectives and framing to address climate vulnerabilities in coastal areas in Indonesia

Demak, Pekalongan, and Semarang are primarily concerned with the day-to-day risks they face—risks that are increasingly exacerbated by climate-related factors. However, these persistent and cumulative threats are often not formally recognised within the disaster management system as events warranting response or intervention. While such conditions may not lead to immediate fatalities, over time, they can contribute to serious mental health and well-being challenges, such as depression, as highlighted by Maharani et al. (2025).

While there have been significant improvements through community-based initiative programmes launched by the government, such as *Kampung Iklim* (Climate Village) and *Desa Tangguh Bencana (*Disaster Resilient Village), across four different research settings, the roundtable discussions underscored the urgent need for disaster management and climate adaptation strategies to incorporate a gender equity perspective. The programme ensures that the specific needs, roles, and voices of women and other marginalised groups are meaningfully considered in both the planning and implementation stages (Prakash et al., 2025; Zaidi & Fordham, 2021). A similar situation was observed in Aceh Province, where the *Destana* programme did not provide equal opportunities for women to participate. As a result, women in the participating communities had comparatively lower levels of knowledge and skills related to disaster preparedness than their male counterparts (Oktari et al., 2021).

Apart from the *Climate Village* and *Disaster Resilient Village* initiatives, several other programmes also contribute to climate adaptation efforts, including urban farming schemes and *waste bank* (*Bank Sampah*) initiatives. In the Tambakrejo Sub-district, Semarang, for example, urban farming has proven effective in raising community awareness about the importance of environmental sustainability while also providing alternative food sources. This initiative forms part of the community's adaptation strategy to address shifts in agricultural patterns resulting from climate uncertainties (Sholeh et al., 2024; Suntoro et al., 2025). Meanwhile, in Jakarta, the Waste Bank initiative not only helps reduce flood risk by decreasing waste in drainage systems and water canals but also provides communities with an opportunity to earn income (Nugraha & Lassa, 2018).

This discussion illustrates that disaster and climate risks are systemic in nature and deeply embedded in the daily lives of affected communities. For those living in coastal areas, coping with disasters and adapting to climate change are not limited to the capacity to respond to acute events such as flooding or storm surges. It also encompasses the ability to meet basic needs, including access to clean drinking water, safe and secure housing, a healthy living environment, fair work compensation, alternative livelihoods, healthcare, and social protection (Maharani et al., 2025; Nurhidayah, 2020; Swarnokar et al., 2025). In this context, disaster risk reduction and climate adaptation are inseparable from broader efforts to ensure livelihoods, well-being, and social equity.

As a way forward, this study offers several key recommendations to strengthen coastal disaster management and climate adaptation efforts.

1. **Promote cross-sectoral integration among community-based resilience programmes.** Enhancing coordination between disaster management, climate change adaptation, and sustainable development initiatives will improve policy coherence, minimise duplication of efforts, and support a more comprehensive and sustainable approach to risk reduction at the community level (Shammin et al., 2022; Swaris et al., 2024).
2. **Incorporate comprehensive health and well-being dimensions** into disaster and climate adaptation policies. As defined by the World Health Organization, health is "a state of complete physical, mental, and social well-being and not merely the absence of disease and infirmity" (Callahan, 1973, p. 77). Recognising and addressing the interconnectedness between climate adaptation and well-being outcomes is essential for promoting transformative adaptation (Heath, 2025). This approach ensures that adaptation efforts not only reduce physical risks but also support mental health, social stability, and overall quality of life, particularly in vulnerable and displaced populations (Aitsi-Selmi et al., 2015).
3. **Strengthen gender equity** by ensuring the active participation of women and marginalised groups in all stages of planning, implementation, and evaluation of

resilience programmes. Inclusive participation not only enhances the relevance and effectiveness of resilience strategies but also helps to address structural inequalities that exacerbate vulnerability. As noted by Aziz and Anjum (2024), gender equity provides a critical foundation for responding to the complex and intersecting impacts of climate-related hazards, thus enabling more just and sustainable adaptation outcomes.

4. **Formally recognise land subsidence and sea-level rise** as slow-onset, anthropogenic, climate-related disasters within disaster management frameworks. These phenomena, as reported by Siriwardane-de Zoysa et al. (2021), are often driven by human activities and exacerbated by climate change, posing complex challenges for policy development, particularly in Southeast Asian countries, including Indonesia. Without formal recognition in disaster management frameworks, efforts to address these hazards may be fragmented and under-resourced. Moreover, a lack of recognition can limit access to sustainable funding mechanisms and institutional support, ultimately hindering long-term resilience and adaptation planning for affected communities (Suprayitno et al., 2024).

Conclusion

This study underscores the critical importance of adopting holistic and context-sensitive approaches to disaster management and climate change adaptation in Indonesia's vulnerable coastal regions. Through comparative analysis of community- and government-led programmes (Kampung Iklim and Desa Tangguh Bencana), as well as in-depth stakeholder roundtables across multiple coastal cities, five core themes emerged: the primacy of relocation as a last resort, the centrality of livelihood vulnerability and adaptive capacity, the need to foreground gender and social equity, the imperative of health and social protection, and the call for truly inclusive decision-making.

The findings reveal persistent gaps between the national disaster management policy and the lived realities of local stakeholders. While government frameworks increasingly acknowledge climate risks, slow-onset hazards, such as land subsidence and sea-level rise, remain underrepresented in both formal recognition and programmatic response. Community voices emphasise that resilience hinges not only on physical infrastructure or institutional mandates but also equally on the preservation of livelihoods, the agency of marginalised groups, social cohesion, and the meaningful participation of those most affected.

Finally, disaster management in coastal Indonesia must move beyond top-down approaches and one-size-fits-all solutions, embracing adaptive and co-produced strategies that are responsive to community needs and grounded in justice and equity. Such a transformation is essential to protect not only the physical environment but also the dignity, livelihoods, and well-being of Indonesia's coastal populations in the face of a changing climate.

References

Aitsi-Selmi, A., Egawa, S., Sasaki, H., Wannous, C., & Murray, V. (2015). The Sendai framework for disaster risk reduction: Renewing the global commitment to people's resilience, health, and well-being. *International Journal of Disaster Risk Science, 6*(2), 164–176.

Aldyan, R. A. (2023). The impact of climate change on water resources and food security in Indonesia. *Journal of Law, Environmental and Justice, 1*(1), 50–63.

Ariyaningsih, Sukmara, R. B., Bachtiar, R., Iriansyah, A. A., & Shaw, R. (2025). Disaster law in Indonesia. In *Disaster law: Implications to governance and implementation* (pp. 183–205). Springer.

Asrofi, A., Giyarsih, S. R., & Hadmoko, D. S. (2024). The impact of tidal floods on poor households in the Sayung Coast, Demak Regency, Indonesia. *The Indonesian Journal of Geography, 56*(3), 456–465.

Aziz, M., & Anjum, G. (2024). Transformative strategies for enhancing women's resilience to climate change: Apolicy perspective for low-and middle-income countries. *Women's Health, 20*, 17455057241302032.

Bott, L.-M., Schöne, T., Illigner, J., Haghighi, M. H., Gisevius, K., & Braun, B. (2021). Land subsidence in Jakarta and Semarang Bay–The relationship between physical processes, risk perception, and household adaptation. *Ocean & Coastal Management, 211*, 105775.

Butt, S. (2013). Disaster management law in Indonesia: From response to preparedness? In *Asia-Pacific Disaster Management: Comparative and socio-legal perspectives* (pp. 183–196). Springer.

Callahan, D. (1973). The WHO definition of 'health'. *Hastings Center Studies*, 77–87.

Djalante, R. (2016). Research trends on hazards, disasters, risk reduction and climate change in Indonesia: A systematic literature review. *Natural Hazards and Earth System Sciences Discussions, 2016*, 1–48.

Djalante, R., & Garschagen, M. (2017). A review of disaster trend and disaster risk governance in Indonesia: 1900–2015. *Disaster risk reduction in Indonesia: Progress, challenges, and issues* (pp. 21–56).

Djalante, R., Garschagen, M., Thomalla, F., & Shaw, R. (2017). *Introduction: Disaster risk reduction in Indonesia: Progress, challenges, and issues*. Springer.

Djalante, R., & Thomalla, F. (2012). Disaster risk reduction and climate change adaptation in Indonesia: Institutional challenges and opportunities for integration. *International Journal of Disaster Resilience in the Built Environment, 3*(2), 166–180.

Dwirahmadi, F., Barnes, P., Wibowo, A., Amri, A., & Chu, C. (2023). Linking disaster risk reduction and climate change adaptation through collaborative governance: Experience from urban flooding in Jakarta. *Geosciences, 13*(11), 353.

Dwirahmadi, F., Rutherford, S., Phung, D., & Chu, C. (2019). Understanding the operational concept of a flood-resilient urban community in Jakarta, Indonesia, from the perspectives of disaster risk reduction, climate change adaptation, and development agencies. *International Journal of Environmental Research and Public Health, 16*(20), 3993.

Esteban, M., Takagi, H., Jamero, L., Chadwick, C., Avelino, J. E., Mikami, T., Fatma, D., Yamamoto, L., Thao, N. D., & Onuki, M. (2020). Adaptation to sea level rise: Learning from present examples of land subsidence. *Ocean & Coastal Management, 189*, 104852.

Fatmawati, F., & Sulistyawati, S. (2019). Climate change and dengue in Indonesia: A systematic review. *Epidemiology and Society Health Review, 1*(1), 29.

Gan, C. C. R., Oktari, R. S., Nguyen, H. X., Yuan, L., Yu, X., Kc, A., Hanh, T. T. T., Phung, D. T., Dwirahmadi, F., & Liu, T. (2021). A scoping review of climate-related disasters in China, Indonesia and Vietnam: Disasters, health impacts, vulnerable populations and adaptation measures. *International Journal of Disaster Risk Reduction, 66*, 102608.

Hadi, S. (2020). Disaster management in the implementation of the 2030 Sustainable Development Goals in Indonesia. *The Journal of Indonesia Sustainable Development Planning, 1*(1), 105–111.

Handayani, W., Chigbu, U. E., Rudiarto, I., & Putri, I. H. S. (2020). Urbanization and increasing flood risk in the Northern Coast of Central Java—Indonesia: An assessment towards better land use policy and flood management. *Land, 9*(10), 343.

Handayani, W., Fisher, M. R., Rudiarto, I., Setyono, J. S., & Foley, D. (2019). Operationalizing resilience: A content analysis of flood disaster planning in two coastal cities in Central Java, Indonesia. *International Journal of Disaster Risk Reduction, 35*, 101073.

Hawken, S., & Sunindijo, R. Y. (2018). City of Kampung: Risk and resilience in the urban communities of Surabaya, Indonesia. *International Journal of Building Pathology and Adaptation, 36*(5), 543–568.

Heath, S. C. (2025). Navigating psychosocial dimensions: Understanding the intersections of adaptation strategies and well-being outcomes in the context of climate change. *Current Opinion in Environmental Sustainability, 72*, 101493.

Heryani, N., Kartiwa, B., Sosiawan, H., Rejekiningrum, P., Adi, S. H., Apriyana, Y., Pramudia, A., Yufdy, M. P., Tafakresnanto, C., & Rivaie, A. A. (2022). Analysis of climate change impacts on agricultural water availability in Cimanuk Watershed, Indonesia. *Sustainability, 14*(23), 16236.

Irda Sari, S. Y., Sunjaya, D. K., Shimizu-Furusawa, H., Watanabe, C., & Raksanagara, A. S. (2018). Water sources quality in urban slum settlement along the contaminated river basin in Indonesia: Application of quantitative microbial risk assessment. *Journal of Environmental and Public Health, 2018*(1), 3806537.

Karlina, T., & Johan, W. (2020). Sea level rise in Indonesia: The drivers and the combined impacts from land subsidence. *ASEAN Journal on Science and Technology for Development, 37*(3), 3.

Krippendorff, K. (2018). *Content analysis: An introduction to its methodology*. Sage Publications.

Maharani, A., Sujarwoto, S., Susanti, H., Brooks, H., & Bee, P. (2025). Association between climate related hazards and depression among coastal communities in Indonesia. *Scientific Reports, 15*(1), 6998.

Mardiah, A. N., Lovett, J. C., & Evanty, N. (2017). Toward integrated and inclusive disaster risk reduction in Indonesia: Review of regulatory frameworks and institutional networks. *Disaster risk reduction in Indonesia: Progress, challenges, and issues* (pp. 57–84).

Mardiasmo, A., & Barnes, P. (2015). Community response to disasters in Indonesia: Gotong Royong; a double edged-sword. In *Proceedings of the 9th annual international conference of the international institute for infrastructure renewal and reconstruction.*

Neuendorf, K. A. (2018). Content analysis and thematic analysis. In *Advanced research methods for applied psychology* (pp. 211–223). Routledge.

Nugraha, E., & Lassa, J. A. (2018). Towards endogenous disasters and climate adaptation policy making in Indonesia. *Disaster Prevention and Management: An International Journal, 27*(2), 228–242.

Nurhidayah, L. (2020). Sea-level rise (slr) and its implication on human security and human rights in Indonesia: A legal analysis. In *Climate change research, policy and actions in Indonesia: Science, adaptation and mitigation* (pp. 33–52). Springer.

Nurhidayah, L., Davies, P., Alam, S., Saintilan, N., & Triyanti, A. (2022). Responding to sea level rise: Challenges and opportunities to govern coastal adaptation strategies in Indonesia. *Maritime Studies, 21*(3), 339–352.

Oktari, R. S., Kamaruzzaman, S., Fatimahsyam, F., Sofia, S., & Sari, D. K. (2021). Gender mainstreaming in a Disaster-Resilient Village Programme in Aceh Province, Indonesia: Towards disaster preparedness enhancement via an equal opportunity policy. *International Journal of Disaster Risk Reduction, 52*, 101974.

Otsuka, Y., Agestika, L., Sintawardani, N., & Yamauchi, T. (2019). Risk factors for undernutrition and diarrhea prevalence in an urban slum in Indonesia: Focus on water, sanitation, and hygiene. *The American Journal of Tropical Medicine and Hygiene, 100*(3), 727.

Pawitan, H. (2018). Climate change impacts on availability and vulnerability of Indonesia water resources. In *IOP Conference Series: Earth and Environmental Science.*

Pereira, J. J., Pulhin, J., Chhinh, N., Trong, T. D., & Satari, S. K. (2019). Appraising slow onset hazards for loss and damage: Case studies in Southeast Asia. *APN Science Bulletin, 9*(1), 45–51.

Prakash, A., Totin, E., Kemp, G., Kerr, R. B., & Roberts, D. (2025). Bridging the gap: Promoting gender equity in climate change adaptation in the global south. *PLOS Climate, 4*(6), e0000556.

Purnomo, A. H., Kurniawan, T., Farandy, A. R., Apriliani, T., Imron, M., & Sajise, A. J. (2024). Revisiting the climate change adaptation strategy of Jakarta's coastal communities. *Ocean & Coastal Management, 253*, 107112.

Roitman, S., & Rukmana, D. (2022). Urban Indonesia: Challenges and opportunities. In *Routledge handbook of Urban Indonesia* (pp. 3–14).

Saputra, E., Reinhart, H., Musthofa, A., Rofi, A., Nu'mah, A. K., & Saiddinullah, A. (2025). The political ecology of disasters: The impact of knowledge/power on the responses to urban coastal disasters in Pekalongan, Indonesia. *Geo: Geography and Environment, 12*(1), e70006.

Shammin, M. R., Haque, A. E., & Faisal, I. M. (2022). A framework for climate resilient community-based adaptation. In *Climate change and community resilience* (pp. 11–30).

Shaw, R., & Cui, Y. (2025). Disaster law: An overview on its history, potentials and challenges. In *Disaster law: Implications to governance and implementation* (pp. 1–14).

Sholeh, M., Naryatmojo, D. L., Putro, S., & Simbolon, A. S. (2024). Climate change adaptation through urban agriculture food barn in Tambakrejo, Semarang City. *Jurnal Abdimas, 28*(2), 273–278.

Siriwardane-de Zoysa, R., Schöne, T., Herbeck, J., Illigner, J., Haghighi, M., Simarmata, H., Porio, E., Rovere, A., & Hornidge, A.-K. (2021). The 'wickedness' of governing land subsidence: Policy perspectives from urban Southeast Asia. *PLoS One, 16*(6), e0250208.

Sufri, S., & Lassa, J. A. (2024). Integration of Disaster Risk Reduction (DRR) and Climate Change Adaptation (CCA) in Aceh province: Progress and challenges after 20 years of Indian Ocean Tsunamis. *International Journal of Disaster Risk Reduction*, 104894.

Suhartini, N., & Jones, P. (2023). Urbanization and the development of the Kampung in Indonesia. In *Beyond the informal: understanding self-organized Kampungs in Indonesia* (pp. 57–77). Springer.

Suntoro, A., Marbun, A. A. Y., & Sayekti, C. W. (2025). Climate change challenge: Coastal community adaptation towards achieving sustainable community resilience in Central Java, Indonesia. In *Climate change: Conflict and resilience in the age of anthropocene* (pp. 235–255). Springer.

Suprayitno, D., Iskandar, S., Dahurandi, K., Hendarto, T., & Rumambi, F. J. (2024). Public policy in the era of climate change: Adapting strategies for sustainable futures. *Migration Letters, 21*(S6), 945–958.

Swaris, N., Halwatura, R. U., & Amaratunga, D. (2024). Policy coherence for resilience in Sri Lanka coherence of climate change adaptation (CCA) disaster risk reduction (DRR) and sustainable development goals (SDGs). *International Journal of Disaster Resilience in the Built Environment, 15*(3), 450–473.

Swarnokar, S. C., Mou, S. I., Sharmi, S. D., Iftikhar, A., & Jesmin, S. (2025). Climate-induced risks, adaptation, and mitigation responses: A comparative study on climate-stressed coastal communities. *Frontiers in Climate, 7*, 1553579.

Takagi, H., Esteban, M., Mikami, T., & Fujii, D. (2016). Projection of coastal floods in 2050 Jakarta. *Urban Climate, 17*, 135–145.

Wijaya, N., Bisri, M. B. F., Aritenang, A. F., & Mariany, A. (2017). Spatial planning, disaster risk reduction, and climate change adaptation integration in Indonesia: Progress, challenges, and approach. In *Disaster risk reduction in Indonesia: Progress, challenges, and issues* (pp. 235–252). Springer.

Yamamoto, K., Sayama, T., & Apip. (2021). Impact of climate change on flood inundation in a tropical river basin in Indonesia. *Progress in Earth and Planetary Science, 8*, 1–15.

Yoseph-Paulus, R., & Hindmarsh, R. (2018). Addressing inadequacies of sectoral coordination and local capacity building in Indonesia for effective climate change adaptation. *Climate and Development, 10*(1), 35–48.

Zaidi, R. Z., & Fordham, M. (2021). The missing half of the Sendai framework: Gender and women in the implementation of global disaster risk reduction policy. *Progress in Disaster Science, 10*, 100170.

Zhang, Y., & Wildemuth, B. M. (2009). Qualitative analysis of content. *Applications of social research methods to questions in information and library science, 308*(319), 1–12.

Zikra, M. (2015). Climate change impacts on Indonesian coastal areas. *Procedia Earth and Planetary Science, 14*, 57–63.

Chapter 11
Climate Change and the Paradox of Stunting Cases in North Coastal Area of Java: A Study in Demak and Semarang

Aditya Rahmadhony, Oktriyanto, Resti Pujihasvuty, Diah Puspita Sari, Amanda Rose Collins, and Diana Nur Afifah

Introduction

Climate change is an urgent global issue affecting many aspects of human life, making it a major concern for local and global environmental policymakers (Abbass et al., 2022; Cadez et al., 2019; Klenert et al., 2020). Indonesia has more than 17,500 islands and a coastline stretching over 108,000 kilometres. As such, the country's territory is dominated by water, being about 70% ocean and 30% land. These circumstances make Indonesia particularly vulnerable to climate change impacts (Kurniawan et al., 2024), especially sea level rise and erosion. Other impacts of climate change seen in Indonesia are rising temperatures, heavy rainfall, floods, droughts, and forest fires (Nashier & Lakra, 2020). In addition to these environmental changes and their consequences for biodiversity, climate change also impacts the economic, social, health, and food security experiences of people in affected regions (Abbass et al., 2022; Gal, 2020; Nashier & Lakra, 2020; Sui et al., 2020). As such, the effects of climate change pose significant challenges to the health and well-being of Indonesians (Nur Amalia et al., 2024).

A. Rahmadhony (✉)
Research Center for Law, National Research and Innovation Agency (BRIN), Jakarta, Indonesia
e-mail: adit030@brin.go.id

Oktriyanto · R. Pujihasvuty · D. P. Sari
Research Center for Population, National Research and Innovation Agency (BRIN), Jakarta, Indonesia

A. R. Collins
School of Health Sciences and Social Work, Griffith University, Brisbane, QLD, Australia

D. N. Afifah
Diponegoro University, Semaramg, Indonesia

L. Nurhidayah et al. (eds.), *Climate Change, Labour and Migration in Indonesia*, Springer Climate, https://doi.org/10.1007/978-3-032-15334-0_11

Children are particularly vulnerable to the health problems caused by climate change (Helldén et al., 2021; Proulx et al., 2024; UNICEF, 2023, 2024). One of these health problems is stunting, defined as a condition of poor physical growth and development in children caused by chronic undernutrition, and manifesting as a height-for-age measurement below −2 standard deviations (SD) from the median of the World Health Organization (WHO) Child Growth Standards (Perumal et al., 2018). Stunting is a temporary loss in growth and a cumulative effect of sustained dietary shortfalls and frequent illnesses, particularly during the first 1000 days of life (Handayani et al., 2024). In addition, United Nations Children's Fund (UNICEF) defines stunting as a condition of failure to thrive in toddlers (under five years old), which is characterised by a height lower than the standard for their age (UNICEF et al., 2020). Stunted children show poor health, cognitive, motor, and psychometric functions, as well as decreased social skills (Phalkey et al., 2015). Climate change and food security are often identified as causes of stunting. Agostoni et al. (2023) found that climate change was a cause of a decrease in the nutritional value of food, resulting in climate change-induced chronic undernutrition, which in turn can lead to stunting. While emphasising the impact of climate change on malnutrition, Agostoni et al. (2023) also identified the underlying causes of stunting—namely, household food security, dietary diversity, nutritional quality of food, and access to health facilities. Furthermore, the social, economic, and political factors determining food and nutrition security (livelihoods, income, infrastructure resources, and political context) were articulated. These findings reinforce knowledge of how climate change can lead to stunting, especially in vulnerable areas.

Other studies have also revealed that increased temperature, extreme heat, decreased rainfall, and drought caused by climate change can lead to low birth weight (LBW), premature births, malnutrition, inhibited growth, and stunting of children (Haines & Ebi, 2019; Ngwira, 2020; Meherali et al., 2024; van der Merwe et al., 2022). It may cause severe food shortages due to these events, especially in environmental conditions primarily influenced by seasonal and annual climate variability (Kinyoki et al., 2016). In contrast, Ngwira (2020) found that increased rainfall will also increase the incidence of stunting. As such, both an increase and decrease in rainfall may lead to malnutrition and cases of stunting, demonstrating not only the extent to which climate change impacts children's health but also the delicacy of maintaining an ecosystem that supports healthy food production.

Climate change in coastal areas is a serious issue because these areas are very vulnerable to climate change impacts. Heavy rainfall, tidal flooding, sea level rise, land subsidence due to excessive groundwater use, coastal erosion, heat waves, and seawater intrusion may all be experienced in these regions (Nicholls & Cazenave, 2010; Zamrsky et al., 2024). The impacts are exacerbated by human activities such as pollution and habitat destruction (Asensio-Montesinos et al., 2024; Marfai et al., 2008; Roy et al., 2023), which disrupt the sustainability of ecosystems and human life. Many Indonesians live in coastal areas, the centre of economic activity. In general, coastal communities are highly dependent on marine products, with the fishing sector being the main livelihood for most residents of coastal areas. Coastal areas prone to extreme weather tend to experience more significant economic losses due to the disruption of fishing activities and reduced fishing catches. This creates fluctuations in income, limited economic conditions, and subsequent low purchasing power (Kapantow et al., 2023). The negative impact of climate change on economic conditions means that coastal communities often cannot utilise the abundant marine resources that should be available to them to improve their nutritional intake. One example is the nutrition from animal protein derived from marine products. Despite its availability, access to this protein source is difficult for most locals due to their low income levels (Has et al., 2020). In addition, rising living costs and climate change threaten the health, well-being, and livelihoods of millions of people in coastal areas by damaging infrastructure, disrupting fish stocks, and weakening human and natural resilience (Posen et al., 2023; Ulfa, 2018; Wu & Barrett, 2022). While coastal areas geographically have direct access to marine resources, climate change, along with various socioeconomic and infrastructural factors, is a significant obstacle to their optimal utilisation for local welfare (Kapantow et al., 2023). In summary, people's dependence on an unstable fisheries sector creates structural poverty (Ulfa, 2018), and risk factors for stunting are higher in low-income families (Yuliantini et al., 2022). This socioeconomic inequality is the root cause of stunting in coastal communities (Hallegatte et al., 2018).

Rising temperatures, changing rainfall patterns, and extreme weather events can negatively impact agricultural productivity and land-based food availability (Agostoni et al., 2023; Praveen & Sharma, 2019). Agricultural production's quantity and nutritional quality depend on the balance of appropriate biophysical resources, including soil quality, water availability, CO_2, sunlight, and temperature suitability (Myers et al., 2017). Seawater intrusion in the agricultural sector can damage farmland and disrupt food yields (Kapantow et al., 2023), negatively affecting health determinants (Agostoni et al., 2023). Seawater intrusion, tidal flooding, and limited access to clean water also increase health risks, such as diarrhoea, dengue fever, malaria, and other infectious diseases, which directly affect the nutritional status of children and lead to stunting (Haines & Ebi, 2019; Rohmah & Mardiana, 2023; UNICEF, 2023; WHO, 2014). As such, it is evident why Haines and Ebi (2019) concluded that the instability of food and nutrition created by climate change has led to crises in vulnerable communities.

Over a decade, from 2013 to 2023, Indonesia succeeded in reducing the prevalence of stunting from 37.6 to 21.5% (Kemenkes, 2024). However, the prevalence of stunting cases in Indonesia remains relatively high. Of the 38 provinces in Indonesia, 15 have prevalence rates below the national rate. The disparity in stunting cases in Indonesia is enormous, with the lowest stunting prevalence being 7.2% in Bali and the highest 39.4% in Central Papua (Kemenkes, 2024). Central Java still has a high prevalence of stunting at 20.7% as of 2023 (Kemenkes, 2024). Of the two study sites in our research, Semarang City experienced an increase in stunting prevalence—from 10.4% in 2022 to 15.7% in 2023—while Demak decreased in prevalence—from 16.2% in 2022 to 9.5% in 2023 (Dinkes Jateng, 2024). The downward trend in stunting cases in Indonesia is still quite far from the target mandated by Presidential Regulation Number 72 of 2021 concerning the Acceleration of Stunting Reduction, which was 14% by 2024. Furthermore, the national rate of 21.5% in

2023 is still relatively high in global terms and tends to experience stagnation (Kemenkes, 2024). The global situation illustrates that the disparity in stunting prevalence experienced in Indonesia remains a serious problem that requires greater attention (Meher et al., 2023).

The impact of climate change in the long term will affect the sustainability of the ecosystem and continue to impact local food security that is sourced from fisheries and agricultural catches. Sea level rise and seawater intrusion will continue negatively impacting many coastal populations relying heavily on freshwater (Nicholls & Cazenave, 2010; Zamrsky et al., 2024). It is increasingly evident that these issues must be addressed, as when obtaining nutritious food becomes difficult, malnutrition and stunting of children become more prevalent (Kapantow et al., 2023). For these reasons, our study sought to understand the relationship between climate change and stunting prevalence in the northern coastal region of Java, focusing on Demak and Semarang. In addition, our research evaluates local context-based policies, particularly measuring the effectiveness of policy implementation in addressing the challenges of climate change-induced stunting across this region. Hopefully, these much-needed strategies to mitigate and adapt communities to climate change vulnerabilities that impact stunting in children can be developed.

Methodology

This study adopted a qualitative approach with an exploratory focus to explore the relationship dynamics between climate change and stunting prevalence in the northern coastal region of Java, specifically in Demak Regency and Semarang City. A non-probability purposive sampling technique was used to select ten key informants, consisting of three groups: (1) coastal communities, especially parents of stunted children, to explore the social, economic, and environmental factors that influence parenting, food access, and child health; (2) health workers, such as midwives and *puskesmas* (community health centre) officers, to obtain information related to maternal and child health status, nutrition services, and available health interventions; and (3) policymakers at the local level and related organisations, to understand the policies and strategies implemented in stunting prevention and climate change impact mitigation. In-depth interviews were the primary data collection method to capture community perceptions, experiences, and adaptive responses to climate risks and their impacts on child health.

Socioeconomic Vulnerabilities and Community Responses to Stunting in Coastal Demak and Semarang

Stunting cases in Tambaklorok Village, Semarang, typically occur because mothers lose their jobs or livelihoods as factory workers due to termination of employment, limiting their ability to meet their children's nutritional needs, resulting in their children suffering from stunting. In addition to factory work, most of the residents in Tambaklorok work as fishermen. According to the head of the Family Empowerment and Welfare Movement (PKK) *Rukun Warga* (RW) (community associations) 15 (in-depth interview, 11 November 2024), most children at risk of stunting have parents who work as fishermen.

Cases of stunting due to economic factors were also found in Sayung Village, Demak. One example is a housewife who had to quit her job at a snack factory in Semarang after losing her mother, who had previously helped with childcare. This decision impacted the family's income, since her husband works as a labourer in a car repair shop in Kendal, Central Java. In addition, the family's living conditions were poor, with a house with almost no doors or windows and a dirt floor. The home was also often affected by flooding during the rainy season and tidal floods.

> I used to work in a ciki (snack) factory in the Semarang area, sir, but because my mother died, I finally resigned to take care of my child. At that time, the economy was still good; for my first child alone, I gave Bebelac brand milk. I only found out that my second child was stunted when I went to the *posyandu* [integrated health post]. (Housewife, 36 years old, 2024, in-depth interview)

One of the causes of stunting is suboptimal parenting, often because parents work and leave parenting to others. In Kalisari Village, Demak, even where families are economically well-off children still experience stunting because both parents work as factory labourers with busy schedules. Childcare is left to grandparents, and the food is monotonous, with only sweetened condensed milk as a nutritional supplement (Grandfather, 68 years old, 2024, in-depth interview).

Another finding is that fisher families themselves rarely consume protein-rich marine catches, with most of the fish caught sold at the market. The money earned from selling the fish is used to buy diesel as boat fuel and rice as staple food, repair damaged nets, and meet other household needs.

Child marriage is another of the contributing factors to stunting cases, especially in coastal areas such as Tambaklorok, and this is reflected in the increasing number of recent stunting cases, in which most of the mothers were married at a young age. In-depth interviews with the head of Tambaklorok Family Empowerment and Welfare in November 2024 revealed that fathers' preferences also significantly influence children's diets in coastal families.

Most mothers who work as factory labourers or shellers leave the responsibility of childcare to grandmothers or other family members. After work, mothers often feel exhausted and prefer to rest or use their mobile phones rather than interact directly with their children. In addition, their participation in *posyandu* activities to monitor their children's health and development is low. Urgent economic needs mean that participating in *posyandu* activities is frequently neglected. The head of PKK Tambaklorok conveyed this:

> Many parents, especially mothers, prefer shelling shells to attend the *posyandu* because they can earn around IDR50,000 per day. This difficult economic situation often forces them to neglect their children's nutritional needs. The women prefer jobs that make money quickly, such as shelling shells. Low education levels and limited access to information on child nutrition exacerbate this condition. (Head of Family Empowerment and Welfare RW 15, 2024, in-depth interview)

Interviews with the chairperson of the PKK in Tambaklorok, Semarang (in-depth interview, 11 November 2024), and the Health Officer of Puskesmas Sayung 1, Demak Regency (in-depth interview, 13 November 2024), revealed that there is a tendency among parents who have stunted toddlers not to feel worried about the condition. They accept the stunting condition of their children with resignation and feel happy because they receive assistance to meet their children's nutritional needs. This phenomenon reveals an interesting social paradox that needs to be studied more deeply: malnutrition in children that results in stunting is no longer a taboo and embarrassing thing among specific communities. On the contrary, this condition is perceived as an opportunity to receive assistance from government programmes and through community initiatives. This may indicate a shift in social norms or perceptions related to the problem of malnutrition and stunting, which can potentially influence efforts to prevent and address this problem sustainably.

Based on an in-depth interview with the head of the Family Planning Division of Demak District (12 November 2024), Demak District's achievement in reducing the stunting rate to 9.5% by 2023 was the result of anticipatory efforts and the implementation of sustainable programmes. One of the significant findings in this study is the role of the *Cengkraman Mata Elang* (CME) programme. Originally designed as an application to monitor the health of pregnant women and infants to reduce maternal and infant mortality, the programme has evolved into a tool with early warning system features to detect toddlers at risk of stunting. This programme has won various awards, both at the regional and national levels, as a form of innovation in public services. CME is one of the key factors in efforts to reduce the prevalence of stunting in the Demak District. Figure 11.1 depicts stunting information available to the community.

Based on an interview with the head of PKK Tambaklorok RW 15 (10 November 2024), stunting remains one of the community's main challenges. However, it has been effectively addressed through collaboration between the community, government, and related institutions. The implementation of programmes such as The Friday Blessing Program (*Jumat Berkah*) and assistance to mothers with children at risk of stunting have shown significant results. The Friday Blessing Program (*Jumat Berkah*) is an initiative launched by KWT *Cantik Bahari*, held every Friday, consisting of social activities aimed at sharing, giving alms, and strengthening community bonds for those in need. With this programme, the stunting rate, which initially stood at 46 children in 2019, declined to 30 children in 2020, 25 in 2021, 20 in 2022, 15 in 2023, and just seven in 2024.

Fig. 11.1 Cengkraman Mata Elang application. (Source: https://cme.demakkab.go.id/)

The *Cantik Bahari* Women Farmers Group (KWT—*Cantik Bahari*) runs a nutritious food provision programme using local ingredients, such as moringa leaves and Brazilian spinach, which are processed into vegetable *bobor* (porridge) using coconut milk. Using coconut milk as an additional ingredient aims to increase the calorie content of food to help toddlers gain weight. In addition, KWT *Cantik Bahari* actively visits and assists mothers with toddlers under five who cannot attend a *posyandu* to monitor the development of children at risk of stunting. Funding for this programme comes from community self-help through selling plant seeds cultivated by the KWT. According to the head of Sayung 1 Health Centre, Demak Regency (in-depth interview 11 November 2024) also has various locally based innovation programmes that involve the community and are supported by the local government. One of these programmes is *Satu Hari Ikan, Satu Hari Telur* or SIST ('One Day Fish, One Day Eggs'), an initiative of Penggerak.

In Bonang District, an innovative programme, *Sukodono Peduli Stunting dan Gizi* or SUSI ('Sukodono Cares for Stunting and Nutrition'), has been developed to utilise local resources from coastal areas. This programme is designed to process local raw materials, especially fish, into value-added food products, such as nuggets, meatballs, and other similar products. This approach aims to increase the attractiveness of consumption and food preferences in children to support the fulfilment of optimal nutritional needs and contribute to efforts to overcome stunting.

The role of the community cannot be denied in overcoming stunting cases in Semarang and Demak. Utilising local resources, the community in both areas has successfully created various innovative and strategic programmes. Tambaklorok Village, North Semarang District, is an example of a self-help programme's success in reducing stunting cases in its area. Groups such as *Cantik Bahari* in Tambaklorok are the driving force behind women's empowerment. They are involved in educating and producing nutritious food and urban agricultural activities such as cultivating Brazilian spinach and moringa leaves. This group also sells their harvest to support the *Jumat Berkah* Program.

One of the key findings of this study is the existence of the resource paradox: although coastal areas are rich in marine resources, stunting prevalence rates remain high.

Climate Change, Socioeconomic Challenges, and Policy Responses to Stunting in Java's Northern Coastal Areas

Climate Change Impacts in the North Coastal Region of Java

The northern coastal areas of Java, especially in the Demak and Semarang regions, experience very complex conditions due to a range of natural and non-natural factors (Ervita & Marfai, 2017). Natural factors such as sea level rise, tidal flooding, erosion, and land subsidence have caused communities in the Demak and Semarang regions to lose their homes, land, and livelihoods. This phenomenon has caused significant coastal erosion in Semarang, threatening infrastructure and settlements along the coastline (Bisyri et al., 2020; Wahyudi et al., 2014). Demak is experiencing similar problems, with people often forced to adapt to increasingly unstable environmental conditions due to climate change (Sunarti & Apriliasari, 2015). The situation is exacerbated by non-natural factors, such as excessive groundwater use for households and industries, which accelerates land subsidence. Research results show that significant land subsidence due to anthropogenic factors or factors caused by rapid population growth and economic activity is the main driving force, which manifests as pressure exerted from overexploitation of groundwater, increased building and infrastructure loads, and subsidence of undeveloped areas (Sarah et al., 2020, 2022).

Climate change and land subsidence in coastal areas are detrimental physically and in terms of infrastructure. However, climate change has exacerbated existing vulnerabilities such as poverty, threatening the food security of people living in coastal areas (Nurhidayah, 2021). Another phenomenon is tidal flooding, which often occurs in coastal areas. This phenomenon threatens settlements and disrupts the community's economic activities, especially fishing and farming (Choirunnisa et al., 2022). In addition, increasing temperatures and changing rainfall patterns impact the agricultural sector and food security, especially for crops that depend on stable rainfall, such as rice (Kurniadi et al., 2023). Farmers often face difficulties determining the right planting time in coastal areas due to weather uncertainty, which can reduce crop yields (Harvian & Yuhan, 2021; Nuraisah & Budi Kusumo, 2019).

Other studies have also shown that climate change can increase the risk of crop failure, especially for rice crops highly dependent on stable rainfall patterns (Nuraisah & Budi Kusumo, 2019; Ruminta, 2016). In addition, fishermen experience difficulties determining the time and location of fishing due to erratic changes in sea conditions, ultimately reducing catches (Madaual et al., 2023). Studies have also shown that climate change can increase crop pests and diseases, further worsening food security in the region (Adriana et al., 2017; Harvian & Yuhan, 2021). The result is a decrease in agricultural productivity, the primary food source for residents in the region (Alfarizi & Ngatindriatun, 2023).

This study demonstrates that threats to food security which affect people experiencing poverty contribute to child malnutrition. Research shows that reduced agricultural yields due to climate change can lead to malnutrition, especially among children (Harini et al., 2022). Climate change negatively affects children in coastal areas, food security, and food diversity, including malnutrition such as stunting, wasting, and underweight (Hastuti et al., 2024). Food supply instability caused by crop failures and declining fish catches can exacerbate malnutrition among children, which in turn can affect long-term health (Alfarizi & Ngatindriatun, 2023; Harvian & Yuhan, 2021). As was expressed by participants, this results in children from low-income families being particularly at risk, as they do not have access to enough nutritious food to meet their needs. It poses a significant challenge for communities that depend on agriculture and fisheries as their primary source of livelihood.

Stunting in Java's North Coastal Region: The Case of Demak and Semarang

Although stunting is not directly caused by climate change, climate change can be a significant driver of the increase in stunting cases through several causal factors (Ruhyana et al., 2021). Climate change resulting in tidal flooding and seawater intrusion reduces the availability of clean water and increases pollution and the risk of diarrhoeal diseases (Abedin et al., 2019) due to poor sanitation conditions. Children living in affected areas in Demak and Semarang experience recurrent diseases such as gastrointestinal infections, which result in the child suffering from inadequate nutrition. WASH (water, sanitation, and hygiene) practices and poor household food security affect the nutritional status of children under five and lead to insufficient food intake (Anggraheny et al., 2024). To address this, the government has issued Minister of Health Regulation Number 35/2012 on Guidelines for the Identification of Health Risk Factors Due to Climate Change, which aims to provide a reference for health officers/apparatus within the Ministry of Health, provinces, and districts/cities, as well as climate change and health observers, to identify health risk factors caused by climate change.

Poor sanitation and limited access to clean water in coastal areas contribute significantly to child malnutrition and stunting, as children living in such environments are more prone to infections like diarrhoea, which hinder nutrient absorption and growth (Permana et al., 2023; Sefrina et al., 2022; Sopianti et al., 2024; Sukmawati et al., 2021; Widyastuti et al., 2023). These issues are often linked to poverty, with low-income families lacking the resources for proper facilities, and are further exacerbated by the mental and environmental stress brought about by climate change (Lawrance et al., 2022). The high rate of stunting in coastal areas is also influenced by poverty. Poverty contributes to families' inability to provide adequate and nutritious food, which is a significant factor in stunting in children (Najib et al., 2023; Yanti et al., 2023). Crop failure due to pond damage caused by tidal floods, such as those in the Demak coastal area, has disrupted food security (Asrofi et al., 2017; Claus et al., 2024) and resulted in economic losses for the affected farmers. Climate change has also led to a decline in fishermen's catches (Chan et al., 2023; Sari & Hadi, 2018), similar to the impact observed among Togo fishermen due to extreme weather such as rising sea levels, strong winds, and unpredictable rainfall that disrupted economic activities (N'Souvi et al., 2024). This situation resulted in decreased family income, meaning children's basic needs and nutritious food went unmet.

Low-income families often cannot afford nutritious food, so they rely on cheap and less healthy food (Yanti et al., 2023). This lack of dietary diversity causes children not to get the nutrients needed for optimal growth (Ahmad et al., 2018; Damanik et al., 2020; Reyes et al., 2004). Other studies have found that poor economic conditions among families are associated with a high risk of stunting (Handayani et al., 2022). Our findings demonstrated that these conditions resulted in reliance on family members to provide childcare; if both parents worked, childrearing was often left to grandparents, who may not have the best understanding of nutrition. This was seen in Demak, when a child was fed sweetened condensed milk. This reliance on childcare provided by grandparents also reduced the security of employment, with mothers having to give up work when family members passed away. Najib et al. (2023) found that the risk of stunting in children from food-insecure families was 6.9 times that of children from food-secure families. The practice of providing non-diverse foods was also found in toddlers in Jakarta (Damanik et al., 2020). In summary, it can be concluded that food insecurity in the household is directly related to the nutritional status of children, with households experiencing food insecurity having a higher probability of having stunted children (Sari et al., 2023). This was the case in both Demak and Semarang.

The relationship between parenting practices and stunting in Demak and Semarang reflects a complex interplay between parental knowledge, educational background, and socioeconomic factors that shape children's nutritional outcomes (Manggala et al., 2018; Rahayuwati et al., 2023). In these regions, parental behaviours—from feeding practices to healthcare decisions—play a vital role in reducing or increasing stunting among children. Studies indicate that maternal and paternal influences are significant (Manggala et al., 2018; Rahayuwati et al., 2023).

Other disruptions to parenting may contribute to higher rates of stunting. Findings suggest that mothers feeling exhausted from work and requiring rest, low participation in *posyandu* activities, and parental focus on urgent economic needs may all be contributing factors. Parents who work outside the area, leaving their children with their grandparents, have also been found to contribute to the high rate of stunting in the area (Azhari et al., 2024). Several studies have shown that child marriage has a significant influence on the incidence of stunting (Hidayah et al., 2024; Restiana &

Fadilah, 2023). The lack of physical and mental readiness of mothers who experience pregnancy as a result of child marriage puts their children at increased risk of stunting.

Often, pregnant mothers have limited access to nutrition and healthcare, along with a reluctance to attend maternity classes that could help improve maternal and child health due to their circumstances. Pregnant mothers in Demak and Semarang experience a variety of obstacles in terms of proper nutrition, access to healthcare, and participation in maternity classes, all of which are inextricably linked to socioeconomic limits and inadequate health education. Evidence demonstrates that social and economic factors—such as low educational attainment and suboptimal income levels—not only reduce these women's ability to acquire nutritious food and quality healthcare but also limit their exposure to vital health education programmes (Ramadhaniyati et al., 2025; Wijayanti et al., 2024). Wijayanti et al. (2024) present an empirical foundation by establishing that lower educational levels are associated with lower income, limiting pregnant women's access to decent nutrition and adequate healthcare. Consequently, pregnant women's lived experiences with obtaining basic healthcare services make them more vulnerable to stunting and other adverse effects. It emphasises the critical need for health policies prioritising nutrition education and economic empowerment, particularly among low-income families in Indonesia. As a result, socioeconomic deprivation makes it difficult to acquire food and medical services and engage in preventative health behaviours such as attending maternity classes, which can help improve mother and child health outcomes.

The Resource Paradox and Aid Dependency

Coastal areas should have great potential as a source of protein through abundant marine products, which can support children's growth needs. However, seafood is often only seen as a commodity traded to fulfil daily needs, not as a resource that can ensure the nutritional well-being of the community, especially children. The poverty that still plagues coastal communities is a significant obstacle to providing nutritious food for children due to low income and limited purchasing power to access healthy food (Has et al., 2020). Research in Buton Regency in 2024 also showed that one of the factors causing stunting is the family's socioeconomic situation, especially in terms of employment and income levels (Azhari et al., 2024). Low levels of protein adequacy were shown to be the most substantial risk factor associated with stunting in coastal areas (Rihi Leo et al., 2018), while food security had a significant positive association with stunting (Yuliantini et al., 2022). This situation illustrates the contradiction between coastal areas rich in protein sources and that protein inadequacy among coastal communities is a significant risk factor for stunting in children.

Fathers play a crucial role in regulating children's nutrition, through providing a healthy diet, taking children to health facilities, and fulfilling vaccination schedules (Prianto et al., 2025); the reality is that children—especially toddlers, who are in a critical growth phase—often only eat the same food as their fathers. This shows that fathers, as the primary breadwinners, play a role in determining the food distribution in the family, where they are considered worthy of receiving the best portion of food. In contrast, children's nutritional needs are often considered less important. This pattern reflects the dominance of a patriarchal culture that influences food distribution decisions in coastal families, where wives/mothers tend to prioritise fathers' preferences without considering children's nutritional needs. This patriarchal culture has the potential to harm children's welfare, especially in terms of their nutritional fulfilment (Islam et al., 2024).

In coastal areas of Demak and Semarang, where many families depend on fish catches as a source of income, selling fish was a top priority. While fishing was a primary source of income for families carried out by men, it was found that families rarely consumed their catch. In many cases, the revenue from fish sales is the primary source of family income, so fishers feel forced to sell fish they could have consumed (Tria et al., 2022). This is related to the high poverty rate, where the proceeds from selling fish are usually used to fulfil other basic needs such as food, education, and health, as determined by fathers. However, other factors, such as fishers' lack of knowledge about the nutritional value of fish, also contributed to the decision to sell the fish they caught. Many fishers and their family members are unaware of the health benefits of consuming fish, which is rich in protein and other essential nutrients (Muzdalifah & Nilamsari, 2021). Nearby shops and street vendors contribute to children in coastal areas preferring fast food or less nutritious snacks, leading to an unbalanced diet (Purwestri et al., 2018). On the other hand, fishing families are also vulnerable to changes in market prices and catches as they depend on fish sales, with the sale of fish being a primary source of income for many participants. Fishing families face significant challenges meeting their nutritional needs when catches decline or fish prices drop.

Findings suggest that this resource paradox has led to aid dependency. Stunting of children meant that families gained access to programmes and initiatives to reduce stunting, thus alleviating the pressure families were facing to meet nutritional needs. This phenomenon had social implications wherein the taboo associated with stunting decreased and became more accepted as a social norm and means of access. As such, programmes and initiatives developed to reduce stunting must take this social phenomenon into account to ensure that the livelihood of families is adequately being met so that malnutrition of children is not a primary means of access to support.

Policy and Case Analysis of Stunting in the North Coastal Region of Java

Presidential Regulation Number 72 of 2021 concerning the Acceleration of Stunting Reduction aims to reduce stunting prevalence to 14% by 2024. Stunting reduction is also one of the targets in the 2020–2024 National Medium-Term Development Plan (RPJMN). Likewise, the United Nations Sustainable Development Goals (SDGs) aim to eliminate all forms of malnutrition, including stunting, by 2030.

Figure 11.2 illustrates the trend in stunting prevalence in Indonesia from 2013 to the planned target of 2024. The graph presents national stunting prevalence data, which shows a gradual decline. In 2013, the stunting prevalence was recorded at 37.6%, decreasing to 30.8% in 2018, 27.7% in 2019, and 24.4% in 2021. Since the issuance of Presidential Regulation Number 72 of 2021 (Perpres 72/2021), the prevalence of stunting has decreased further, reaching 21.6% in 2022 and 21.5% in 2023. This graph reflects the success of the government's efforts to reduce stunting through various strategic policies and nutrition interventions formulated in Presidential Regulation 72/2021. However, although progress has been significant, challenges remain to meet the ambitious target set in 2024.

Nationally, the stunting prevalence rate has been consistently decreasing but was still 21.5% as of 2023. A reduction in stunting prevalence of 7.5% is therefore needed to reach the target of 14% by 2024. The dynamics at the local level, as seen in Semarang, provide a more detailed picture of the challenges and opportunities in addressing stunting. Figure 11.3 illustrates the trend in stunting prevalence in Semarang from 2019 to 2023, showing the dynamics of change over the period. In 2019, the prevalence of stunting was recorded at 26.0%, gradually declining to 21.3% in 2021 and reaching its lowest point of 10.4% in 2022. However, in 2023, the prevalence increased again to 15.7%.

The Semarang City Government has a specific policy related to accelerating stunting reduction, with the issuance of Mayor's Regulation Number 27 of 2022 on the Acceleration of Stunting Reduction in Semarang City. The Semarang City Government, which previously aimed to decrease stunting by 4% by 2026, brought

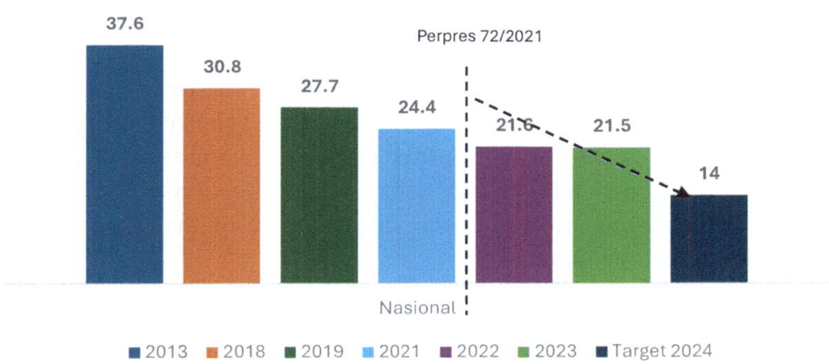

Fig. 11.2 Stunting prevalence trends in Indonesia. (Source: Ministry of Health, 2024)

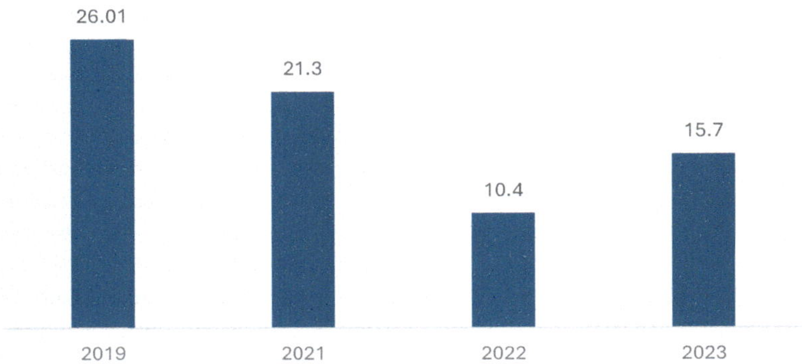

Fig. 11.3 Stunting prevalence trends in Semarang. (Source: Ministry of Health, 2023)

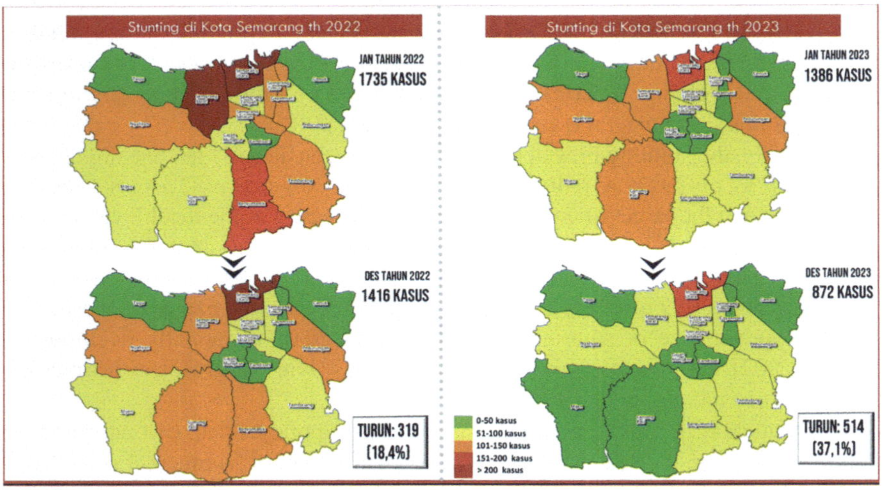

Fig. 11.4 Stunting case data in Semarang. (Source: Semarang City Health Department, 2024)

forward this target to 2024 based on Mayor Regulation 45 of 2023. The change in target to 4% decrease by 2024 represents a strong commitment of the Semarang City Government to reducing stunting prevalence. However, there was an anomaly in the trend after the policy was issued, with stunting prevalence in 2023 increasing to 15.7% from 10.4% in 2022. Thus, despite early success in reducing stunting the increase in 2023 indicates the need for a thorough evaluation of the policies and interventions implemented to ensure the sustainability of the trend in reducing stunting prevalence in the future.

Figure 11.4 shows that West Semarang and North Semarang, which are both coastal areas, have relatively high stunting rates of over 200 cases compared to other regions of Semarang City. It can be seen in the 2022 and 2023 maps that both West Semarang and North Semarang still have high stunting rates (marked in orange and red, respectively) despite the overall decline. However, more intensive intervention is needed in North Semarang, which has consistently been one of the centres of high stunting prevalence. A study conducted by Laode Wado et al. (2019) in the North Semarang area showed that family food security, father's occupation, family income, and respondents' lack of nutritional knowledge had a significant relationship with stunting.

Based on data from the Semarang City Health Office, the majority of stunted children are between 24 and 59 months of age. The data show that risk factors that may cause the high incidence of stunting in this age group included low birth weight (less than 2500 grams) in 17.9% of children under five and birth length less than 47 cm in 21.3% of children under five (Dinkes, 2024). One of the leading causes is the mother's pregnancy history; only 8.6% of mothers had more than six prenatal check-ups, and 27.4% were anaemic during pregnancy. With 67.6% of mothers not working and 52.4% of families earning less than the Regional Minimum Wage, family economic conditions also contribute to stunting (Dinkes, 2024). Only 18% of children receive age-appropriate complementary foods, and 29.1% are not exclusively breastfed. In addition, health problems such as Acute Respiratory Tract Infection (9.3%) and diarrhoea (7.4%) worsen nutrient absorption (Dinkes, 2024). The prevalence of stunting in this age group raises concerns since it represents not only the children's current nutritional state but also the accumulated harmful exposures and health disparities they encountered during their early years.

The Semarang City Government has made various efforts to tackle stunting cases through programme innovations such as establishing the *Rumah Pelita/Rumah Penanganan Stunting Lintas Sektor Bagi Baduta* ('Cross-Sector Stunting Handling House for Under-fives') and *Rumah Pelangi Nusantara* ('Rainbow House Nusantara'), which involve cross-sectoral input to overcome malnutrition and stunting. In addition, there is a programme that involves the private sector by collaborating with entrepreneurs in the hospitality sector and the Indonesian Food Service Providers Association (PPJI)/*Perkumpulan Penyelenggara Jasa Boga Indonesia* called the CEMPAKA Program/*Cegah Stunting Bersama Pengusaha di Kota Semarang* (Prevent Stunting with Entrepreneurs in Semarang City), as well as many other programmes intended to monitor child development and growth, promotive interventions for pregnant women, and nutrition education programmes.

The situation in Semarang City is rather different. Demak experienced a consistent downward trend in stunting prevalence, as shown in Fig. 11.5, from 2019 to 2023. In 2019, the stunting prevalence was recorded at 35.76%; this decreased to 25.5% in 2021, then continued to decline until it reached 9.5% in 2023. This significant decline reflects the success of various interventions carried out in the region and shows consistent progress in efforts to achieve the national target of reducing stunting prevalence.

Demak's success in reducing stunting rates is most likely due to the government taking anticipatory steps as early as possible to manage children at risk of stunting. Demak has issued a policy in the form of Demak Regent Regulation Number 11 of 2019 concerning Reducing the Prevalence of Stunting in Demak Regency, long before the national policy, namely the issuance of Presidential Regulation Number

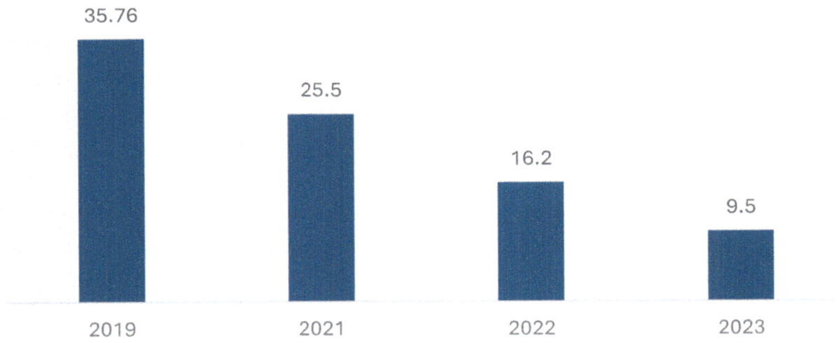

Fig. 11.5 Stunting prevalence trends in Demak. (Source: Ministry of Health, 2023)

72 of 2021 concerning the Acceleration of Stunting Reduction. Demak Regent Regulation Number 11 of 2021 regulates through four approaches, including (1) the Family Independence approach, a health and nutrition education strategy carried out with promotive and preventive efforts; (2) the Healthy Living Community Movement, synergising the actions of promotive and preventive efforts to prevent stunting problems and increase community productivity; (3) the First Thousand Days of Life Movement, which mobilises the participation and concern of stakeholders for the nutritional needs of foetuses and toddlers in the first 1000 days of life; and (4) conducting education, training, and nutrition counselling to the community.

Multi-sectoral involvement in managing stunting in Demak Regency is essential. To coordinate efforts, the Demak Government issued Regent Regulation Number 29 of 2019 concerning the Regional Action Plan for Stunting Prevention in Demak Regency 2019–2021. This policy was issued due to the need to synchronise programmes and activities aimed at reducing and preventing stunting across various sectors. The first phase of the Regional Action Plan was conducted in ten villages targeted for sensitive and specific nutrition interventions. These were implemented over two years, 2020 and 2021, in five target villages with high stunting rates.

In 2024, the Demak District Government issued a new policy, Regent Regulation Number 8 of 2024 on Accelerating Stunting Reduction in the Region, which replaced Regulation Number 11 of 2019. The new policy focuses more on increasing commitment and coordination across sectors, improving behaviour change communication, strengthening specific and sensitive interventions, and food and nutrition security in stunting prevention. Although Demak has successfully reduced the stunting prevalence rate to 9.5% by 2023, Regent Regulation Number 8 of 2024 sets an intermediate target of 14% by 2024 to achieve the 2030 SDG target of eliminating all forms of malnutrition, including stunting.

Figure 11.6 shows the distribution map of stunting cases in Demak in 2023 by subdistrict. There are still areas with high stunting cases, such as Kecamatan Karangtengah and Kecamatan Guntur, and some in the medium category, such as Kecamatan Bonang, Sayung, Dempet, and Kebonagung; the rest, shown in green,

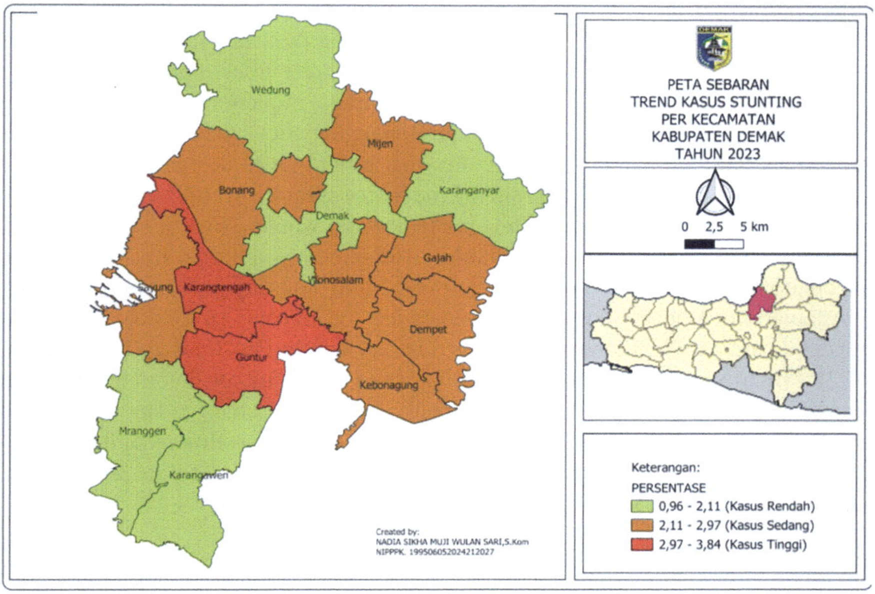

Fig. 11.6 Map of distribution of stunting cases. (Source: Demak Regency Health Department, 2024)

have low stunting rates. Coastal areas such as Sayung, Bonang, and parts of Karangtengah still tend to fall into the medium and high categories.

Successful Stunting Management: The Importance of Community Participation

National and local policies are essential in creating an environment that supports stunting management at the macrosystem level. The government's commitment to addressing stunting is demonstrated by policies such as Presidential Regulation Number 72 of 2021, Legislation at the regional level as seen in Demak and Semarang, and innovative local-based programmes such as SIST and SUSI. Although these policies do not specifically regulate the handling of stunting due to climate change or specifically for affected coastal communities, there is already a firm commitment from the government to handling stunting sustainably. Regulation of the Minister of Health Number 35 of 2012 concerning Guidelines for Identification of Health Risk Factors Due to Climate Change needs to be particularly appreciated as it provides a guideline for identifying health risk factors caused by climate change, one of which is malnutrition in children.

The success of programmes in the Semarang and Demak areas such as *Cantik Bahari*, *Jumat Berkah*, SIST, and SUSI indicates the importance of close collaboration between the community, government, and related institutions in efforts to address stunting. This approach shows that the challenge of stunting can be overcome gradually through utilising local resources and implementing community-based innovation. The programme implementation model in Tambaklorok Village, Sayung District, and Bonang District can be an inspiration for other areas with similar problems. Innovative programmes designed to empower local communities contribute to reducing the risk of stunting in children and increase the community's social and economic resilience as a whole. This community-based approach confirms that active community involvement, supported by commitment and creativity, plays a significant role in overcoming stunting problems.

Adaptation of health policies must consider the impacts of climate change to prevent an increase in malnutrition cases that risk stunting in children (Lieber et al., 2022). Providing education and public awareness involving local communities can strengthen the community's capacity to deal with climate change. The widespread threat to health posed by climate change demands decisive action from health professionals and governments to protect the health of current and future generations, especially in reducing stunting rates (Haines & Ebi, 2019).

Analysis Based on Brofenbrenner's Ecological Systems Theory Concept

The ecological systems theory of Brofenbrenner (1979) provides a solid theoretical framework for examining several factors that influence stunting cases in the northern coastal areas of Java across the micro, meso, exo, and macro layers. The micro or individual layer can be seen in practising parenting, the mother's education level, and family habits in providing nutritious food. Parental care and family habits significantly affect child development. Parents being ill-equipped or educated causes most cases of malnutrition in children. Lack of adequate nutritional intake—for example, giving large amounts of condensed milk as a substitute for nutritious milk—is a major contributing factor. Mothers who work to meet the family's economic needs lack sufficient free time to provide optimal childcare. Diarrhoea is a recurrent infectious disease in coastal areas due to poor sanitation and significantly affects children's health and nutrition. At the meso layer, family access to health services, such as *posyandu* and *puskesmas*, affects the nutritional status of children. Due to economic pressures, mothers prioritise work over participating in integrated health post activities. This results in a lack of regular monitoring of child growth and health. However, as in the case of the *Cantik Bahari* Women Farmers Group in Tambaklorok, community empowerment programmes show that collaboration between families, communities, and health institutions can raise awareness of the importance of child nutrition.

At the exo layer community, environmental factors such as seawater intrusion and tidal flooding affect food security, finance, and health, contributing to child

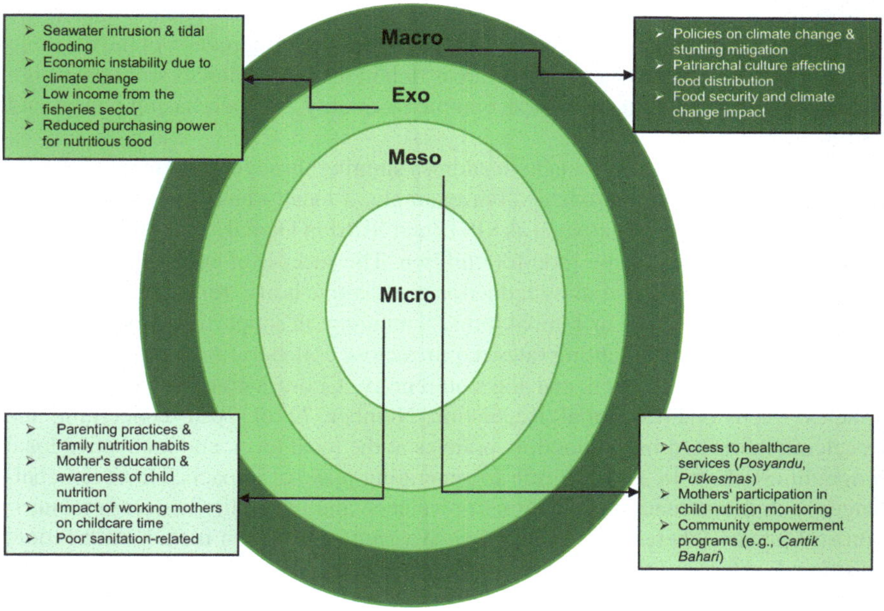

Fig. 11.7 Interaction system model: Climate change and stunting in coastal areas. (Source: modification of Bronfenbrenner's ecological systems theory, 1979)

stunting. Economic instability caused by climate change further exacerbates this issue. People's purchasing power is reduced due to low income from the unstable fisheries sector and unfavourable environmental conditions, limiting their ability to provide nutritious food. The effectiveness of policies related to climate change and stunting mitigation at a community and macro level determines families' access to support and the prevalence of stunting. Finally, a macro issue is the patriarchal culture, with the need for fathers to earn a living leading to them being given priority in the distribution of family food, reducing children's access to good nutrition. This impact is also illustrated at the micro level, with children's diets determined by their fathers' food preferences. Furthermore, climate change can worsen ecosystems supporting food security and public health, potentially increasing the number of stunted children (WHO, 2014). Figure 11.7 depicts an interaction system model.

Conclusion

Managing stunting in the northern coastal areas of Java, especially in Semarang and Demak, is a complex challenge influenced by various ecological, social, economic, and environmental factors. The impact of climate change worsens this situation, with multiple dimensions affecting sanitation, fishermen's catches, and the availability of clean water. If not addressed, this problem has the potential to increase the risk of malnutrition in children in the long term. The phenomenon of the resource paradox is also evident, where coastal areas rich in marine resources still face a high prevalence of stunting due to socioeconomic inequality, low purchasing power, and limited infrastructure.

Various efforts have been made to address stunting through government policies and community-based approaches. For example, a locally based approach can be seen in creative programmes such as SIST (and SUSI in Demak), which utilise local resources to provide healthy food for children. The success of community empowerment in supporting the stunting reduction programme is also reflected in the active role of the *Cantik Bahari* in Tambaklorok, Semarang, in supporting these efforts.

Success in reducing stunting rates requires close collaboration between the government, the private sector, and the community. Coordination between national policies, such as Presidential Regulation Number 72 of 2021 concerning the Acceleration of Stunting Reduction, policies at the local level, and innovative local programmes plays an essential role in generating positive impacts. However, challenges such as fluctuations in stunting prevalence and inequality in access to nutrition still persist. Therefore, a comprehensive approach, which includes education, nutritional interventions, technology-based monitoring, and mitigation of the impacts of climate change, is needed to ensure the sustainability of stunting reduction.

Recommendations

Efforts to overcome stunting on the north coast of Java require a holistic approach and involve various sectors. The involvement of local groups, such as the Women Farmers Group, as agents of change is crucial. At the same time, community education must be strengthened, emphasising the importance of fulfilling nutrition, proper parenting patterns, and good sanitation. In addition, there is a need to increase access to health services through integrated health posts and community health centres, accompanied by implementing outreach programmes for families with children at risk of stunting.

Strengthening local food security through processing seafood into nutritious food and providing subsidies to low-income families can be economically beneficial. Sustainable programmes require multi-sector collaboration between the government, private sector, and local communities. The development of infrastructure to deal with tidal flooding and the use of digital technology, such as applications to monitor stunting, must be top priorities to reduce the impact of climate change.

The government must ensure that policies such as Presidential Regulation Number 72 of 2021 are consistently followed while also assessing their effectiveness. In addition, to achieve more significant results, women's empowerment through local programmes such as the *Cantik Bahari* KWT must continue to be supported. Hopefully, this can accelerate the reduction in stunting rates in a substantive and sustainable manner.

References

Abbass, K., Qasim, M. Z., Song, H., Murshed, M., Mahmood, H., & Younis, I. (2022). A review of the global climate change impacts, adaptation, and sustainable mitigation measures. *Environmental Science and Pollution Research, 29*(28), 42539–42559. https://doi.org/10.1007/s11356-022-19718-6

Abedin, M. A., Collins, A. E., Habiba, U., & Shaw, R. (2019). Climate change, water scarcity, and health adaptation in Southwestern Coastal Bangladesh. *International Journal of Disaster Risk Science, 10*(1), 28–42. https://doi.org/10.1007/s13753-018-0211-8

Adriana, G., Pandjaitan, N. K., & Dharmawan, A. H. (2017). Kohesivitas komunitas nelayan dalam menghadapi perubahan iklim di Pesisir Jawa Barat. *Sodality Jurnal Sosiologi Pedesaan, 5*(1), 67–74.

Agostoni, C., Baglioni, M., Vecchia, A. La, Molari, G., & Berti, C. (2023). Interlinkages between climate change and food systems: the impact on child malnutrition—narrative review. In *Nutrients*. mdpi.com. https://www.mdpi.com/2072-6643/15/2/416

Ahmad, I., Khalique, N., Khalil, S., Urfi, & Maroof, M. (2018). Dietary diversity and stunting among infants and young children: A cross-sectional study in Aligarh Istiyaq. *Indian Journal of Community Medicine, 43*(1), 34–36. https://doi.org/10.4103/ijcm.IJCM

Alfarizi, M., & Ngatindriatun. (2023). Perubahan iklim, covid-19, perang Ukraina: Ancaman ketahanan pangan indonesia dan mitigasinya. *Jurnal Pertahanan & Bela Negara, 13*(1), 64–87. https://doi.org/10.33172/jpbh.v13i1.2035

Anggraheny, H. D., Setiawan, M. R., Widiasih, E., & Dewi, E. C. (2024). Household food security and WASH practices in children with undernutrition: a case study in low middle income area in Semarang, Indonesia. *BKM Public Health and Community Medicine, 40*(5), e12356. https://doi.org/10.22146/bkm.v40i05.12356

Asensio-Montesinos, F., Molina, R., Anfuso, G., Manno, G., & Lo Re, C. (2024). Natural and human impacts on Coastal Areas. *Journal of Marine Science and Engineering, 12*(11), 1–10. https://doi.org/10.3390/jmse12112017

Asrofi, A., Hardoyo, S. R., & Sri Hadmoko, D. (2017). Strategi adaptasi masyarakat pesisir dalam penanganan bencana banjir rob dan implikasinya terhadap ketahanan wilayah (studi di Desa Bedono Kecamatan Sayung Kabupaten Demak Jawa Tengah). *Jurnal Ketahanan Nasional, 23*(2), 1. https://doi.org/10.22146/jkn.26257

Azhari, Dedihasriadi, & Muis. (2024). Problematika faktor stunting pada wilayah pesisir (studi kebijakan pada Kabupaten Buton Tengah tahun 2022). *Journal Publicuho, 7*(1), 128–139. https://doi.org/10.35817/publicuho.v7i1.330

Bisyri, F. A., Satriadi, A., & Purwanto, P. (2020). Studi muka air laut rencana dan elevasi puncak breakwater di Wilayah Pesisir Kecamatan Tugu, Kota Semarang. *Indonesian Journal of Oceanography, 2*(2), 129–136. https://doi.org/10.14710/ijoce.v2i2.7987

Brofenbrenner, U. (1979). *The Ecology of Human Development: Experiments by nature and design*. The president and fellows of Harvard College All.

Cadez, S., Czerny, A., & Letmathe, P. (2019). Stakeholder pressures and corporate climate change mitigation strategies. *Business Strategy and the Environment, 28*(1), 1–14. https://doi.org/10.1002/bse.2070

Chan, S., Haridhi, H. A., Damora, A., Aprilla, R. M., Rahmah, A., & Asni, K. (2023). Economic loss as the impact of climate change on tuna fishermen in Northern Indonesian waters. *IOP Conference Series: Earth and Environmental Science, 1221*(1). https://doi.org/10.1088/1755-1315/1221/1/012072

Choirunnisa, L. A. D., Purwaningsih, Y., & Prasetyani, D. (2022). Adaptasi nelayan pesisir Kabupaten Pacitan akibat perubahan iklim. *Jurnal Wilayah Dan Lingkungan, 10*(2), 166–181. https://doi.org/10.14710/jwl.10.2.166-181

Claus, B., Nugraha, R. B. A., Gultom, I. S., Anggoro, T. D., Jayawiguna, M. H., Setiyono, H., & Wirasatriya, A. (2024). Analisis model rendaman rob di Desa Tambakbulusan Kabupaten Demak sebagai evaluasi rencana pengembangan kawasan wisata bahari. *Jurnal Kelautan Nasional, 19*(2), 149. https://doi.org/10.15578/jkn.v19i2.14000

Damanik, S. M., Wanda, D., & Hayati, H. (2020). Feeding practices for toddlers with stunting in Jakarta: A case study. *Pediatric Reports, 12*. https://doi.org/10.4081/pr.2020.8695

Dinkes Jateng. (2024). *Laporan akhir: Analisis faktor determinan penanganan stunting Provinsi Jawa Tengah tahun 2024*.

Dinkes. (2024). *Upaya percepatan penurunan stunting di Kota Semarang*. Dinas Kesehatan Kota Semarang.

Ervita, K., & Marfai, M. A. (2017). Shoreline change analysis in Demak, Indonesia. *Journal of Environmental Protection, 08*(08), 940–955. https://doi.org/10.4236/jep.2017.88059

Gal, A. M. (2020). Konsep pemberdayaan nelayan pesisir Kota Surabaya sebagai bentuk adaptasi perubahan iklim berbasis sustainable livelihood. *Jurnal Planoearth, 5*(1), 45–51.

Haines, A., & Ebi, K. (2019). The imperative for climate action to protect health. *New England Journal of Medicine, 380*(3), 263–273. https://doi.org/10.1056/nejmra1807873

Hallegatte, S., Fay, M., & Barbier, E. B. (2018). Poverty and climate change: Introduction. *Environment and Development Economics, 23*(3), 217–233. https://doi.org/10.1017/S1355770X18000141

Handayani, D., Ardhiyanto, O., Kusumah, M., Asghor, A., Dwidayanto, A., Kurniawan, K., Nafianto, D., Fadilah, F., Pamungkas, A., Fadhillah, M., & Zahir, M. (2024). Penerapan aplikasi stunting pada posyandu kenanga sebagai gerakan masyarakat sadar stunting. *Journal of Computer Science Contributions (JUCOSCO), 4*(1), 77–86. https://doi.org/10.31599/jucosco.v4i1.3676

Handayani, D., Kusuma, E., Puspitasari, H., & Nastiti, A. D. (2022). The factors affecting stunting on toddlers in Coastal Areas. *Jurnal Aisyah: Jurnal Ilmu Kesehatan, 7*(3), 755–764. https://doi.org/10.30604/jika.v7i3.967

Harini, R. H., Ariani, R. D., & Yulianda, Y. (2022). Strategi adaptasi ketahanan pangan terhadap perubahan iklim di pinggiran Kota Yogyakarta. *Majalah Geografi Indonesia, 36*(1), 1. https://doi.org/10.22146/mgi.60245

Harvian, K. A., & Yuhan, R. J. (2021). Kajian perubahan iklim terhadap ketahanan pangan. *Seminar Nasional Official Statistics, 2020*(1), 1052–1061. https://doi.org/10.34123/semnasoffstat.v2020i1.593

Has, E. M. M., Prakhasita, R. C., Krisnana, I., Wahyuni, S. D., Arief, Y. S., & Nursalam. (2020). Risk factors for stunting and severe stunting among children (12-59 months old) in the Coastal Area of Surabaya, East Java. *Indonesia. International Journal of Psychosocial Rehabilitation, 24*(April), 9020–9027.

Hastuti, V. N., Afifah, D. N., Sugianto, D. N., Anjani, G., & Noer, E. R. (2024). Socio-demographics, dietary diversity score, and nutritional status of children aged 2–5 years: A cross-sectional study of Indonesian coastal areas. *Clinical Epidemiology and Global Health, 27*(September 2023). https://doi.org/10.1016/j.cegh.2024.101599

Helldén, D., Andersson, C., Nilsson, M., Ebi, K. L., et al. (2021). Climate change and child health: a scoping review and an expanded conceptual framework. In *The Lancet Planetary* …. thelancet.com. https://www.thelancet.com/journals/lanplh/article/PIIS2542-5196(20)30274-6/fulltext

Hidayah, I., Suryahadi, A., Palmisano, F., & Kiefte-de Jong, J. C. (2024). The role of parental child marriage in children's food security and nutritional status: a prospective cohort study in Indonesia. *Frontiers in Public Health, 12*(2). https://doi.org/10.3389/fpubh.2024.1469483

Islam, T. M. T., Kearns, M., & Mitra, S. (2024). Patriarchy and child well-being: evidence from India. *Journal of Economics, Race, and Policy, 7*(4), 294–308. https://doi.org/10.1007/s41996-024-00152-4

Kapantow, N. H., Sanggelorang, Y., & Adam, H. (2023). Non-exclusive breastfeeding as the risk factor of stunting among Indonesian toddler in coastal areas. *International Journal of Community Medicine And Public Health, 10*(2), 581–585. https://doi.org/10.18203/2394-6040.ijcmph20230208

Kemenkes. (2024). *Laporan tematik survei kesehatan indonesia tahun 2023*. Kementerian Kesehatan Republik Indonesia.

Kinyoki, D. K., Berkley, J. A., Moloney, G. M., Odundo, E. O., Kandala, N. B., & Noor, A. M. (2016). Environmental predictors of stunting among children under-five in Somalia: Cross-sectional studies from 2007 to 2010. *BMC Public Health, 16*(1), 1–9. https://doi.org/10.1186/s12889-016-3320-6

Klenert, D., Funke, F., Mattauch, L., & O'Callaghan, B. (2020). Five lessons from covid-19 for advancing climate change mitigation. *Environmental and Resource Economics, 76*(4), 751–778. https://doi.org/10.1007/s10640-020-00453-w

Kurniadi, B., Minsas, S., & Helena, S. (2023). Sosialisasi dan edukasi mitigasi bencana perubahan iklim bagi masyarakat pesisir Batu Ampar. *Lumbung Inovasi: Jurnal Pengabdian Kepada Masyarakat, 8*(4), 791–796. https://doi.org/10.36312/linov.v8i4.1486

Kurniawan, T. A., Bandala, E. R., Othman, M. H. D., Goh, H. H., Anouzla, A., Chew, K. W., Aziz, F., Al-Hazmi, H. E., & Khoir, A. N. (2024). Implications of climate change on water quality and sanitation in climate hotspot locations: A case study in Indonesia. *Water Supply, 24*(2), 517–542. https://doi.org/10.2166/ws.2024.008

Laode Wado, L. A., Sudargo, T., & Armawi, A. (2019). Sosio demografi ketahanan pangan keluarga dalam hubungannya dengan kejadian stunting pada anak usia 1 – 5 tahun (studi di wilayah kerja Puskesmas Bandarharjo Kelurahan Tanjung Mas, Kecamatan Semarang Utara, Kotamadya Semarang, Provinsi Jawa Tengah). *Jurnal Ketahanan Nasional, 25*(2), 178. https://doi.org/10.22146/jkn.45707

Lawrance, E. L., Thompson, R., Newberry Le Vay, J., Page, L., & Jennings, N. (2022). The impact of climate change on mental health and emotional wellbeing: A narrative review of current evidence, and its implications. *International Review of Psychiatry, 34*(5), 443–498. https://doi.org/10.1080/09540261.2022.2128725

Lieber, M., Chin-Hong, P., Kelly, K., Dandu, M., & Weiser, S. D. (2022). A systematic review and meta-analysis assessing the impact of droughts, flooding, and climate variability on malnutrition. *Global Public Health, 17*(1), 68–82. https://doi.org/10.1080/17441692.2020.1860247

Madaual, R. A., Ibal, L., & Abubakar, E. (2023). Strategi penghidupan masyarakat desa pesisir Kabupaten Seram Bagian Timur dalam menghadapi perubahan iklim. *Jurnal Perencanaan Wilayah, 8*(1), 80–92. https://doi.org/10.33772/jpw.v8i1.373

Manggala, A. K., Kenwa, K. W. M., Kenwa, M. M. L., Sakti, A. A. G. D. P. J., & Sawitri, A. A. S. (2018). Risk factors of stunting in children aged 24-59 months. *Paediatrica Indonesiana, 58*(5), 205–212. https://doi.org/10.14238/pi58.5.2018.205-12

Marfai, M. A., King, L., Sartohadi, J., Sudrajat, S., Budiani, S. R., & Yulianto, F. (2008). The impact of tidal flooding on a coastal community in Semarang, Indonesia. *Environmentalist, 28*(3), 237–248. https://doi.org/10.1007/s10669-007-9134-4

Meher, C., Zaluchu, F., & Eyanoer, P. C. (2023). Local approaches and ineffectivity in reducing stunting in children: A case study of policy in Indonesia. *F1000Research, 12*, 217. https://doi.org/10.12688/f1000research.130902.1

Meherali, S et al., (2024). Impact of climate change on child outcomes: an evidence gap map review. *BMJ Paediatrics Open, 8*(1), e002592.

Merwe, E. Van der, et al. (2022). *Climate change and child malnutrition: a Nigerian perspective. Food Policy*. https://www.sciencedirect.com/science/article/pii/S0306919222000604

Muzdalifah, L., & Nilamsari, W. (2021). Peran istri nelayan dalam meningkatkan perekonomian keluarga di Pulau Tidung. *Jurnal Pemberdayaan Komunitas MH Thamrin, 3*(2), 121–136. https://doi.org/10.37012/jpkmht.v3i2.670

Myers, S. S., Smith, M. R., Guth, S., Golden, C. D., Vaitla, B., Mueller, N. D., Dangour, A. D., & Huybers, P. (2017). Climate change and global food systems: Potential impacts on food security and undernutrition. *Annual Review of Public Health, 38*(April), 259–277. https://doi.org/10.1146/annurev-publhealth-031816-044356

N'Souvi, K., Adjakpenou, A., Sun, C., & Ayisi, C. L. (2024). Climate change perceptions, impacts on the catches, and adaptation practices of the small-scale fishermen in Togo's coastal area. *Environmental Development, 49*(May 2023), 100957. https://doi.org/10.1016/j.envdev.2023.100957

Najib, Rum Giyarsih, S., Listyaningsih, U., & Nawawi. (2023). Analysis of feeding behavior and family food security as a stunting risk factor in Semarang City. *Kemas, 19*(2), 270–277. https://doi.org/10.15294/kemas.v19i2.43144

Nashier, I., & Lakra, P. (2020). Global climate change and its effects. *Integrated Journal of Social Sciences, 2020*(1), 14–23.

Ngwira, A. (2020). Climate and location as determinants of childhood stunting, wasting, and overweight: An application of semiparametric multivariate probit model. In *Nutrition*. Elsevier. https://www.sciencedirect.com/science/article/pii/S2665902620300054

Nicholls, R. J., & Cazenave, A. (2010). Sea-level rise and its impact on coastal zones. *Science, 328*(5985), 1517–1520. https://doi.org/10.1126/science.1185782

Nur Amalia, R., Windusari, Y., Sari, N., Alam Fajar, N., & Rahmiwati, A. (2024). Strategi promosi kesehatan dalam menghadapi dampak perubahan iklim bagi kesehatan masyarakat: Systematic literature review. *Media Publikasi Promosi Kesehatan Indonesia (MPPKI), 7*(1), 51–57. https://doi.org/10.56338/mppki.v7i1.4346

Nuraisah, G., & Budi Kusumo, R. A. (2019). Dampak perubahan iklim terhadap usahatani padi Di Desa Wanguk Kecamatan Anjatan Kabupaten Indramayu. *MIMBAR AGRIBISNIS: Jurnal Pemikiran Masyarakat Ilmiah Berwawasan Agribisnis, 5*(1), 60. https://doi.org/10.25157/ma.v5i1.1639

Nurhidayah, L. (2021). Sea-Level Rise (SLR) and its implication on human security and human rights in Indonesia: A legal analysis. In R. Djalante, J. Jupesta, & E. Aldrian (Eds.), *Springer Climate. Springer* (pp. 33–52). Springer International Publishing. https://doi.org/10.1007/978-3-030-55536-8_3

Permana, I., Firlianty, & Mentari, T. (2023). Gerakan pencegahan stunting melalui edukasi pada masyarakat. *Pengabdian Kampus: Jurnal Informasi Kegiatan Pengabdian Pada Masyarakat, 10*(1), 65–69. https://doi.org/10.52850/jpmupr.v10i1.8606

Perumal, N., Bassani, D. G., & Roth, D. E. (2018). Use and misuse of stunting as a measure of child health. *Journal of Nutrition, 148*(3), 311–315. https://doi.org/10.1093/jn/nxx064

Phalkey, R. K., Aranda-Jan, C., Marx, S., Höfle, B., & Sauerborn, R. (2015). Systematic review of current efforts to quantify the impacts of climate change on undernutrition. *Proceedings of the National Academy of Sciences of the United States of America, 112*(33), E4522–E4529. https://doi.org/10.1073/pnas.1409769112

Posen, P. E., Beraud, C., Jones, C. H., Tyllianakis, E., Joseph-Witzig, A., & Louis, A. S. (2023). Vulnerability of coastal infrastructure and communities to grenada and its dependencies. *Land, 12*, 1418.

Praveen, B., & Sharma, P. (2019). Climate change and its impacts on Indian agriculture: An econometric analysis. *Journal of Public Affairs, 20*(1), e1972.

Prianto, B., Molyo, P. D., Widayati, S., Setiyaningsih, L. A., & Nuswantari, S. A. (2025). Reducing stigma toward mothers in stunting incidence by increasing fathers' participation in raising children. *Journal of Population and Social Studies, 33*, 159–174. https://doi.org/10.25133/JPSSV332025.0009

Proulx, K., Daelmans, B., Baltag, V., et al. (2024). Climate change impacts on child and adolescent health and well-being: A narrative review. In *Journal of Global Health*. ncbi.nlm.nih.gov. https://www.ncbi.nlm.nih.gov/pmc/articles/PMC11115477/

Purwestri, R. C., Barati, Z., Wirawan, N. N., Fahmi, I., Lauvai, J., & Scherbaum, V. (2018). 7. What explains stunting among children living in a rice surplus area in Central Java, Indonesia? In *BRILL* (pp. 137–151). Wageningen Academic. https://doi.org/10.3920/978-90-8686-864-3_7

Rahayuwati, L., Komariah, M., Hendrawati, S., Sari, C. W. M., Yani, D. I., Setiawan, A. S., Ibrahim, K., Maulana, S., & Hastuti, H. (2023). Exploring the relationship between maternal education, parenting practice, and stunting among children under five: Findings from a cross-sectional study in Indonesia. *F1000Research, 12*, 722. https://doi.org/10.12688/f1000research.133916.1

Ramadhaniyati, Lestari, L., Saito, R., & Tsuda, A. (2025). Exploring mothers' experiences in accessing primary healthcare service facilities in rural Indonesia: Implication of Stunting Prevention. *Journal of Rural Community Nursing Practice (JRCNP), 3*(1), 129–149.

Restiana, R. P., & Fadilah, T. F. (2023). Relationship between early marriage and incidence of stunting in children aged 24-59 months. In *Proceedings of the 3rd Borobudur International Symposium on Humanities and Social Science 2021 (BIS-HSS 2021), November 2019*, pp. 924–929. https://doi.org/10.2991/978-2-494069-49-7_156

Reyes, H., Pérez-Cuevas, R., Sandoval, A., Castillo, R., Santos, J. I., Doubova, S. V., & Gutiérrez, G. (2004). The family as a determinant of stunting in children living in conditions of extreme poverty: A case-control study. *BMC Public Health, 4*, 1–10. https://doi.org/10.1186/1471-2458-4-57

Rihi Leo, A., Subagyo, H. W., & Kastasurya, M. I. (2018). Faktor risiko stunting pada anak usia 2-5 tahun di wilayah gunung dan pesisir pantai. *Jurnal Gizi Dan Pangan Soedirman, 2*(1), 51. https://doi.org/10.20884/1.jgps.2018.2.1.512

Rohmah, A. N., & Mardiana, M. (2023). Hubungan tingkat kecukupan protein, zink, pendapatan perkapita, dan sanitasi lingkungan dengan kejadian stunting pada balita di Kelurahan Tanjung Mas Kota Semarang. *Nutrizione (Nutrition Research and Development Journal), 3*(3), 11–24.

Roy, P., Pal, S. C., Chakrabortty, R., Chowdhuri, I., Saha, A., & Shit, M. (2023). Effects of climate change and sea-level rise on coastal habitat: Vulnerability assessment, adaptation strategies and policy recommendations. *Journal of Environmental Management, 330*(August 2022), 117187. https://doi.org/10.1016/j.jenvman.2022.117187

Ruhyana, N. F., Nurfindarti, E., & Essa, W. Y. (2021). Studi prioritas lokus penanganan stunting Kabupaten Sumedang dengan pendekatan kajian resiko adaptasi perubahan iklim. *Jurnal Borneo Administrator, 17*(1), 65–88. https://doi.org/10.24258/jba.v17i1.768

Ruminta, R. (2016). Analisis penurunan produksi tanaman padi akibat perubahan iklim di Kabupaten Bandung Jawa Barat. *Kultivasi, 15*(1), 37–45. https://doi.org/10.24198/kultivasi.v15i1.12006

Sarah, D., Hutasoit, L. M., Delinom, R. M., & Sadisun, I. A. (2020). Natural compaction of Semarang-Demak alluvial plain and its relationship to the present land subsidence. *Indonesian Journal on Geoscience, 7*(3), 273–289. https://doi.org/10.17014/ijog.7.3.273-289

Sarah, D., Mulyono, A., Satriyo, N. A., Soebowo, E., & Wirabuana, T. (2022). Towards sustainable land subsidence mitigation in Semarang and Demak, Central Java: Analysis using DPSIR Framework. *Journal of Water and Land Development, 55*, 150–165. https://doi.org/10.24425/jwld.2022.142317

Sari, I. K. W., & Hadi, S. P. (2018). Climate change anticipation on supporting capacity of fishing environment in the coastal area of Tanjungmas Semarang City. *E3S Web of Conferences, 31*, 1–7. https://doi.org/10.1051/e3sconf/20183109016

Sari, I. P., Ningsih, W. I. F., & Sari, D. M. (2023). The correlation between the household food security and the incidence of stunting in toddlers 6-59 months in Seberang Ulu I Palembang. *Jurnal Ilmu Kesehatan Masyarakat, 14*(2), 198–209. https://doi.org/10.26553/jikm.2023.14.2.198-209

Sefrina, L. R., Kurniasari, R., Elvandari, M., Utami, A. R., Aiboga, A. M., & Ulya, N. (2022). The interaction of three types of undernutrition and health insurance subscriptions at the children in the Karawang Coastal Area, Indonesia. *Amerta. Nutrition, 6*(1SP), 19–24. https://doi.org/10.20473/amnt.v6i1sp.2022.19-24

Sopianti, M., Alam Fajar, N., Sunarsih, E., Windusari, Y., & Novrikasari. (2024). Air bersih dan jamban sehat terhadap kejadian stunting di negara berkembang: Literature review. *Media Publikasi Promosi Kesehatan Indonesia (MPPKI), 7*(1), 8–14. https://doi.org/10.56338/mppki.v7i1.4211

Sui, L., Wang, J., Yang, X., & Wang, Z. (2020). Spatial-temporal characteristics of coastline changes in Indonesia from 1990 to 2018. *Sustainability (Switzerland), 12*(8), 1–28. https://doi.org/10.3390/SU12083242

Sukmawati, Abidin, U. W., & Hasmia. (2021). Hubungan hygiene dan sanitasi lingkungan terhadap kejadian stunting pada balita di Desa Kurma. *Journal Peqguruang: Conference Series, 3*(2), 495–502.

Sunarti, & Apriliasari, N. K. (2015). Climate change impact toward to the coastal settlements In Demaan Sub-District, Jepara Regency. *Jurnal Tataloka, 17*(4), 248–256.

Tria, R., Ramayani, I., & Laba, M. S. (2022). The innovation of coastal community fish catching tools in efforts to increase fisherman income in Polewali Mandar Regency, West Sulawesi. *Jurnal Perikanan Unram, 12*(4), 660–669.

Ulfa, M. (2018). Persepsi masyarakat nelayan dalam menghadapi perubahan iklim (ditinjau dalam aspek sosial ekonomi). *Jurnal Pendidikan Geografi: Kajian, Teori, Dan Praktik Dalam Bidang Pendidikan Dan Ilmu Geografi, 23*(1), 41–49. https://doi.org/10.36349/easjacc.2024.v06i01.002

UNICEF. (2023). *The climate-changed child: a children's climate risk index supplement.* UNICEF. https://doi.org/10.7927/h49c6vhw

UNICEF. (2024). A threat to progress. In *UNICEF*. UNICEF. https://doi.org/10.18356/9789280655797

UNICEF, WHO, & Bank, W. (2020). Levels and trends in child malnutrition, joint child malnutrition estimates 2020 edition. In *World Health Organization*. World Health Organization.

Wahyudi, A., Hendrarto, B., & Hartoko, A. (2014). Penilaian kerentanan habitat mangrove di Kelurahan Mangunharjo, Kecamatan Tugu, Kota Semarang Terhadap Variabel Oseonografi Berdasarkan Metode CVI (Coastal Vulnerability Index). *Diponegoro Journal of Maquares, 3*(1), 89–98.

WHO. (2014). Quantitative risk assessment of the effects of climate change on selected causes of death, 2030s and 2050s. *WHO Press*. http://www.who.int/globalchange/publications/quantitative-risk-assessment/en/

Widyastuti, D., Jamaluddin, H. N., Arisanti, R., & Kartiasih, F. (2023). Analisis pengaruh faktor sosial ekonomi terhadap akses sanitasi layak di Indonesia tahun 2021. *Seminar Nasional Official Statistics, 2023*(1), 105–116. https://doi.org/10.34123/semnasoffstat.v2023i1.1853

Wijayanti, E. E., Anggoro, Y., Hidayati, N., & Munawarah, Z. (2024). Social and economic factors affecting the incidence of anemia in pregnant women. *Oshada: Jurnal Kesehatan, 1*(December), 1–16.

Wu, T., & Barrett, J. (2022). Coastal land use management methodologies under pressure from climate change and population growth. *Environmental Management, 70*(5), 827–839. https://doi.org/10.1007/s00267-022-01705-9

Yanti, D. S., Sumardiyono, S., & Kusnandar, K. (2023). The relationship between household good security and incidence of stunting in toddlers during the new normal: A systematic review. *Epidemiology and Society Health Review (ESHR), 5*(1), 51–59. https://doi.org/10.26555/eshr.v5i1.7279

Yuliantini, E., Sukiyono, K., Yuliarso, M. Z., & Sulistyo, B. (2022). Food security and stunting incidences in the coastal areas of Indonesia. *Open Access Macedonian Journal of Medical Sciences, 10*(F), 454–461. https://doi.org/10.3889/oamjms.2022.9335

Zamrsky, D., Oude Essink, G. H. P., & Bierkens, M. F. P. (2024). Global impact of sea level rise on coastal fresh groundwater resources. *Earth's Future, 12*(1). https://doi.org/10.1029/2023EF003581

Part II
Labour

Chapter 12
Discourse on Labour in the Fisheries Sector of Jakarta Bay: Vulnerabilities, Challenges, and Strategies in Response to Climate Change

Vera Bararah Barid, Ahmad Fathony, Nurrahman Aji Utomo, and Intan Hapsari Surya Putri

Introduction

In Indonesia, the high proportion of workers in the informal sector, the mismatch between workforce skills and industry demands, and elevated unemployment rates among young people are key issues that require serious attention. According to data from the BPS (Indonesian Central Bureau of Statistics), as of February 2024 the total labour force was approximately 144 million people, with around 60% being workers in the informal sector.[1] Particularly in coastal regions such as Jakarta Province, Pekalongan, and Demak, many traditional fishers face seasonal disruptions in their livelihood due to the west monsoon (December–February), which brings strong winds and high waves that make fishing dangerous or impossible, and to tidal flooding (locally known as *rob*), which damages boats, floods coastal infrastructure, and impedes access to the sea. As a result, many fishers are unable to go to sea and are forced to seek alternative sources of income during these periods.

The small-scale fisheries sector, which constitutes a significant portion of informal employment, plays a crucial role in sustaining livelihoods, particularly in

[1] Badan Pusat Statistik. (2024, June 7). *Labor Force Situation in Indonesia February 2024*. BPS-Statistics Indonesia. Retrieved from https://www.bps.go.id/en/publication/2024/06/07/112a10c79b8cfa70eec9f6f3/labor-force-situation-in-indonesia-february-2024.html.

V. B. Barid (✉) · A. Fathony · N. A. Utomo
Research Center for Law, National Research and Innovation Agency (BRIN), Jakarta, Indonesia

I. H. S. Putri
Center for Urban and Regional Resilience Research (CURE) – Faculty of Engineering, Diponegoro University, Semarang, Indonesia

L. Nurhidayah et al. (eds.), *Climate Change, Labour and Migration in Indonesia*, Springer Climate, https://doi.org/10.1007/978-3-032-15334-0_12

developing countries. However, workers in this sector face substantial challenges as a result of climate change, including rising sea temperatures, altered rainfall patterns, and sea-level rise, all of which directly impact fish availability, ecosystem stability, and income reliability. Barange et al. (2018) note that climate-induced shifts in fish distribution led to uncertainty in catch rates and can undermine the economic security of small-scale fishers.

The fisheries sector in Jakarta Bay plays a pivotal role in sustaining the livelihoods of coastal communities, particularly in densely populated neighbourhoods such as Kalibaru, North Jakarta. However, this sector is increasingly vulnerable to intersecting environmental and socio-economic stressors. Climate change, marked by rising sea levels, coastal erosion, and more erratic weather patterns, disproportionately affects small-scale fishers who often operate without adequate social protection (ADB, 2021; ILO, 2020). As a result, many fishers face job insecurity and income loss, contributing to heightened social vulnerability in coastal communities (Adi Renaldi, 2024).

Recent studies reveal that overfishing and ecological degradation have severely impacted marine biodiversity in Jakarta Bay, forcing traditional fishing families into poverty and, in some cases, pushing them towards harmful coping strategies such as early marriage (Mongabay, 2024). Furthermore, limited access to technology, capital, and institutional support makes informal fishing communities particularly vulnerable to environmental stressors. In this context, community-based adaptation strategies and enhanced local capacities are essential for strengthening the resilience and long-term sustainability of informal fisheries in the face of ongoing climate challenges.

Alongside ecological concerns, labour conditions in the fisheries sector remain precarious. The workforce, largely informal, gendered, and comprised of migrant labour, is often subjected to exploitative practices including forced labour and human trafficking, especially in major fishing hubs like Muara Angke (Indonesia Business Post, 2025). These structural inequalities are emblematic of a broader transformation in global labour systems, wherein material production is increasingly intertwined with immaterial forms of labour, including the production of knowledge, communication, and social relations (Hardt & Negri, 2000). This shift has marginalised traditional sectors like fisheries and has pushed workers towards insecure and informal economies.

In *Empire*, Hardt and Negri argue that the global labour system has undergone a fundamental transformation. Labour is no longer limited to the production of physical goods but increasingly involves the production of meanings, relationships, and services which they term immaterial labour. One notable consequence of this transformation is the widespread shifting of labour, especially from traditional sectors such as fisheries to informal or service-based occupations.

Given these overlapping pressures, the discourse on labour in the fisheries sector of Jakarta Bay demands urgent attention. Workers in this sector face intersecting vulnerabilities arising from environmental disruptions and declining fish stocks, precarious working conditions and limited access to social protection. These challenges are further complicated by structural inequalities, informal labour arrangements, and shifting socio-economic dynamics driven by climate change and urban development. In light of this, a nuanced understanding of labour discourse is essential, not only to examine how labour is organised, valued, and contested but also to explore the power relations and narratives that shape policy and practice.

This chapter explores these dynamics by investigating the vulnerabilities and challenges faced by fisheries labourers in Jakarta Bay, while also identifying and analysing strategies, both formal and community-based, employed to adapt and resist. In doing so, it seeks to provide a deeper understanding of the complex interplay between labour dynamics, environmental change, and governance in shaping livelihood strategies and resilience within coastal communities situated in rapidly evolving urban maritime contexts.

Methodology

This study employs a qualitative research approach to explore the discourse on labour in the fisheries sector of Jakarta Bay, focusing on the vulnerabilities, challenges, and adaptive strategies of fishing communities in the face of climate change. The research was conducted in three coastal subdistricts of North Jakarta: Kalibaru, Muara Angke, and Marunda. These areas were selected due to their strategic roles in small-scale fisheries and their increasing exposure to climate-related stressors such as tidal flooding, sea-level rise, and seasonal disruptions (Wannewitz & Garschagen, 2024; Prana et al., 2024).

A snowball sampling technique was utilised to identify key informants, resulting in a total of 23 respondents. Participants included small-scale fishers, local government officials, community leaders, and women leaders actively involved in fisheries-related livelihoods. This diverse representation aimed to capture multiple perspectives on labour dynamics, environmental vulnerability, and grassroots adaptation efforts.

Data were collected through in-depth interviews and focus group discussions conducted across the three study locations between January, July, and September 2024, and again during February to March 2025. The qualitative data were then analysed thematically to uncover recurring patterns in labour conditions, coping strategies, and governance responses. This methodological framework allows for a nuanced understanding of how labour in the fisheries sector is shaped by both ecological pressures and socio-political structures in a rapidly urbanising coastal context. This chapter outlines detail and provides insight about the discourse on labour in the fisheries sector of Jakarta Bay, and specifically its vulnerabilities, challenges, and strategies in response to climate change.

Literature Review

The concept of livelihood resilience is central to understanding how communities, particularly those in vulnerable sectors such as small-scale fisheries, respond to the multifaceted challenges posed by climate change. Livelihood resilience refers to the ability of individuals or households to maintain or restore their means of living in the face of external shocks, including environmental degradation, tidal flooding, and extreme weather events (Nurzaman et al., 2020). In coastal communities like Kalibaru, Muara Angke, and Marunda, where economic activities are heavily reliant on fisheries, such resilience is being increasingly tested by climate-induced disruptions. These impacts not only affect the availability of marine resources but also disrupt the labour patterns and social organisation that sustain fishing livelihoods (Kristiningrum et al., 2023).

Within this context, labour in the fisheries sector has become more precarious. Changing seasonal patterns, sea-level rise, and pollution have altered fishing cycles, forcing many communities to diversify income sources or adopt new forms of employment. This reflects a shift in labour dynamics from traditional fishing to broader informal economic activities. The Sustainable Livelihoods Approach (SLA) offers a useful framework to examine this transition by highlighting the intersection of human, social, and natural capital in coastal adaptation strategies (Allison & Ellis, 2001). The capacity of fishing households to access alternative livelihoods, such as aquaculture, eco-tourism, or coastal conservation work, has become a key determinant of labour resilience (Prado et al., 2015; Susiloningtyas, 2019).

Importantly, this diversification is not merely an economic decision but a restructuring of labour relations within coastal communities. Historical research in climate-affected fishing villages, such as those in Bangladesh after recent cyclones, shows that community-wide shifts to agriculture or tourism result in new labour hierarchies and a redefinition of social roles (Islam & Walkerden, 2022; Uddin et al., 2021). Similar patterns can be observed in North Jakarta, where informal labour, including women's contributions as shellfish peelers or fish processors during the productive monsoon season, plays an essential role in household survival and adaptive capacity.

Furthermore, the literature highlights the significance of adaptive co-management in strengthening labour resilience. This approach involves shared governance between communities and institutions, fostering local empowerment and flexibility in resource use (Roscher et al., 2022; Backler et al., 2025). In Jakarta Bay, where formal governance mechanisms often overlook informal workers, participatory management models can enable fishing communities to gain greater control over their resources while adapting their labour to environmentally sustainable practices (Hsu & Peng, 2023).

Finally, the evolving nature of labour in the fisheries sector underscores the importance of inclusive and adaptive policy interventions. Labour resilience cannot be achieved solely by creating new jobs; it also requires that emerging livelihoods are culturally appropriate, socially equitable, and environmentally sustainable. Without such considerations, climate adaptation may deepen inequality or marginalise already vulnerable groups (Uy et al., 2011; Uddin et al., 2021). Recognising the discourse of labour as part of a broader immaterial transformation, where livelihoods are no longer defined only by physical production but by relational, communicative, and social dimensions (Hardt & Negri, 2000), is essential for building effective responses to climate change in coastal regions like Jakarta Bay.

Study Participants

Table 12.1 below outlines the demographics of the participants in this research. There were 9 female and 14 male informants. Most of them (6) were fishermen or civil servants working in subdistrict government offices (11), with the remainder comprising community leaders/activists (3), cooperative association staff (2), and one being a housewife. Most had informal jobs and had several side jobs as clam peelers, small traders, and green mussels' farmers. Most of the informants were married and were between 30 and 50 years old. Most were residents of the four subdistricts in North Jakarta.

Findings and Discussion

The northern coast of Jakarta Province hosts diverse labour patterns, predominantly centred around the fisheries sector. In areas like Muara Angke, fishers typically operate larger vessels, engaging in offshore fishing and wholesale fish trading. Conversely, communities in Kalibaru and Marunda primarily consist of small-scale

Table 12.1 Demographic information of informants

Characteristics of informants	Respondents (n = 23)
Gender	
Female	9
Male	14
Occupation	
Fisherman	6
Civil servant	11
Community activist/leaders	3
Cooperative association	2
Housewife	1
Residence	
Muara Angke	6
Kalibaru	3
Cilincing	3
Marunda	1
Jakarta Province	10

Source: modified from data participants

fishers utilising boats under two gross tons, limiting their fishing range and catch volume. This disparity in vessel size significantly influences income levels, with larger vessels correlating with higher earnings (Hasiholan et al., 2023).

Seasonal monsoons further impact fishing activities. The east monsoon offers favourable conditions for fishing, whereas the west monsoon presents challenges due to rough seas and reduced fish availability. During these periods, many fishers seek alternative income sources, and women often engage in shellfish processing to supplement household earnings. However, in Marunda such supplementary labour is less prevalent, with many women relying solely on their husband's income and government assistance. In Kalibaru, the fishing community grappled with declining fish stocks and increasing pollution levels, compelling fishers to venture further offshore, thereby incurring higher fuel costs and greater risks. These hardships have led some families to resort to early marriages for their daughters as a coping mechanism for economic strain (citiespossibilities.com). Muara Angke, a densely populated fishing hub, continues to be a central point for fish landing and distribution. Despite infrastructural developments, such as the modernisation of the fish market, fishers reported hazardous working conditions, including excessive working hours and inadequate safety measures (ILO, 2025). In Marunda, the community faced similar adversities. Environmental degradation and overfishing have led to reduced catches, impacting the livelihoods of small-scale fishers. The community also contended with pollution from industrial activities, further compromising their fishing grounds. Marunda's fishers face reduced catches due to environmental degradation and industrial pollution, severely impacting their livelihoods (Verité, 2012).

Collectively, these communities exemplify the vulnerabilities of small-scale fishers in the face of climate change, environmental degradation, and socio-economic pressures. Addressing these challenges necessitates integrated strategies that encompass environmental sustainability, labour rights protection, and community resilience building. The following section presents an overview of the labour discourse in each of the three locations, highlighting the specific strategies, challenges, and the ways in which fishers adapt and sustain their livelihoods.

Labour Conditions in Kalibaru

Kalibaru Subdistrict, located along the northern coastline of Jakarta, is one of the prominent coastal settlements in the region. Covering an area of 246.70 hectares, it consists of 172 neighbourhood units (RT) and 14 community units (RW). The subdistrict is home to approximately 82,203 residents across 36,169 households (BPS Kota Jakarta Utara, 2024). Due to its coastal location, Kalibaru possesses distinct topographical characteristics, with a significant portion of its population engaged in maritime-related sectors such as fishing, shipping, and port services. However, the area also contends with various environmental and infrastructural challenges, particularly in flood control and coastal conservation efforts.[2] In addition to its geographic importance, Kalibaru is recognised as a developing urban area, experiencing ongoing progress in both its economic and social sectors.

As part of densely populated urban-peripheral communities, many inhabitants rely on maritime-based livelihoods such as small-scale fishing, shipping, and port services. However, the subdistrict faces recurring environmental and infrastructural challenges—including tidal flooding, coastal erosion, and inadequate flood controls—linked to its low-lying geography and subsiding land. Studies show that Kalibaru ranks among the most flood-vulnerable coastal settlements in North Jakarta, with moderate to high exposure levels stemming from combined physical and social risk factors. Moreover, research on participatory mapping highlights complex land uses—spanning residential, industrial, and public spaces—underscoring the need for integrated coastal conservation efforts. With its strategic role in supporting North Jakarta's port activities, Kalibaru plays a significant part in the local economy, especially in sectors like maritime logistics, commerce, and small-scale industry. Yet, its susceptibility to environmental hazards and the lack of robust infrastructure demand targeted resilience planning.

Almost all inhabitants around Kalibaru's shore work as fishermen, with 80% also farming fish. Their fishing boats are modest and operated by daily fishermen who depart and return every day. As a result, these fishermen rely on the east monsoon wind to fish because catches are better during this season and fishermen are more likely to be safe. When there is a west wind, they choose to rest as the conditions are not conducive to sailing. This is also a time to repair their equipment and inspect their ships.

The west monsoon normally arrives in early December, but because of natural variation—and perhaps due to the impacts of climate change—it can begin in October. The risk to smaller fishing boats is strongly influenced by the direction of this westerly wind, although some will still choose regardless.

During the west monsoon, some fishermen work as motorcycle taxi drivers, some support their wife's business (if she has one), some become construction workers or project workers, and some participate in Cangkring, a community-based organisation which coordinates the production of a variety of products from green mussel waste, such as paving blocks, toilets, accessories, coffee tables, and other items. This shift in employment while fishermen are unable to go to sea allows them to make money in other ways. When the east monsoon arrives, the fishermen resume sailing as normal.

Some fishermen also decide not to work during the west monsoon since the pay is low, instead choosing to rest at home and not seek additional revenue. However, many are obliged to work in order to meet the needs of their families. Temporary work as a labourer or construction worker is less lucrative than fishing. For example, a fisherman who typically earns IDR 400,000 per day will only earn a fifth of that as a labourer or project worker. The bulk of fishermen's lifestyles are characterised

[2] https://utara.jakarta.go.id/kelurahan/kelurahan-kalibaru/profil.

by poor financial management; when they receive money, it usually disappears quickly because they believe they will be able to recoup it readily. This becomes a difficulty in itself when financial conditions are tight and the west monsoon arrives, so they must find a way to meet their living needs.

Good financial management in a fisherman's household will influence the urgency of the requirement for extra work. This means that it typically comes down to the wife's role: if the budget is managed properly, income from the east season will be set aside to cover expenses during the west wind season. However, if the wife does not manage the household finances efficiently and the money runs out, a shift to casual labour in the off-season will become necessary.

The life of a fisherman is actually quite varied. In terms of finance there are couples who work together, such as when the husband is a fisherman and the wife works in a firm, becomes a teacher, owns livestock, or sells daily necessities. There are also those who operate stalls at home, and those who become clam peelers. Revenue fluctuates: usually, if the wife works for a corporation or in a textile factory, the family will be financially well-off and the west season has little affect. However, families who rely only on their husband's income will struggle during the west season and will usually seek loans from cooperatives and mobile banks. The income they get is only enough to pay off debts as a means of survive. Approximately half the fishermen on the Kalibaru coast have wives who earn money separately.

At least two Kalibaru residents have created microbusinesses, manufacturing shell crackers; however, their sale model is traditional and does not use e-commerce platforms such as Shopee or Tokopedia. Local agencies and governments provided no support for marketing through online channels and, as a result, these businesses are limited to selling via small stalls and setting up stands at exhibitions, or accepting orders via WhatsApp. The potential of these ventures could be realised even further by leveraging digital and other social media technology, but the training provided by the provincial government largely focuses on how to create processed fish products, with little attention to packaging or product marketing. This is one of the shortcomings of the recently announced government program.

The emergence of Cangkring as a community venture for utilising shellfish waste has the potential to create income for Kalibaru fishermen and their wives. Shellfish peelers in Cangkring mostly have low-income husbands. Cangkring adds value by repurposing a by-product of their livelihood that would otherwise go to waste, processing it and transforming it into usable and sellable items to boost the finances of the neighbouring fishermen, particularly during the west season. Cangkring was founded in 2019 with ten active members and now provides opportunities for fishermen—including those no longer strong enough to sail—their wives, and the elderly who want to be productive and earn income by recycling green mussel waste.

The Cangkring community also opens up employment opportunities for other residents to earn income. During the east season, green mussels will be abundant and require additional labour to peel them. However, the proportion is low (only about 10%) and they only come when there are a lot of green mussels. Even though the distance from where they live is far, they still come because transportation is easy and cheap. However, during the west season, the amount of mussels is lower and does not require labour from outside the region.

Labour Conditions in Marunda

The circumstances at Marunda are nearly identical to those in Kalibaru, with daily fishermen numbering 200 in total. Fishermen in the Marunda area catch more shrimp, although some prefer to catch fish. Pollution in the Marunda area has limited catches, with fish and shrimp dying as a result of industry waste. This was reported to the appropriate agency in 2015, but no firm action was taken. The presence of this pollution has a significant impact on fishermen, forcing them to go farther out to sea. There are various factories in this area, including oil, palm oil, paper, and paint factories, and the waste from Marunda companies affects not only jobs but also health.

Fishermen's wives in Marunda are mostly housewives and do not work, meaning household finances are solely dependent on fishing income. Unlike Kalibaru, there is no shellfish peeling industry because the fishermen's catch consists almost wholly of fish and shrimp. Only a few women work, even in the informal sector. Women's empowerment has not been fully realised, especially in terms of supporting the family economy. Fishermen in Marunda desperately need social support during the west season or when there is rob flooding, and many people are in debt or have taken out internet loans. When fishermen are unable to go to sea at all, local government support and social aid provided to local inhabitants can help them survive.

Cooperatives existed in this area for a period, but have mostly ceased to operate. Cooperative membership provides a range of benefits, with government social assistance usually received through the cooperative. However, the monthly donations expected from each member are not reliable; some members pay on time, while others do not. Several years ago, prior to the COVID-19 pandemic, the subdepartment provided social aid in the form of rice and basic necessities in December, but this no longer happens. The program now in operation is the Family Hope Program (PKH) from the Local Department of Social Affairs, which is available only to families who meet specific criteria and is subject to a range of rules and provisions.

The Marunda residents have limited freedom of movement. Some fishermen who attempted to raise livestock locally were ultimately unsuccessful because the animals died as a result of pollution. Furthermore, since the establishment of the Marunda Centre Port, fishermen from Marunda have not been permitted to approach and fish in the region; this forces them to travel further, which naturally demands more fuel and increases costs, as well as only being possible when the weather is good.

Labour Conditions in Muara Angke

Muara Angke is a delta surrounded by the Asin River in the east, Adem River in the west, and Jakarta Bay in the north. It includes Pluit Village and Penjaringan Subdistrict. The fishing port of Muara Angke was built in 1977 and the region was developed to accommodate the fisheries that spread over several locations in Jakarta. Muara Angke has been designated as a centre for fisheries development in Jakarta since 1990. Initially the region had an area of 62 hectares, but the Jakarta Government expanded this through land reclamation from 2006 to around 71 hectares at present. The region comprises fishermen housing (21.2 ha), fish landing bases (5 ha), mangrove forests (8 ha), fish processing (5 ha), docks (1.4 ha), vacant land (6.7 ha), markets (1 ha), terminals (2.6 ha), soccer fields (1 ha), and ferry ports to Pulau Seribu (2 ha). Fishermen housing was designed to be inhabited by owners of fishing boats, crews, fishing workers, salted fish labourers, and fish traders (UPT PPKP and PPI Muara Angke, 2011).

The availability of jobs in Muara Angke is quite similar to other coastal areas, with the main difference being that large vessels dominate while smaller boats are very limited in number. Large vessels tend to support more stable fishing activities, typically lasting from one to three months at sea. Women living along this coast also work as green mussel peelers, and some run small food businesses, such as selling rice, to supplement the household income. This is necessary because relying solely on their husbands' income is often insufficient, especially during the west monsoon season, which can be unpredictable in duration. Women have become active community agents by initiating the formation of neighbourhood units (RT), demonstrating strong female agency and participation in the community. In certain situations, these women also help the community access government aid and assistance from other institutions.

Aside from fishermen, there are other 3D (dirty, dangerous, and difficult) jobs available in this area, particularly those referred to as TKBM (steward unloaders). Approximately 90% of TKBM workers come from Banten Province, while the rest are local residents. Wages for this work are calculated based on the number of loads carried and delivered from sellers to buyers. This job requires physical strength, as it involves transporting heavy goods or fish. Becoming a TKBM worker is relatively easy—one simply needs to offer their services to fish sellers or buyers who are unable to make direct purchases at the market. These workers also offer personal shopping services, fulfilling specific requests from buyers to earn extra income.

Muara Angke tends to have greater job stability compared to Marunda and Kalibaru. Fishermen with large vessels and TKBM workers almost never shift to other jobs during the west and east monsoon seasons. Fishermen generally prefer to rest during the west monsoon period, and since most of them come from outside the city, they return to their hometowns to rest and spend time with their families. As for the TKBM workers, they rarely take time off and are not dependent on seasonal winds. The only factor that affects them is the availability of fish in the market—if the supply decreases, their wages are affected accordingly. Nevertheless, they can continue working as long as the fish market remains open.

Long-settled large-vessel fishermen living in Muara Angke may face financial difficulties if the household relies solely on the husband's income, especially when the wife does not work and the family lacks financial management skills. This situation does occur, and some eventually take on temporary jobs as construction labourers, online drivers, or other types of gig work. Job shifting in Muara Angke is influenced by several factors: the number of family members, the number of income-earning family members, and how household finances are managed.

Fisheries Sector Strategies Due to Climate Change Impact in the Coastal Area Jakarta

The fisheries sector in Jakarta Bay faces a multitude of vulnerabilities due to the impacts of climate change, overfishing, and environmental degradation. This situation is particularly challenging for fishers who depend on the marine ecosystem for their livelihoods. Employment opportunities in the coastal regions of Jakarta Province are predominantly concentrated in the fisheries sector and within the informal labour market. The challenges faced by coastal communities arise during the west monsoon season or when the sea recedes, both of which significantly affect the livelihoods of daily fishermen and green mussel peelers, an occupation primarily undertaken by coastal women. Additionally, climate change phenomena such as tidal flooding and sea level rise further disrupt the ability of residents to engage in their economic activities. Areas most affected by these environmental changes include Cilincing, Marunda, Kalibaru, and the Muara Angke coastline.

Tidal flooding and rising sea levels do not disrupt activities at the Muara Angke Fish Market. As a result, the role of TKBM (steward unloaders) tends to remain stable. TKBM work constitutes one of the most accessible forms of informal employment, offering widespread opportunities for income generation. However, Jakarta residents are often reluctant to pursue this type of work due to its relatively low remuneration and its classification as a 3D job. Approximately 90% of TKBM workers originate from Banten Province, particularly from Serang, Lebak, and surrounding areas. This occupational path is often inherited, passed down through family members, neighbours, and local networks. TKBM work has the potential to offer an alternative livelihood for fishermen during periods when fishing is not feasible, such as during the west monsoon season. Its accessibility and demand, particularly in fish markets like Muara Angke, make it a viable temporary employment option for those affected by seasonal and climate impact. Even so, Muara Angke fishermen, who struggle with the effects of climate change, are not exactly jumping at this chance.

As a strategy and form of adaptation, livelihood diversification is one of the most effective strategies for reducing vulnerability and increasing resilience in the fisheries sector. Jakarta Bay fishers, like many coastal communities, rely heavily on fishing as their primary income source. However, climate change is making traditional fishing practices increasingly unreliable. Shifting fish stocks, rising sea temperatures, and increased storm frequency are undermining the sustainability of fishing labour in the region. To adapt, fishers are diversifying their livelihoods to reduce dependence on fishing alone.

Challenges related to income sustainability emerge due to the growing prevalence of green mussel cultivation and peeling processes. The decrease in green mussel prices affects the wages for peeling, while efforts at mangrove cultivation have yet to sustain routine production. However, the success of livelihood diversification depends on local socio-economic contexts. In Jakarta Bay, policies must ensure that alternative livelihood strategies are tailored to the local environment and are compatible with the social structure of the fishing community. Furthermore, the shift towards diversified livelihoods requires policy interventions that provide training, capital, and infrastructure support to make these transitions sustainable (Uddin et al., 2021).

Income stability within fishermen households is a critical factor in adapting to the socio-economic impacts of climate change. However, the concept of asset and financial management is still not well understood among the fishing communities in Jakarta Bay. This lack of knowledge impedes their ability to secure resources for future resilience, making them more vulnerable to income loss during times of resource depletion or extreme weather. To address this, capacity-building initiatives are needed to educate fishers on asset management principles, such as resource monitoring, financial literacy, and risk mitigation strategies. Strengthening these areas can improve their ability to manage their livelihoods and support the sustainable use of marine resources (Susiloningtyas, 2019).

The intersection of livelihood resilience and labour discourse in the context of coastal fisheries is key to understanding how fishing communities can adapt to climate change. By emphasising livelihood diversification and adaptive co-management, it is possible to create a more resilient workforce that can shift from traditional forms of fishing labour to more diversified and sustainable livelihood strategies. This approach not only addresses labour market transitions but also builds a robust foundation for long-term economic stability in coastal areas vulnerable to climate change impacts. The resilience of labour in coastal fisheries is not just about survival but about transforming the way labour is structured to better cope with the changing environmental and socio-economic conditions brought about by climate change (Balgos, 2005; Kristiningrum et al., 2023; Uy et al., 2011).

Strategies for Resilience in the Fisheries Sector of Jakarta Bay

The fisheries sector in Jakarta Bay faces a multitude of vulnerabilities due to the impacts of climate change, overfishing, and environmental degradation. This situation is particularly challenging for fishers who depend on the marine ecosystem for their livelihoods. The combination of declining fish stocks, pollution, and sea level rise in Jakarta Bay presents serious challenges to both the labour force in the fisheries sector and the sustainability of the industry itself. This discussion explores key strategies for resilience in the fisheries sector of Jakarta Bay, focusing on asset management, livelihood diversification, and social safety nets. These strategies are essential for both addressing the immediate challenges faced by fishers and ensuring long-term sustainability of fisheries as a critical source of livelihood.

Asset Management: Securing Resources for Sustainable Labour

Effective asset management is essential to ensure the resilience of Jakarta Bay's fisheries sector amidst climate change pressures. Natural resources, such as fish stocks and coastal ecosystems, face degradation due to overexploitation and environmental changes. Sustainable management of these resources is vital for the livelihoods of fishers, who rely on healthy marine environments like mangroves, coral reefs, and seagrass beds. These ecosystems not only support fish populations but also act as natural barriers against extreme weather events. Additionally, human capital, such as skills and knowledge, is crucial for resilience. Many fishers in Jakarta Bay are adopting sustainable fishing practices and engaging in community-based resource management, which can enhance long-term sustainability (Uy et al., 2011).

However, the concept of asset management is still not well understood among the fishing communities in Jakarta Bay. Many fishers lack understanding of how to manage both natural and financial assets effectively. This lack of knowledge impedes their ability to secure resources for future resilience, making them more vulnerable to income loss during times of resource depletion or extreme weather. To address this, capacity-building initiatives are needed to educate fishers on asset management principles, such as resource monitoring, financial literacy, and risk mitigation strategies. Strengthening these areas can improve their ability to manage their livelihoods and support the sustainable use of marine resources (Susiloningtyas, 2019).

Livelihood Diversification: Mitigating the Vulnerabilities of Fishing Labour

Livelihood diversification is one of the most effective strategies for reducing vulnerability and increasing resilience in the fisheries sector. For Jakarta Bay fishers, like many coastal communities, climate change is making traditional fishing practices increasingly unreliable. Shifting fish stocks, rising sea temperatures, and increased storm frequency are undermining the sustainability of fishing labour in the region. To adapt, fishers are diversifying their livelihoods to reduce dependence on fishing alone. Aquaculture, eco-tourism, and small-scale businesses such as handicrafts and local trade have become alternative income sources. For example, some fishers in Jakarta Bay have shifted to fish farming to supplement their catch, while others have explored eco-tourism opportunities that capitalise on the region's unique coastal ecosystems.

Livelihood diversification not only provides a buffer against resource scarcity but also helps alleviate income inequality within communities. For instance, women in fishing households have increasingly become involved in seaweed farming or local crafts, thus creating economic opportunities for marginalised groups and enhancing the social resilience of the community. By enabling households to access alternative sources of income, diversification can protect fishers from economic shocks and support their adaptation to changing environmental conditions. Furthermore, the shift towards diversified livelihoods requires policy interventions that provide training, capital, and infrastructure support to make these transitions sustainable (Uddin et al., 2021).

However, the success of livelihood diversification depends on local socio-economic contexts. In Jakarta Bay, policies must ensure that alternative livelihood strategies are tailored to the local environment and are compatible with the social structure of the fishing community. For example, some fishers may prefer to stay within the traditional fishing sector, while others may choose transition to non-fishing activities. Community-led initiatives, where local preferences and traditional knowledge are considered in diversification plans, can improve the overall success and sustainability of these strategies.

Social Safety Nets: Supporting Vulnerable Fisheries Labour

Social safety nets are crucial for enhancing the resilience of fisheries communities in Jakarta Bay, especially for those whose livelihoods are vulnerable to environmental shocks. These mechanisms can be divided into formal and informal systems. Formal safety nets, such as government disaster relief funds, insurance schemes, and social protection programs like PKH (Program Keluarga Harapan), provide immediate financial support during crises. These programs are essential for preventing vulnerable populations, such as small-scale fishers, from falling deeper into poverty during difficult times. However, these formal systems may be slow to implement and might not always accommodate the seasonal income fluctuations typical in the fishing industry.

In contrast, informal community-based safety nets like cooperatives, savings groups, and mutual aid societies are more flexible and responsive to the immediate needs of local communities. By pooling resources and sharing knowledge, fishers can support each other during income loss or off-season downturns. Strengthening these informal networks is vital, as they foster social cohesion and offer quicker, more accessible support compared to formal systems.

To improve overall resilience, expanding government programs like PKH to include more fishing families and supporting the formation of local cooperatives or insurance schemes can help communities better manage financial risks and sustain their livelihoods through challenging periods. In addition, local governments and NGOs can play a role in bridging the gap between formal and informal safety nets by providing training and capacity-building programs to strengthen local institutions. These efforts could include educating fishers on accessing formal financial support or formalising informal groups to increase their impact and sustainability in the long term.

Policy Implications: Environmentally and Socially Adaptive Safety Nets for Climate Change

The policy implications for the fisheries sector in Jakarta Bay include the importance of prioritising both environmental sustainability and social protection to build long-term resilience against climate change. From an environmental perspective, policies should focus on strengthening coastal ecosystem protection, such as through the establishment of marine protected areas and the promotion of sustainable fishing practices. These policies can help maintain fish stocks, safeguard marine biodiversity, and reduce the pressure on overexploited resources. Additionally, adaptive resource management should be a central aspect of policy, where fisheries management is flexible enough to respond to shifting environmental conditions like changing sea temperatures and fish migration patterns. Supporting integrated coastal zone management (ICZM) can enhance collaboration between stakeholders, ensuring a sustainable balance between economic activity and environmental conservation, thus securing the foundation of the fisheries sector for future generations.

On the social side, policies should ensure that vulnerable fishers and their families are protected through adaptive social safety nets that respond to climate change impacts. These safety nets can include climate-responsive insurance programs, cash transfers, and social welfare schemes designed to mitigate income loss during extreme weather events or declining fish stocks. By providing targeted financial support, these programs will allow marginalised groups, such as small-scale fishers and women, to cope with the disruptions to their livelihoods. Moreover, livelihood diversification policies must be promoted to create alternative income sources, reducing the dependency on the fisheries sector. Investing in community-based programs and capacity-building initiatives will empower local communities to adopt new economic activities and ensure their resilience. Thus, social protection policies must be integrated into the broader climate adaptation strategy, ensuring that fishers in Jakarta Bay have the tools and support to navigate the challenges posed by climate change while maintaining their economic stability.

Recommendations

Acknowledge Climate-Induced Labour Shifts as Structural, Not Temporary

Climate change and monsoon seasonality significantly impact fishing schedules, especially for daily fishers who rely on the west monsoon for sailing. These disruptions have led to unavoidable shifts in labour patterns. Labour shifts in this context should not be seen as incidental or seasonal but as structural changes requiring long-term planning and support mechanisms. To ensure effective and evidence-based responses to labour shifts in fishing communities, it is essential to conduct regular assessments of labour trends. These assessments should capture not only changes in employment patterns but also the underlying drivers such as seasonality, resource availability, and socio-economic pressures. Furthermore, integrating climate vulnerability data into local labour planning frameworks is equally important. By aligning labour policies with environmental data, local governments can anticipate the impacts of climate change on employment and proactively design interventions that support sustainable livelihoods for coastal populations.

Strengthen Household Economic Resilience Through Skill Diversification

To strengthen household resilience in coastal communities, local governments should offer vocational training and microbusiness support for fishermen and their wives. These programs could include online marketing modules for fish-based products, developed in partnership with private platforms or cooperatives. Encouraging small home industries such as smoked fish, salted fish, and seafood snacks can also help families generate additional income, especially during periods when fishing is not possible.

Institutionalise Multi-Stakeholder Collaboration for Economic Alternatives

To support long-term adaptation in coastal communities, it is important to establish community resilience centres in fishing areas. These centres can function as hubs for vocational training, access to market networks, and emergency assistance during climate-related disruptions. In parallel, a public–private partnership framework should be developed to mobilise funding and resources for alternative employment programs, ensuring broader support and sustainability of livelihood initiatives in coastal regions.

Extend Labour Rights and Social Protection to Informal and Gig Workers

Most fishers fall into informal or seasonal labour categories. Existing labour laws predominantly protect formal sector workers, leaving fishers and temporary workers without access to social security, healthcare, or minimum wage guarantees. With climate-induced job shifts increasing, legal clarity and protection for these categories are urgently needed. The central government should establish a regulatory framework that formally recognises gig workers, including seasonal fishers, as legitimate categories of labour entitled to basic protections. These should include access to health insurance through social security for health, occupational accident protection via social security for employment, and the enforcement of minimum income standards. In addition, portable social protection schemes should be introduced to ensure that workers maintain continuous coverage even as they move between different types of employment. This approach would provide greater security and equity for informal and mobile labourers affected by climate-related disruptions.

Strengthen Local Governance and Data Collection on Labour Vulnerabilities

To design effective and inclusive policies, governments must rely on accurate and up-to-date data that captures labour transitions within fishing communities. Disaggregated data based on gender, age, education, and income is particularly important to ensure that interventions address the diverse realities of coastal populations. In this context, conducting regular vulnerability assessments in coastal areas becomes essential to identify specific risks and the adaptive capacities of households affected by climate change. At the same time, developing a comprehensive labour database to monitor informal and seasonal employment trends, especially those driven by environmental factors, will provide critical insights for both local and national governments. Together, these efforts will enable the formulation of more responsive, targeted, and sustainable policies that support the evolving nature of work in climate-vulnerable coastal regions.

Conclusion

The labour discourse in the fisheries sector along the north coast of Jakarta Province reflects a dynamic interaction between environmental pressures, socio-economic conditions, and structural changes in employment. Several interrelated factors shape this reality, including the seasonal nature of fishing activities governed by the west and east monsoons, the increasing unpredictability of climate change, and the gradual impact of sea level rise. These forces have altered the traditional rhythms of work in fishing communities, compelling many fishers to seek alternative forms of livelihood during periods when they are unable to go to sea.

As a survival strategy, many fishermen have transitioned into informal and seasonal jobs, ranging from construction work and transportation services to small-scale home industries. In Kalibaru, innovative community-based initiatives such as Cangkring have demonstrated the potential of empowering coastal women by transforming shellfish waste into eco-friendly products such as paving blocks and household items. This not only generates economic value but also promotes circular economy practices. In Muara Angke, where larger fishing vessels dominate, fishers and crew members often remain inactive during idle periods, while steward unloaders generally continue their routines except during major holidays. In Marunda Cilincing, fishers commonly take on temporary work in construction or as taxi drivers during the west monsoon season when sailing is not feasible.

These adaptive responses demonstrate the resilience of coastal communities but also highlight the precarious nature of their livelihoods. Climate change has intensified these vulnerabilities, creating conditions in which labour shifts are no longer optional but essential for daily survival. Despite some localised innovations, most of these responses remain reactive and lack institutional support.

Therefore, addressing the impacts of climate-induced labour shifts requires coordinated and sustained efforts from multiple stakeholders. Capacity building, skill diversification, legal protection for informal workers, and inclusive policies that recognise the realities of small-scale fishers are all crucial components of a just and sustainable adaptation strategy. Only through such integrated efforts can coastal communities in Jakarta Bay navigate the growing challenges of climate change while maintaining their dignity, productivity, and economic well-being.

References

Adi Renaldi. (2024). *Climate change and overfishing hit Indonesia's tuna catch*. Retrieved from https://www.context.news/nature/climate-change-and-overfishing-hit-indonesia-tuna-catch

Allison, E. H., & Ellis, F. (2001). The livelihoods approach and management of small-scale fisheries. In *Marine Policy* (Vol. 25). http://www.fao.org/fi/projects/sflp/index.html

Asian Development Bank. (2021). *Climate Risk Country Profile: Indonesia*. ADB.

Backler, S., Davies-Vollum, K. S., & Raha, D. (2025). An evaluation of coastal resilience policies in Anglophone West Africa through a regional comparative analysis. *Marine Policy, 176*. https://doi.org/10.1016/j.marpol.2025.106648

Balgos, M. C. (2005). Integrated coastal management and marine protected areas in the Philippines: Concurrent developments. *Ocean and Coastal Management, 48*(11–12), 972–995. https://doi.org/10.1016/j.ocecoaman.2005.03.003

Barange, M., Bahri, T., Beveridge, M. C. M., Cochrane, K. L., Funge-Smith, S., & Poulain, F. (2018). *Impacts of climate change on fisheries and aquaculture: Synthesis of current knowledge, adaptation and mitigation options*. FAO Fisheries and Aquaculture Technical Paper No. 627. FAO. https://openknowledge.fao.org/handle/20.500.14283/i9705en

Hardt, M., & Negri, A. (2000). *Empire*. Harvard University Press. https://www.hup.harvard.edu/books/9780674006713

Hasiholan, V. M., Fauzi, A., & Kinseng, R. A. (2023). The effect of fishermen's income factors on fishermen's poverty on The North Coast of Jakarta. *Jurnal EMT KITA, 7*(2), 482–495. https://doi.org/10.35870/emt.v7i2.1161

Hsu, K., & Peng, L. P. (2023). Understanding vulnerability and sustainable livelihood factors from coastal residents in Taiwan. *Marine Policy, 155*. https://doi.org/10.1016/j.marpol.2023.105793

ILO. (2020). *Social insurance and climate change in Indonesia: Implications for adaptive social protection*. International Labour Organization.

ILO and PRK-BRIN, Understanding working conditions of fishers in Indonesia: Evidence from the 2024 Survey on Decent Work in Marine Fishing, Geneva: International Labour Office, 2025. ILO. https://doi.org/10.54394/HTBY3881

Indonesia Business Post. (2025, January 3). *Systemic exploitation of fisheries workers in Indonesia persists*. Retrieved from https://indonesiabusinesspost.com/insight/systemic-exploitation-of-fisheries-workers-in-indonesia-persists/

Islam, R., & Walkerden, G. (2022). Livelihood assets, mutual support and disaster resilience in coastal Bangladesh. *International Journal of Disaster Risk Reduction, 78*. https://doi.org/10.1016/j.ijdrr.2022.103148

Kristine Sabillo. (2024). *Climate change forces Jakarta fishing families to marry off young daughters*. Retrieved from https://news.mongabay.com/short-article/climate-change-forces-jakarta-fishing-families-to-marry-off-young-daughters/

Kristiningrum, R., Sari, W. I. R., & Chaniago, H. D. (2023). Performance of sustainable livelihoods assets in coastal communities in dealing with climate change: A case study at Kuaro District, Paser Regency. *IOP Conference Series: Earth and Environmental Science, 1282*(1). https://doi.org/10.1088/1755-1315/1282/1/012012

Nurzaman, A., Shaw, R., & Roychansyah, M. S. (2020). Measuring community resilience against coastal hazards: Case study in Baron Beach, Gunungkidul Regency. *Progress in Disaster Science, 5*. https://doi.org/10.1016/j.pdisas.2020.100067

Prado, D. S., Seixas, C. S., & Berkes, F. (2015). Looking back and looking forward: Exploring livelihood change and resilience building in a Brazilian coastal community. *Ocean and Coastal Management, 113*, 29–37. https://doi.org/10.1016/j.ocecoaman.2015.05.018

Prana, A. M., Dionisio, R., Curl, A., Hart, D., Gomez, C., Apriyanto, H., & Prasetya, H. (2024). Informal Adaptation to Flooding in North Jakarta, Indonesia. *Progress in Planning, 186*, 100851. https://doi.org/10.1016/j.progress.2024.100851

Roscher, M. B., Eriksson, H., Harohau, D., Mauli, S., Kaltavara, J., Boonstra, W. J., & van der Ploeg, J. (2022). Unpacking pathways to diversified livelihoods from projects in Pacific Island coastal fisheries. *Ambio, 51*(10), 2107–2117. https://doi.org/10.1007/s13280-022-01727-x

Susiloningtyas, D. (2019). Coastal livelihood sustainability analysis of migration activity at Tanjung Luar and Salura Island Indonesia. *IOP Conference Series: Earth and Environmental Science, 338*(1). https://doi.org/10.1088/1755-1315/338/1/012032

Uddin, M. S., Haque, C. E., Khan, M. N., Doberstein, B., & Cox, R. S. (2021). 'Disasters threaten livelihoods, and people cope, adapt and make transformational changes': Community resilience and livelihoods reconstruction in coastal communities of Bangladesh. *International Journal of Disaster Risk Reduction, 63*. https://doi.org/10.1016/j.ijdrr.2021.102444

Uy, N., Takeuchi, Y., & Shaw, R. (2011). Local adaptation for livelihood resilience in Albay, Philippines. *Environmental Hazards, 10*(2), 139–153. https://doi.org/10.1080/17477891.2011.579338

Verité. (2012). *Research on indicators of forced labor in the supply chain of fish in Indonesia: Platform (jermal) fishing, small-boat anchovy fishing, and blast fishing*. Verité. Retrieved from https://verite.org/wp-content/uploads/2016/11/Research-on-Indicators-of-Forced-Labor-in-the-Indonesian-Fishing-Sector__9.16.pdf

Wannewitz, M., & Garschagen, M. (2024). The role of social identities for collective adaptation capacities–general considerations and lessons from Jakarta, Indonesia. *International Journal of Disaster Risk Reduction, 100*, 104194. https://doi.org/10.1016/j.ijdrr.2023.104194

Chapter 13
Strengthening Social Protection Among Informal Sector Workers Affected by Climate Change Impact: Pekalongan Case Study

Josefhin Mareta, Nawawi, Dewi Analis Indriyani, and Donna McAuliffe

Introduction

Climate change is anticipated to have a transient adverse effect on agricultural productivity, diminish the capacity of forests to facilitate livelihood diversification using timber harvesting and the collection of non-timber forest products, and influence the distribution and abundance of fish and other marine resources (Call & Sellers, 2019). Due to land subsidence and climate change, Pekalongan City and Regency in Central Java have been consistently affected by tidal floods for several years. The current situation in these areas is a cause for concern. The wider Pekalongan area on Java's north coast, encompassing Pekalongan City and Regency, is frequently subject to severe coastal and tidal floods. Consequently, the 1.2 million inhabitants of the region suffer the loss of their means of living, belongings, and fertile land (Syam et al., 2021).

The impact permeates into other growth sectors, influencing industries linked to the coastline, such as tourism and fisheries. Over time, tidal floods have submerged residential neighbourhoods and productive agricultural districts in approximately one third of Pekalongan City in Central Java. The ongoing floods have had a detrimental impact on the economic endeavours of the residents. In addition, the persistent tidal inundation has destroyed hundreds of hectares of rice fields and jasmine flowers, rendering them unsuitable for cultivation. These two commodities were the

J. Mareta (✉) · D. A. Indriyani
Research Center for Law, National Research and Innovation Agency (BRIN), Jakarta, Indonesia
e-mail: jose003@brin.go.id

Nawawi
Research Center for Population, National Research and Innovation Agency, Jakarta, Indonesia

D. McAuliffe
School of Health Sciences and Social Work, Griffith University, Brisbane, QLD, Australia

L. Nurhidayah et al. (eds.), *Climate Change, Labour and Migration in Indonesia*, Springer Climate, https://doi.org/10.1007/978-3-032-15334-0_13

primary income source for most farmers in the coastal areas of Pekalongan. In the case of Jeruksari Village in Pekalongan Regency, the villagers, who were previously engaged in farming, are now largely unemployed and seek work in different occupations due to the inundation of their farmlands by tidal floods. Their welfare has declined compared to before the tidal floods became a significant problem. In addition, climate change has adversely impacted fishing catches, causing loss of income for fishing-dependent communities and families.

An observed social consequence within the communities is a shift in how men and women do their everyday duties. Women are assuming more significant leadership roles in the public sector, which men historically dominated. Due to tidal floods, males residing in coastal areas primarily employed in the fishing sector have been compelled to switch professions or face unemployment. Women, including spouses and female offspring, also work to meet the requirements of their families. Each household member must be willing to generate income to fulfil the family's essential requirements, regardless of the amount earned. Coastal women can sustain themselves and generate revenue by engaging in informal sector activities along the coast.

The literature provides numerous definitions for the informal economy. The 'informal economy' refers to all economic activities that contribute to the publicly calculated national revenue but are not formally registered. The informal economy is also denoted by synonymous phrases such as 'informal sector' or 'informal employment'. In this chapter, we use the terms 'informal sector' and 'informal employment' interchangeably, despite acknowledging the distinction made by the International Labour Organization (ILO) in 2018. Approximately 85% of all employment is attributed to the informal sector, while households and the official sector constitute the remaining 15% (Elgin & Elveren, 2021).

Women in low-income nations have been disproportionately employed in the informal economy under the neoliberal paradigm. The ILO has determined that a significant proportion of employed women in low-income countries—92.1%—are engaged in informal labour (ILO, 2018). Companies prefer to hire women for two reasons: first, due to their transient employment status, women initially exhibit higher productivity levels; and, second, women are more willing to embrace lower income and precarious working conditions because flexible and home-based employment aligns more closely with their preferences (Elgin & Elveren, 2021).

In the absence of a formal employment agreement, informal sector workers are vulnerable due to the lack of legal labour protections, such as minimum wage, severance pay, overtime pay, old-age insurance, and health insurance. Unlike formal labour, informal employment lacks sufficient regulation at the legislative level. It raises various inquiries on the characteristics of casual work and jobs, its boundaries and extent, social security concerns brought about by its extensive use, and future possibilities for formalising casual work (Zanfirova et al., 2023).

The primary aim of our research is to delineate the legal dimensions of safeguarding informal workers, examine the social protection obstacles faced by informal workers, and propose a climate change policy that protects informal labour. We focus in particular on women compelled to be involved in family economic activities, who are often overlooked in government empowerment programs.

Informal Sector: Concept, Definition, and Employment Relationship

The informal sector plays a substantial role in the labour market and is a primary catalyst for production, employment growth, and revenue generation in numerous developing countries. However, individuals who engage in informal work are at a higher risk of experiencing vulnerability and insecurity (Huang et al., 2024; *Statistics on the Informal Economy*, n.d.). Informal employment is typically characterised by lacking recognition, registration, regulation, or protection under labour legislation and social security. Due to its easy exit nature, the informal sector is often the primary exit strategy for those looking for work or who have lost their jobs in the formal sector (Nawawi et al., 2013). 'Working circumstance' is a comprehensive term encompassing income, welfare, security, and the work environment. Informal workers endure severe working circumstances due to a lack of fundamental legal and social protection. They may be terminated without recourse to legal proceedings, their remuneration may be delayed, and they may be deprived of employment perks, sick leave, mandated rest days, and pension plans (Huang et al., 2024).

The ILO defines informal employment as work which, in law or practice, is not covered or is insufficiently covered by formal arrangements (Goodman et al., 2024). The informal sector typically lacks formalised regulations and is characterised by familial employment relationships, limited educational attainment, and a lack of specialised skills (Sukismanto et al., 2023). The ILO Conventions and Recommendations seldom mention workers in the informal sector, either to encompass or exclude them from their applicability. Convention No. 150 and Recommendation No. 169 are two instances of exemptions. Some conventions explicitly state in their language that they were intended to apply to all workers universally. One example of such a convention is the Social Policy Convention of 1962 (No. 117), which requires governments to assist 'independent producers and wage-earners', who likely include workers in the informal sector in urban and rural areas, to improve their living conditions (Schlyter, n.d.).

Direct reasons constitute the principal causes of irregular employment, whereas indirect reasons represent secondary causes that amalgamate with other elements to create informal work. For example, companies use irregular workers to circumvent the expenses associated with tax liabilities, social security contributions, administrative complexities, regulatory inflexibilities, punitive measures, and unemployment expenditures. Cultural acceptability, poverty, state confidence, and employment levels are the elements that compel individuals to refrain from declaring work as indirect causes (Barra & Papaccio, 2024).

In the informal sector, labourers' and employers' identities are ambiguous. Several factors can contribute to the ambiguity of the employment relationship, such as noncompliance with legal regulations and inadequate enforcement, the unclear roles of employers and workers, and uncertainty over workers' rights, and the final responsibility for the workers (Hasbullah & Ibrahim, 2020).

Informal employment is a significant and varied category. Categorising informal employment into self-employment and paid employment and further subdividing these groups into more homogeneous subcategories depending on work status is helpful (Chen, 2016). These categories can be divided as follows. Informal self-employment encompasses: (a) employers in informal firms; (b) own-account workers in informal companies; (c) contributing family workers (in both informal and formal enterprises); and (d) members of informal producers' cooperatives (where applicable). In informal wage employment workers are engaged without social protection contributions by formal or informal firms or as compensated domestic labourers by households. Some categories of paid labour are more prone to informality than others. The categories encompass: (a) employees of informal firms; (b) casual or day labourers; (c) temporary or part-time workers; (d) compensated domestic workers; (e) contract workers; (f) unregistered or undeclared workers; and (g) industrial out-workers, also referred to as home workers.

Informal or precarious work can be characterised using four dimensions established by Ernest Cano (Tomayahu, 2024): (1) lack of job continuity; (2) inadequate wages; (3) deterioration of worker–employer relationships and worker vulnerability; and (4) weakening of social protection for workers.

In the Indonesian context, the use of the informal sector category often refers to worker status as used by the Central Statistical Agency (Badan Pusat Statistik/BPS) in the national labour survey known as Sakernas (Survei Angkatan Kerja Nasional). This category includes informal employment, which encompasses self-employment, businesses supported by temporary or unpaid labour, freelance workers, and family members who work without pay. In Indonesia, free labourers, especially in the agriculture sector, are classified as informal workers due to their typically informal characteristics, including low productivity, insufficient remuneration, and comparatively unfavourable working conditions (Sibagariang et al., 2023). As per Law No. 13 of 2003 (Law 13 of 2003) on Manpower, informal workers in Indonesia labour without an employment contract, meaning there is no agreement governing work, wages, and authority (Hamid & Intan, 2024).

In other countries, such as India and Zimbabwe, informal workers already endure perilous living and working conditions that intersect with many climate-related hazards. Although the impacts of climate change do not function independently of other socioeconomic, political, and environmental factors, they do intensify specific concerns and generate new challenges for informal workers. Substandard housing and labour circumstances frequently worsen owing to severe weather occurrences, potentially leading to increased health issues and exacerbated poverty among informal labourers (Dodman et al., 2023).

Climate change has precipitated economic deterioration for coastal inhabitants in Pekalongan, compelling them to engage in the informal sector, particularly women.

In this study, the term 'informal sector worker' denotes an employed individual who lacks a written contract outlining the employer's and worker's duties, rights, and responsibilities.

Social Protection for the Informal Worker

Over the next few decades, cities will face various climate change impacts, with significant implications for informal livelihoods and public health. Paola et al. (2023) assert with a high level of certainty in the Sixth Assessment Report of the Intergovernmental Panel on Climate Change (IPCC) that urban areas will experience a greater frequency of extreme climate events, including heatwaves resulting in increased hot days and nights, rising sea levels, and heightened rainfall and storm surge intensity from tropical cyclones. These factors will significantly increase coastal city flooding risk.

Many individuals residing in low-income and informal settlements are employed in the informal economy, where the lack of regulation in their neighbourhoods and work environments can give rise to various health hazards. They are primarily involved in temporary, seasonal, and transient jobs. In the case of Pekalongan, men who previously worked as farmers are mainly employed in the services, trade, and construction sectors, while women and children, especially boys, are heavily involved in the batik industry, with jobs such as washing batik material using chemicals. Informal settlements and informal employment are interconnected but not synonymous: a significant portion (though not all) of the inhabitants of informal settlements engage in informal economic activities. When precarious living conditions and precarious job conditions coincide, it can worsen the risks experienced by individuals and households (Dodman et al., 2023).

Despite the diverse demographics and working conditions of informal labourers, they often need more fundamental infrastructure and services such as water, sanitation, and rubbish collection. These workers typically have lower wages and receive less training than official jobs; additionally, their access to social safeguards such as healthcare is limited. Women are usually over-represented in low-wage, unstable informal industries such as domestic or home-based jobs (Kadungure et al., 2023).

Such workers are particularly susceptible to the effects of climate change, due to several factors. Firstly, their low incomes make it difficult to afford the necessary measures to protect themselves from climate-related impacts. For instance, many residents along the coast in Pekalongan use a proportion of their income to raise their houses so that they are not affected by ongoing tidal flooding. They admitted that even though their income is low, they are often forced to reduce their daily needs (food, education, and health) so that their house does not sink. Discrimination further exacerbates their vulnerability, with women and ethnic minorities having limited access to resources and opportunities to cope with climate change. Moreover, their location in climate-sensitive areas, such as coastal regions, exposes them to extreme weather events and other climate-related hazards. Lastly, their informal employment status leaves them without the coverage of social protection programs, such as unemployment insurance or disability benefits (Dartanto, n.d.).

In 1952, the ILO established and solidified the concept of social security by adopting Convention No. 102 on Social Security (Minimum Standards). A comprehensive definition was suggested, encompassing nine fundamental factors that result in the termination or significant decrease of income, which should be protected by social security. The benefits encompassed several areas: sickness, maternity, employment accidents, unemployment, invalidity, old age, and death. Additionally, they covered medical care and subsidies for families with children (Canelas & Niño-Zarazúa, 2022; Unni & Rani, 2002).

Social protection measures may significantly contribute to mitigating the socio-economic effects of climate change due to their established and pivotal function in controlling poverty and income uncertainties. Formal social protection is the governmental transfer of resources to people and families to sustain income during adversity or enhance their living conditions. Social protection encompasses social assistance, insurance, and other programs, including employer mandates, universal allowances, active and passive labour market policies, and certain tax benefits (Costella et al., 2023).

Social protection refers to the array of policies and programs to alleviate poverty and vulnerability by fostering efficient labour markets, reducing individuals' exposure to risks, and augmenting their ability to safeguard themselves against hazards and income disruption or loss. Social protection comprises five principal components: (a) labour markets, (b) social insurance, (c) social assistance, (d) micro and area-based initiatives for community protection, and (e) child protection. Social protection is predominantly understood as employment and social insurance. The ILO offers technical assistance mainly in social security systems, social protection administration and coverage expansion, social expenditure management, labour relations, vocational training, employment policy, labour administration, labour law, industrial working conditions, management development, cooperatives, labour statistics and occupational safety and health (Heeswijk, 2003).

In Indonesia, the social assistance program initiated in 1998 is designed for low-income and other at-risk populations. Despite Law No. 40 of 2004 on the National Social Security System being enacted in 2004, the national health insurance system only commenced in January 2014. Currently, it encompasses two thirds of Indonesia's population, with approximately 50% of its expenses subsidised by the government. The employment social insurance system initiated in January 2015 currently encompasses only a small fraction of formal sector workers, government employees, and the military (The National Team for The Acceleration of Poverty Reduction, 2018).

Social security systems are often established through retirement savings, social insurance, and employer responsibility (Perwira, 2003). According to Article 99 of Law 13 of 2003, social security is an entitlement for workers and their families. The article states that 'every worker and their family has the right to receive labour social security.' Social security is a component of social protection, along with social assistance. In Indonesia, regulations mandate the government's responsibility (as a state organiser) to execute social protection initiatives (Faturrachman et al., 2024).

Employees are entitled to a suitable salary and comprehensive social security benefits. Social security encompasses various aspects such as healthcare, workplace injury coverage, unemployment benefits, retirement pensions, survivor benefits, and provisions for old age security. As per Article 1 of the National Social Security System Law, social security is a means of social protection that guarantees the fulfilment of necessities for a satisfactory standard of living for all individuals.

Law No. 2 Year 2022 on Job Creation (Law 2 of 2022) modifies various regulations, covering remuneration, temporary employment agreements, and social security. The Job Creation Law introduced job loss insurance as a new addition to the existing social security policies. According to Article 18 of Law 2 of 2022, social security programs encompass (a) health insurance, (b) work accident insurance, (c) old-age insurance, (d) pension insurance, (e) death insurance, and (f) job loss insurance.

Research Method

The study was carried out in Pekalongan, Central Java Province. This area was selected because Pekalongan has been categorised as a region prone to the impacts of climate change, according to the vulnerability maps. Pekalongan is susceptible to climate hazards such as landslides, floods, and coastal inundation. This study utilised qualitative methods for data collection. To determine the most appropriate district and neighbourhood, we conducted thorough interviews. The informants were chosen according to their position in the community, involvement in the informal sector, and climate change adaptation and comprised representatives from local administrations, non-government organisations (NGOs), and batik women workers.

A total of ten women batik workers were interviewed from five sub-districts in Pekalongan City/Regency, including one from Jeruksari Village, four from Kemplong Village, two from Kauman Village, one from Gamer Village, and two from Panjang Wetan Village. Comprehensive key informant interviews enhanced women's engagement by allowing them to express their thoughts, experiences, and expertise in a variety of ways. Questions addressed several topics related to climate threats, their impacts, and women's capacity to combat climate change in the informal sector.

Participation of Women in the Informal Sector Affected by Climate Change

Pekalongan City experiences recurrent flooding during the rainy season and high tides, with prevalent riverine and tidal inundations. Sea level rise and land subsidence intensify Pekalongan City's susceptibility to flooding, affecting both communities and the environment (Putri et al., 2024). Pekalongan City is low-lying, resulting in annual flooding during intense rainfall. Floods inflict significant damage, destroying land and property. The residents have undertaken various measures to mitigate the impact of tidal floods, yet flooding continues to be severe. The ocean has engulfed many hectares of agricultural land and whitefish, and prawn aquaculture. Similarly, the majority of the residences have inundated and individuals who opt to remain must consistently invest substantial funds to maintain their homes.

Numerous residents have lost their employment as a result of their enterprises and agricultural lands being inundated by seawater. In homes, women are especially vulnerable to the lasting impacts of land loss due to their heightened responsibilities for childcare and family care. Land loss leads to less stability and security for the entire family, with women sometimes bearing psychological and emotional duties and economic difficulties. The impact of losses on women workers is more significant since they lose a source of income and face even greater challenges in securing help and restoring stability for their families than men.

> We regularly rebuilt homes, which were flooded a few times, to ensure we could live in them. Our homes and fields where we grow jasmine, water spinach, and rice were flooded, which meant we lost money from farming. (Woman Batik Worker, Jeruksari Village)

Economically disadvantaged individuals typically employ survival methods to sustain their household livelihood. Usually, informal labour is feasible, such as working in the home batik industry. As an informal sector, the batik industry is regarded as a flexible employment option for women as they can simultaneously fulfil their homemaker roles. Nonetheless, women engaged in batik production are classified as impoverished based on their pre-batik income, which is inadequate to meet their necessities (Hidayaturrahman et al., 2023; Ibrahim & Md. Shah, 2020).

> I have worked in batik for five years, helping my husband, a peddler. From 8 am to 4 pm, I work tracing batik motifs on a cloth and am paid [IDR] 50,000. Though it's not very large, it can meet daily living expenses with three children. (Woman Batik Worker, Kemplong Village)

Furthermore, women workers in this business encounter a range of challenges. These include health risks due to their direct exposure to batik chemicals; our interview results indicate that most women in the batik industry do not understand the long-term risks and dangers of using material liquids, especially when washing batik materials. The flexible working hours also influence household dynamics as women must reconcile their responsibilities across professional, domestic, and social spheres. This study found a link between informal work and stunting among children because their mothers were forced to work long hours and hand over the care of toddlers to older children or other family members. Finally, a distinction exists in the labour roles of male batik artisans, designated for stamping due to their association with heavy machinery, and female artisans, tasked with dyeing.

> While female workers colour-stamped batik or 'Colet,' men labour stamping, canting and cleaning batik or 'Melorod.' Washing batik pays [IDR]. 80,000 daily, colouring pays [IDR] 40,000 to 50,000 daily. (Woman Batik Worker, Panjang Wetan Village)

Women employed in the informal sector exhibit various vulnerabilities, particularly in realising their rights, including menstruation leave and maternity leave, during necessary periods of rest. A further concern pertains to social security and health insurance. Women employed in the informal sector refrain from advocating for more compensation. They often remain silent and accept their low pay, as they primarily function as supplementary economic support, while men remain the principal earners. Labour relations in the informal sector are not precisely articulated in employment contracts.

> The social security program for workers has been directed but has not been implemented because workers are not bound and can suddenly move; there is no work responsibility. Workers have been given one day off a week. If sick, they may not work and are given money for medical treatment. (Batik Entrepreneur, Gamer Village)

Labour is typically conducted by employees within their residences through a system known as the 'Output System' or *borongan*. In this system, workers are classified as homeworkers—defined as individuals who receive tasks from employers without formal employment relationships, complete the work at home without supervision, supply their work resources, assume their production risks, and receive compensation based on output units (*borongan*) determined by the employer's scale (Rahmawati et al., 2014). Wages are disbursed after work is completed as per the terms of the agreement established before the commencement of the task. The remuneration provided pertains solely to their labour, excluding compensation for transportation or additional benefits.

> I also have to support the family since my husband is a tiny fisherman and hardly returns home. I work on batik at the house. Making one Kodi of cloth or twenty pieces every day brings [IDR] 35,000. Depending on the weather, I don't have to work every day; if it floods, I cannot work, so I am not compensated. (Woman Batik Worker, Jeruksari Village)

> Here, the female employees get paid when they fulfil their tasks. Working on hand-drawn batik depends on the degree of design complexity. More complicated designs translate into more pay for the employees. Some undertake work at home for [IDR] 7,000 for each piece, earning [IDR] 50,000–100,000 on average every week. Daily operations are carried out in the plant under a pay of [IDR] 50,000, therefore averaging [IDR] 300,000 every week. (Woman Batik Worker, Kauman Village)

The monthly income of women workers remains far below the regional minimum wage (Upah Minimum Regional/UMR) in Pekalongan City/District of IDR 2.5 million.

> There aren't enough earnings from this business to give the worker a raise. Some people in the batik business find it hard to market their products because they focus on written batik, which takes a long time and costs a lot (IDR 250,000 or more per piece). (Batik Entrepreneur, Gamer Village)

Batik workers experience health issues in conjunction with poor wages. Occupational ailments in the batik business include respiratory disorders, skin irritation, and eye tiredness. Occupational diseases in batik craftspeople may result from the work environment and work practices. The average artisan works 7–7.5 hours per day six days a week, so the average craftsman works 45 hours per week in a dusty work environment, while the dyeing procedure produces smoke that has the potential to cause respiratory difficulties. Moreover, the chemicals used in colouring batik material may lead to skin discomfort, particularly on the hands (Febriana et al., 2023). In addition to occupational disorders, there is a risk of accidents linked to working in the batik business, notably burns. Burns may arise during the canting process due to the utilisation of hot liquid wax and during the boiling of batik cloth, which involves wax melting. Batik artisans are also at risk of direct contact with hot tools, boiling water, and embers (Lestari & Warseno, 2019).

The working conditions fail to comply with health and safety norms. In the batik sector, personnel engaged in dyeing frequently apply chemicals without sufficient protection, thereby jeopardising their long-term health.

> Flooding causes trouble with my batik work at home. When the high tide flooded the house, I couldn't work. I also had to care for the house while my husband worked as a labourer at the market. The task takes a long time if the motif is challenging and the fabric is silk. I often got splashed hot wax when I worked. (Woman Batik Worker, Kauman Village)

Analysis and Policy Implications

In the last 20 years, Indonesia has markedly transformed its social policies. Social protection initiatives, particularly conditional cash transfers and food assistance programs, have been essential to Indonesia's endeavours to mitigate poverty (Nugroho et al., 2021). Nonetheless, the advancement of Indonesia's social protection policies continues to encounter implementation challenges, revealing deficiencies in the existing social protection framework. Policies have failed to account for the impact of natural catastrophes and climate change on poverty in their design, despite Indonesia's susceptibility to such events. The disastrous effects of climate change have also influenced society, with impoverished and vulnerable populations likely to be the most impacted (SMERU Research Institute, n.d.).

Social protection for informal workers has garnered substantial support from the public and stakeholders. NGOs have been promoting the significance of social protection for informal workers by leveraging online platforms as a tool for issue mobilisation. Various stakeholders, including labour unions, the scholarly community, and the media, continue to emphasise the necessity of converting the concept of social security for informal workers into practical public policy implementations. Nevertheless, engaging informal labourers presents a formidable challenge (Madya, 2024).

Social protection is significant for women compelled to engage in the informal sector, such as the batik industry, and is primarily associated with health safeguards and protection against insufficient income. Informants stated that business proprietors fail to offer labour-related rights, such as the risk of work-related injuries and minimum wage. This failure occurs because the relationship between employees and employers is not formally documented in a contract. Thus, even though they occasionally face demands that necessitate more effort (e.g., overtime without compensation), workers continue to endeavour to meet all requests from the business owner, referred to as the employer.

Law 13 of 2003, the Labour Law, contains three pertinent sections that do not address the requirements and safeguarding of informal workers. Article 1 defines 'labourer' as 'every individual capable of performing work to fulfil the necessities of their life and family'. This broad definition does not explicitly encompass informal workers and engenders ambiguity in articulating and asserting their rights. Article 5 delineates the rights and responsibilities of workers, encompassing entitlements to remuneration, leave, and social security. However, these rights typically pertain solely to formal employees with a clearly established employment relationship with an organisation. Informal workers, frequently lacking a written contract or formal employment connection, encounter difficulties securing these rights (Panjaitan & Ritonga, 2024).

Employees in the domestic batik sector, predominantly mothers, exceed the standard working hours. Employees frequently labour for over 12 hours daily to meet the targets set by the employer. Women batik artisans labour late into the night and early morning, rising at dawn to pray and attend to domestic responsibilities, yet they receive no overtime compensation. The Labour Law stipulates that employees working six days a week are limited to a maximum of seven hours per day, while those working five days a week are permitted to work a maximum of eight hours daily. Any work beyond that threshold qualifies for overtime compensation, with a maximum limit of three hours per day.

Workers also require social security in the form of health and life insurance. Engaging in traditional batik production poses significant health hazards. Furthermore, no specialised safety equipment is provided to protect them from the toxic chemicals used. The batik makers perform their tasks wearing only their regular daily attire, without gloves, masks, etc. Women workers often experience dermatological conditions, including skin peeling and stinging, attributed to the chemical dyes used to apply batik motifs to the fabric.

The Pekalongan Employment Social Security Agency (BPJS Ketenagakerjaan Pekalongan) administers social security protection for informal workers through a social insurance system implemented via the National Social Security Program based on Law 40 of 2004 and Law 24 of 2011. In 2025, BPJS Ketenagakerjaan Pekalongan, in collaboration with the Pekalongan City Industry and Manpower Office (Disperinaker/Dinas Perindustrian dan Tenaga Kerja Kota Pekalongan), facilitated a quota for the Worker Social Security Program (Jamsostek) payment assistance, contributing IDR 16,800 per individual through the Joint Poverty Prevention and Reduction Program via the Empowerment and Protection of Vulnerable Workers (Batik Berlian). Registered informal labourers typically engage as batik vendors in markets, fishermen, or mobile traders utilising motorcycles. Informal labourers in vulnerable situations are registered at the local level. Additionally, village officials will furnish data regarding residents categorised as extremely impoverished and employed. The Pekalongan City Government directly disburses Jamsostek payment support for one year. In the subsequent year, it is expected that informal workers benefiting from this assistance will be able to independently remit payments for Work Accident Insurance and Death Insurance (interview with BPJS Ketenagakerjaan Pekalongan).

The quota is contingent upon budget availability. This quota is under the Decree of the Mayor of Pekalongan No. 500.15.14.2/0180 of 2023 concerning the Identification of Vulnerable Workers' Occupations and the Allocation of Participants Per Village for the Empowerment and Protection of Vulnerable Workers Program in Pekalongan City. There are 21 classifications of vulnerable workers, including batik labourers (interview with Disperinaker Kota Pekalongan).

Pekalongan City Regulation No. 10 of 2023 regarding the Employment Social Security Program for Vulnerable Workers in the Informal Sector within Pekalongan City targets vulnerable workers in the informal sector, including non-wage earners, wage earners, and other residents who meet the criteria for participation in the labour social security program. The legislation stipulates that the work accident insurance and life insurance programs safeguard vulnerable workers in the informal sector. The implementation of Jamsostek for vulnerable workers in the informal sector aims to offer social security to those at risk of work-related injuries and fatalities.

Nevertheless, implementation poses challenges, particularly in the informal sector, where ambiguous regulations about informal sector employees complicate oversight and enforcement. The enactment of the Job Creation Law modified at least nine articles of the Labour Law; however, none of these improvements protect informal labourers. The Job Creation Law should ideally enhance the Labour Law by acknowledging informal workers and regulating their employment ties to ensure protection, social security, and workplace safety (Tomayahu, 2024).

Disperinaker Kota Pekalongan described the implementation of Regulation No. 10 of 2023 as challenging due to the distinct nature of the employer–employee relationship in the informal sector, which is heavily reliant on the employer. The batik business operates informally and lacks official employment relationships, which complicates the implementation of assistance programs via BPJS Ketenagakerjaan Pekalongan. The absence of oversight of business entities is a contributing factor, with supervision serving as a mechanism for labour protection. Regulation No. 10 mandates that the mayor establish fostering, supervision, and control.

Adaptive Social Protection for Climate Change

In terms of climate change impacts, informal workers are often unable to work due to extreme weather events, leading to drastically worsened living and working conditions. While social protection initiatives are in place to support Pekalongan's informal workers, access to benefits is uneven. The Ministry of Social Affairs (Kementerian Sosial) provides the Adaptive Family Hope Program (Program Keluarga Harapan Adaptif/PKH Adaptif), a conditional cash assistance scheme aimed at supporting disaster victims in their social recovery and reinforcement, as stipulated in Regulation No. 4 of 2015 concerning Direct Cash Assistance for Disaster Victims.

The Directorate of Social Security (Direktorat Jaminan Sosial) stated that adaptive social protection emphasises implementing social protection for disaster victims. However, this campaign is not explicitly directed towards victims of climate change-related disasters. Households must satisfy the education, health, and social welfare requirements, meaning not all victims of natural catastrophes qualify for financial assistance. Local governments record disaster victims who experience the most substantial losses to secure adaptive social protection assistance (interview with Direktorat Jaminan Sosial Kementerian Sosial).

> My house was surveyed, and the roof was fixed as part of a house rehabilitation project during the 2018 flood. Furthermore, it included social aid [of IDR] 200,000. Still, I have asked the village official to create a Certificate of Insolvency/Poverty (Surat Keterangan Tidak Mampu) to get the social aid. (Woman Batik Worker, Panjang Wetan Village)

Informal workers should enjoy the same entitlement as formal workers to healthy and safe working conditions and access to social protection programs. Social protection is an obligation of those who profit from the labour of workers and the commodities and services they produce. Employers or capital owners are obligated to offer social benefits to their employees and others whose labour indirectly advantages them, irrespective of the status of the employment relationship. An optimal social protection system comprises a combination of public and private provisions, provided that the latter does not undermine the former (Unni & Rani, 2002).

Many responses centred upon local organisations' efforts to address climate-related concerns and strengthen livelihoods. As just two examples, NGO Bintari (Bina Karta Lestari) undertakes initiatives to build resilience, including rehabilitating mangrove and coastal ecosystems; Mercy Corps, in partnership with the Pekalongan City Marine and Fisheries Agency, has provided training on milkfish aquaculture to promote resilient and sustainable livelihood development in coastal communities impacted by floods and tidal surges. Nevertheless, current programs and policies predominantly concentrate on victims of natural disasters in general, neglecting the specific impact on informal labourers. Civil society organisations are expected to advocate for governmental acknowledgement and community access to social protection.

Access to assets can enhance the resilience of informal workers to climate change, particularly when they can leverage social protection and employ various methods simultaneously. Integrating state-sponsored social protection with human responses can markedly enhance the resilience of low-income urban dwellers. Social welfare programs can offer substantial support for informal workers, particularly in the context of climate change (Dodman et al., 2023).

In Pekalongan City, the low participation of informal workers in the Jamsostek is primarily due to inadequate dissemination of information regarding its social security advantages, and to informal workers' lack of knowledge regarding the registration process for the program. Additionally, the Pekalongan community's local culture, characterised by strong religious convictions, views involvement in labour social security as an illicit matter (Ardianingsih et al., n.d.).

BPJS Ketenagakerjaan Pekalongan and the Pekalongan City Government are working to increase the registration quota for Employment Social Security to address this issue. Registration is conducted by proactively identifying potential registrants, aided by 'shield agents' in the field who locate informal labourers eligible for inclusion in the government-assistance-recipient quota. They facilitate socialisation within the community, including informal workers, and provide direct assistance for community members to register for Jamsostek (interview with BPJS Ketenagakerjaan Pekalongan).

The municipal administration of Pekalongan is concentrating on the batik industry to mitigate the effects of climate change by addressing manufacturing waste. Liquid waste from the batik business contains dangerous chemicals capable of causing water pollution. Numerous initiatives have been instituted, including: (1) the continuous application of clean production methods and recycling within the manufacturing process, (2) the establishment of regional associations or institutions to address localised issues, such as in Kauman Batik Village and Pesindon Batik Village, and (3) the design of a batik wastewater treatment facility (Instalasi Pengolah Air Limbah/IPAL) aimed at diminishing pollutant levels in the liquid waste generated by the batik industry through a Matching Fund (Dana Padanan) Program, organised by university and local governments. This program targets small and medium batik industries in Pekalongan City to implement green industry concepts (interview with Disperinaker Kota Pekalongan).

Expanded social protection could address the issues that climate change poses to informal workers, such as by offering pay supplements to people unable to work due to flooding and providing healthcare and social security to meet fundamental living requirements. Collaboration between governmental entities and civil society organisations is essential to mitigate the climate hazards confronting informal labourers. Community organisations and informal worker groups can provide a basis for focused initiatives to identify beneficiaries of social protection programs.

Conclusion

Amid climate change in the Pekalongan region, women are compelled to engage in the informal sector, enduring extended working hours, meagre wages, and a dangerous work environment. Law 13 of 2003 fails to provide legal certainty around, and thus to safeguard, the rights of informal sector workers. The government has established labour social security, a component of social protection under Law 40 of 2004, which encompasses informal workers; however, its implementation remains problematic due to insufficient awareness among informal workers and inadequate oversight of business proprietors.

The national and municipal governments have established social protection programs to assist disaster victims; however, these initiatives do not specifically target women employed in the informal sector. The majority of women involved in the informal sector do not have the social protection necessary to protect themselves from work-related accidents or injury. Most prioritise meeting daily basic needs over paying monthly contributions to the social protection scheme. NGO-facilitated adaptation projects have partly involved women through empowerment initiatives, but they do not directly address workers in the informal sector.

The government must collaborate with community organisations and women's groups representing the most vulnerable informal sector workers to formulate policies and programs responsive to climate change, including providing health insurance and satisfying basic requirements during natural disasters. Solutions are required that draw on alternative schemes, such as village fund allocations and giving priority to free social protection programs from local government, so that women forced to work in the informal sector remain protected.

References

Ardianingsih, A., Langelo, J. S., & Wicaksono, P. (n.d.). Analisis kepesertaan jaminan sosial ketenagakerjaan pada pekerja sektor informal di kota Pekalongan. *Jurnal Litbang Kota Pekalongan, 19*(2), 143–150.

Barra, C., & Papaccio, A. (2024). Does regulatory quality reduce informal economy? A theoretical and empirical framework. *Social Indicators Research, 172*(2), 543–567. https://doi.org/10.1007/s11205-024-03319-6

Call, M., & Sellers, S. (2019). How does gendered vulnerability shape the adoption and impact of sustainable livelihood interventions in an era of global climate change? *Environmental Research Letters, 14*(8), 083005. https://doi.org/10.1088/1748-9326/ab2f57

Canelas, C., & Niño-Zarazúa, M. (2022). *Social protection and the informal economy: What do we know?* (Working Paper Series, pp. 1–45). United Nations ESCAP, Social Development Division. http://www.unescap.org/kp

Chen, M. A. (2016). The informal economy: Recent trends, future directions. *New Solutions: A Journal of Environmental and Occupational Health Policy, 26*(2), 155–172. https://doi.org/10.1177/1048291116652613

Costella, C., Van Aalst, M., Georgiadou, Y., Slater, R., Reilly, R., McCord, A., Holmes, R., Ammoun, J., & Barca, V. (2023). Can social protection tackle emerging risks from climate change, and how? A framework and a critical review. *Climate Risk Management, 40*, 100501. https://doi.org/10.1016/j.crm.2023.100501

Dartanto, T. (n.d.). *Climate change, worker vulnerability and adaptive social protection*. ILO-Adaptive Social Protection Forum.

Dodman, D., Sverdlik, A., Agarwal, S., Kadungure, A., Kothiwal, K., Machemedze, R., & Verma, S. (2023). Climate change and informal workers: Towards an agenda for research and practice. *Urban Climate, 48*.

Elgin, C., & Elveren, A. Y. (2021). Informality, inequality, and feminization of labor. *Women's Studies International Forum, 88*, 102505. https://doi.org/10.1016/j.wsif.2021.102505

Faturrachman, M., Adriaman, M., & Nazar, J. (2024). Perlindungan hak jaminan sosial tenaga kerja bagi pihak yang tidak menyelenggarakan jaminan sosial tenaga kerja. *Yustisi Jurnal Hukum dan Hukum Islam, 11*(2), 273–283.

Febriana, S. A., Ridora, Y., Indrastuti, N., Dewi, K., Erdina, Oginawati, K., Tanziha, I., Prakoeswa, C. R. S., Waskito, F., & Schuttelaar, M.-L. (2023). Hazard identification and the prevalence of occupational skin disease in Indonesian Batik workers. *Scientific Reports, 13*(1), 5231. https://doi.org/10.1038/s41598-022-17890-w

Goodman, J. M., Lara-Mejía, V., Hernández-Cordero, S., & Vilar-Compte, M. (2024). Breastfeeding among women employed in Mexico's informal sector: Strategies to overcome key barriers. *International Journal for Equity in Health, 23*(1), 144. https://doi.org/10.1186/s12939-024-02147-x

Hamid, A., & Intan, A. M. (2024). Legal protection for informal sector workers in employment development in Indonesia: Challenges and opportunities. *Research in Business & Social Science, 13*(5), 880–892. https://doi.org/10.15294/ijicle.v6i1.4717

Hasbullah, N., & Ibrahim, D. (2020). Employment relations in the informal sector: A need for redefinition. *Humanities and Social Sciences Letters, 8*(4), 397–406. https://doi.org/10.18488/journal.73.2020.84.397.406

Heeswijk, J. P. M. V. (2003). *Social protection*. Asian Development Bank.

Hidayaturrahman, M., Resdiana, E., Syafriyani, I., Elazhari, E., & Sugiantiningsih, A. A. P. (2023). Kemiskinan dan peran berlebih perempuan pembatik di Indonesia. *Jurnal Ilmiah Manajemen Publik Dan Kebijakan Sosial, 7*(1), 88–106. https://doi.org/10.25139/jmnegara.v7i1.6094

Huang, G., Cai, B., Liu, S., & Xue, D. (2024). Analysing the heterogeneity in working conditions of migrant informal workers in China: A test of the WIEGO model of informal employment. *Humanities and Social Sciences Communications, 11*(1), 882. https://doi.org/10.1057/s41599-024-03405-7

Ibrahim, D., & Md. Shah, J. (2020). Preliminary research on the survival strategies among pala'o people in Banggi Island. *Community: Pengawas Dinamika Sosial, 6*(1), 1–9. https://doi.org/10.35308/jcpds.v6i1.1859

Kadungure, A., Machemedze, R., Malaya, W., Banda, N., Loewenson, R., & Sverdlik, A. (2023). Towards climate change resilience and informal workers' health in Zimbabwe: An action-research case study. *Cities & Health, 7*(6), 1055–1071. https://doi.org/10.1080/23748834.2023.2261755

Lestari, R., & Warseno, A. (2019). Analisis risiko penyakit dan kecelakaan kerja menggunakan model upaya kesehatan kerja di industri batik rumahan. *Jurnal Wacana Kesehatan, 3*(2), 315–323. https://doi.org/10.52822/jwk.v3i2.78

Madya, S. H. (2024). Jaminan sosial pekerjai informal: Studi webnografi. *Jurnal Jamsostek, 2*(2), 118–140. https://doi.org/10.61626/jamsostek.v2i2.27

Nawawi, N., Nagib L., Sri Soesilawati, E., Asiati D., & Harfina, D. (2013). *Pekerja informal di perkotaan dan aspirasi terhadap pekerjaan layak*. Research Center for Population LIPI.

Nugroho, A., Amir, H., Maududy, I., & Marlina, I. (2021). Poverty eradication programs in Indonesia: Progress, challenges and reforms. *Journal of Policy Modeling, 43*(6), 1204–1224. https://doi.org/10.1016/j.jpolmod.2021.05.002

Panjaitan, S., & Ritonga, R. (2024). Analisis perlindungan hukum bagi pekerja informal dalam perspektif hukum ketenagakerjaan. *Doktrin: Jurnal Dunia Ilmu Hukum dan Politik, 2*(3), 250–262. https://doi.org/10.59581/doktrin.v2i3.3330

Paola, A. A., Bellouin, N., Coppola, E., Jones, Richard. G., Krinner, G., Marotzke, J., Naik, V., Palmer, Matthew. D., Plattner, G.-K., Rogelj, J., Rojas, M., Sillmann, J., Storelvmo, T., Thorne, P. W., & Trewin, B. (2023). Technical Summary. In *Climate Change 2021: The Physical Science Basis. Contribution of Working Group I to the Sixth Assessment Report of the Intergovernmental Panel on Climate Change* (pp. 33–144). Cambridge University Press. https://doi.org/10.1017/9781009157896.002

Perwira, D. (2003). *Perlindungan Tenaga Kerja Melalui Sistem Jaminan Sosial:* Lembaga Penelitian SMERU.

Putri, S. I. P., Syakira, H., Nugroho, E. O., Layaliya, H. N., Soeharno, A. W. H., Kusuma, M. S. B., Farid, M., Kuntoro, A. A., Sidiq, T. P., Nurmalia, S. L., Widigdo, C., & Sadat, S. H. (2024). An overview of flood risk components: The land subsidence case of Pekalongan City. In N. Suprapto, B. K. Prahani, M. Satriawan, H. P. A. Tjahyaningtijas, M. Abdul Ghofur, & S. Andari (Eds.), *E3S Web of Conferences* (Vol. 513, p. 01001). https://doi.org/10.1051/e3sconf/202451301001

Rahmawati, R., Rismawati, S. D., & Zaduqisti, E. (2014). Sistem pengupahan dan pembagian kerja perempuan buruh batik berbasis putting out system di kota Pekalongan (proses dan implikasi marginalisasi terhadap perempuan buruh batik). *Jurnal Penelitian, 10*(2), 274–293. https://doi.org/10.28918/jupe.v10i2.365

Schlyter, C. (n.d.). *International Labour Standards and the Informal Sector: Developments and Dilemmas.*

Sibagariang, F. A., Mauboy, L. M., Erviana, R., & Kartiasih, F. (2023). Gambaran pekerja informal dan faktor-faktor yang memengaruhinya di Indonesia tahun 2022. *Seminar Nasional Official Statistics, 2023*, 151–160. https://doi.org/10.34123/semnasoffstat.v2023i1.1892

SMERU Research Institute. (n.d.). *Protecting the most vulnerable in Rural Indonesia: How can we develop more adaptive social protection.* Retrieved January 17, 2025, from https://smeru.or.id/en/research?field_research_expertise_tid=66&page=1

Statistics on the informal economy. (n.d.). ILOSTAT. Retrieved August 24, 2024, from https://ilostat.ilo.org/topics/informality/

Sukismanto, S., Hartono, H., Sumardiyono, S., & Andayani, T. R. (2023). Family support to improve knowledge, attitude, and practices of occupational health and safety (OHS) in the informal sector. *Malaysian Journal of Medicine and Health Sciences, 19*(2), 175–181. https://doi.org/10.47836/mjmhs.19.2.26

Syam, D. A., Wengi, K. R. L., & Gandapurnama, A. (2021). *Climate risk and impact assessment of Pekalongan, Indonesia.* Mercy Corps Indonesia.

The National Team for The Acceleration of Poverty Reduction. (2018). *The future of the social protection system In Indonesia: Social protection for all* (First). Office of the Vice President of the Republic of Indonesia.

Tomayahu, N. S. (2024). *Legal implications for home-based workers following the judicial review decision of the constitutional court of the republic of Indonesia. 6.*

Unni, J., & Rani, U. (2002). *Social protection for informal workers: Insecurities, instruments and institutional mechanisms* (1st ed.). International Labour Office.

Zanfirova, T. A., Kuzmenko, V. V., Yehorova-Lutsenko, T. P., Voytsyshena, O. V., & Chernysh, A. R. (2023). Legal and social aspects of informal labour formalisation: Prospects and challenges. *Preliminary Communication, x*(2).

Chapter 14
Women's Resilience Strategies in Responding to Climate Change Impacts in the North Coastal Area of Semarang: Economy, Environment, and Empowerment

Agus Suntoro, Ade Angelia Yusniar Marbun, Laely Nurhidayah, and Lauren Terzis

Introduction

The World Bank warns of the potential uncertainty of Indonesia's future, given its heavy reliance on the marine sector. Indonesia comprises 17,504 islands with a total coastline of 108,000 kilometres, and three quarters of its territory is ocean. The influence of climate change is a challenge, with increases in sea level rise, water temperature, storms, and ecosystem changes having implications for communities that depend on the sea for their livelihoods (Kaczan et al., 2022). Research conducted by Partnerships in Environmental Management for the Seas of East Asia (PEMSEA) in 2018 showed that Indonesia's dependence on the sea and coast was estimated at US$182.54 billion, or 28% of gross domestic product (PEMSEA, 2019).

In its 2009 Report on Climate Change, the United Nations warned that climate change will have a significant impact on women. The Office of the United Nations High Commissioner for Human Rights (OHCHR) notes that women constitute the majority of the world's poor and often directly depend on natural resources and the environment as their primary source of food and income. As such, they are more likely to suffer the impacts of environmental degradation. OHCHR documents five dominant aspects of climate change impacts on women: food security, health,

A. Suntoro (✉) · A. A. Y. Marbun · L. Nurhidayah
Research Center for Law, National Research and Innovation Agency (BRIN), Jakarta, Indonesia
e-mail: agus154@brin.go.id

L. Terzis
The School of Health Sciences and Social Work, Griffith University, Brisbane, QLD, Australia

L. Nurhidayah et al. (eds.), *Climate Change, Labour and Migration in Indonesia*, Springer Climate, https://doi.org/10.1007/978-3-032-15334-0_14

gender-based violence, decent work, and women's environmental human rights defenders (OHCHR, 2020). Balgis Osman-Elasha, lead author of the Fourth Assessment Report of the Intergovernmental Panel on Climate Change, has stated that the impacts of climate change on gender are not equal. Women are characteristically more vulnerable than men because they represent the majority of the world's poor and are more dependent on natural resources. Other factors increasing women's vulnerability include their more limited responses, roles, responsibilities, and decision-making skills compared to men in the context of climate change adaptation (Osman-Elasha, 2009).

The impact of climate change on women is particularly significant on the north coast of Java, especially around Semarang City and Demak Regency, Central Java. Research by Kesemat, a civil society organisation focusing on mangrove rehabilitation, found that women and children suffer the most from climate change. Therefore, interventions and coaching for adaptation to the impacts are carried out by rehabilitating mangroves and producing processed mangrove products in the form of snacks, batik, and coffee. Even in the Karimunjawa area many children and women are 'forced to work' because their parents are poor due to the impact of climate change (Kesemat, 2024). The Indonesia–German Partnership Policy Advice for Climate Change study also found the double burden of women affected by climate change, especially for low-income women in Semarang City. Women have limited access to natural resources and control over public policy; thus, gender mainstreaming into climate change urgently needs to be implemented in Semarang City, Central Java (Rusmadi et al., 2017).

Research by Griffith University, the National Research and Innovation Agency (BRIN), and Diponegoro University (UNDIP) in Semarang City and Demak confirms that women and children are affected by climate change. In several study areas, such as Tambaklorok, Tanjung Mas Village, Karangrejo Village, Mangkang Kulon and Wetan villages, and Mangunharjo Village, Semarang City, the role of women was found to be essential, providing the backbone of families affected by climate change. Economic diversification through new jobs as shrimp paste makers, green mussel peelers, smoked fish processors, cracker makers, mangrove coffee processors, mangrove batik, and low-level factory workers are among the efforts to maintain the economy of families affected by climate change (Young et al., 2024).

This chapter will discuss women's resilience in coastal areas, especially the city of Semarang, as a form of adaptation to climate change. Several previous studies related to coastal issues in Semarang City have been conducted; however, this collaborative research between Griffith University, BRIN, and UNDIP makes a novel contribution through its methodological approach, involving a series of in-depth interviews with affected women, roundtables with both women and policymakers, and direct field observations for more than 20 days. Likewise, the research results have also been packaged as a policy brief directly discussed with national and regional policymakers to encourage policy updates and legislation in efforts to adapt and mitigate the effects of climate change.

Women, Climate Change, and Resilience

Climate change, including extreme weather events, rainfall variability, and rising temperatures, has implications for basic physical and mental health determinants, such as clean air, clean water, food security, and the destruction of livelihoods and homes. A person's gender, age, race, economic status, education, and other social characteristics are factors that determine their vulnerability and capacity to adapt to climate change—with women among those suffering the most (Wigley, 2021). Although women are vulnerable to the impacts of climate change, however, they are also effective agents of change in mitigation and adaptation. Women have adaptive characteristics and knowledge that are valuable in climate change mitigation, disaster reduction, and adaptation strategies. Experience in household responsibilities as resource managers positions them to contribute to livelihood strategies tailored to their social and environmental realities (UN Women, 2009).

Women's movement to become essential to climate change mitigation and adaptation actions has been a long process, starting with the first Conference on Women in Mexico in 1975, the Nairobi conference in 1985, and the Earth Summit in 1992 in Rio de Janeiro. Women's awareness and the impetus of the movement increased at the 1992 Earth Summit, emphasising that women are victims of climate change and the main agents for adaptation, mitigation, and recovery of places and communities affected by climate. This movement emphasised women's strategic position to combat the impacts of climate change, especially about access to resources, livelihoods, fuel, and medicines (Resurrección, 2013).

At the Seventh Conference of the Parties (COP) to the United Nations Framework Convention on Climate Change (UNFCCC) in 2001, it was decided to increase women's participation in climate change adaptation and mitigation under the UNFCCC and the Kyoto Protocol. This policy was a historic decision: for the first time, the COP discussed the importance of increasing women's participation in handling climate change. However, although COP-7 encouraged Parties to involve women in decision-making mechanisms addressing climate change, the challenges faced by women have continue to be under-represented in negotiations and discussions (Mayesa Alam et al., 2015). To protect and recognise women in climate change adaptation and mitigation under the international framework, the 2030 Agenda relates gender, the rule of law, and climate change to the United Nations Sustainable Development Goals (SDGs)—particularly, SDG 5 (on gender equality and women's empowerment), SDG 13 (on climate action), and SDG 16 (on peaceful, just, and inclusive societies). These obligations require states to promote capacity-building mechanisms for climate change planning and management—including a focus on women—especially in developing and island countries most affected by climate change, (IDLO, 2022).

Natural disasters and climate change cause losses to humans and impact the economy and welfare. The consequences are disruption of the community's economic system due to loss of assets, production factors, output, employment, or consumption (Hallegatte, 2014). In this context, there are two approaches to the resilience of women and communities in areas affected by climate change: one is related to ecological resilience, and the other to economic resilience. Walker in 'Resilience, Adaptability and Transformability in Social-Ecological Systems' describes resilience as the ability of a system to remain stable so that its function, structure, and identity do not change when experiencing severe danger (Brian Walker et al., 2004). The Factory Mutual Global Resilience Index 2015 emphasises the importance of resilience in ecosystem recovery by outlining four main factors: political risk, infrastructure quality, natural hazards, and commitment to risk management (Moghim & Garna, 2019).

Carl's (2006) view is that ecological resilience is a vital characteristic of the objective existence of environmental systems, the evolutionary balance of their current conditions, and directly relates to the value-based needs of healthy human survival and development. Li et al. (2023) assess that ecological resilience is related not only to the impacts generated by socio-economic activities and nature but also to environmental recovery and management that reflect defensive resilience, adaptive recovery, and sustainable ecosystem transformation. Simmie (2009) argue that economic resilience is the adaptive capacity—that is, the differential reliability—of a region or area to adapt to changes and challenges that shape the dynamics and trajectory of the regional or local economy. This economic resilience, called 'engineering resilience', emphasises stability or balance against disturbances and the speed of returning to the previous balance.

Case Study Context and Method

Case Study

Semarang City—the capital city of Central Java Province, as established by Law No. 16 of 1950 and based on Government Regulation No. 50 of 1992—comprises 16 subdistricts. Geographically, it is located between 6° 50′–7° 10′ south and 109° 50′–110° 35′ east and covers an area of 373.70 kilometres, with its administrative boundaries being the Java Sea to the north, Semarang Regency to the south, Kendal Regency to the west, and Demak Regency to the east (Rahayu, 2014).

The Semarang City Regional Development Planning Agency's Regional Action Plan for Climate Change Adaptation in Semarang City notes that, based on the results of a study by the Asian Cities Climate Change Resilience Network (ACCCRN), Diponegoro University (UNDIP), and Bogor Agricultural Institute study, the region has experienced a ten-degree increase in air temperature during the dry season (June, July, August), from 25.3 to 26.3°C. Likewise, the sea level rise in between 1985 and 1998 was 58.2 cm; this was also influenced by soil characteristics and land subsidence, with the average sea level rise in Semarang City reaching 4.4 cm/year. The Semarang City Government identified four macro-impacts due to climate change with rising sea levels, namely (a) ecosystem impacts in the form of expanding tidal flooding (also known as rob flooding), decreasing groundwater quality, and decreasing biodiversity; (b) impacts on humans via the emergence of disease outbreaks, social symptoms (e.g., the presence of climate refugees), and migration of coastal communities, with most disaster victims being women; (c) economic impacts due to increased risk of losses due to inundation, as well as reduced land productivity value for economic activities such as agriculture; and (d) damage to infrastructure (Semarang City Regional Development Planning Agency, 2019).

Methodology

A qualitative approach was utilised in this research to understand the unique experiences of women in Semarang, with methods including a literature review, policy analysis, and data collection. The data collection process was carried out for almost 20 days with a focus on four aspects, namely:

(a) Interviews with women who focus on climate change adaptation, and how it is impacting economic sectors such as low-level textile factory workers in Tambakrejo Village, shrimp paste workers in Tambakrejo Village, fish smokers in Tambakrejo and Tanjung Mas villages, and fishermen's wives in Tanjung Mas, Mangkang Wetan, and Mangkang Kulon villages.
(b) Interviews with women who are active in communities adapting to climate change, including in Tambaklorok, Tanjung Mas Village, which focuses on urban farming and stunting eradication, and mangrove managers in Mangunharjo Village.
(c) Government institutions, both in Central Java Province and Semarang City, which are the main parties responsible for climate change mitigation and adaptation at the regional level.
(d) Business representatives, who also have an obligation to protect workers affected by climate change and who ensure that corporate governance operates in accordance with labour standards.

This qualitative research seeks to find explanations for social phenomena—the social world humans live in and why they think and act in the way they do. Qualitative research is valuable in that it (a) explores or identifies concepts or views of society, (b) explores implementation, and (c) understands the context of real life (LIPI, 2017). This research is related to the social aspects of human and seeks to answer questions about their adaptation to the impacts of climate change due to rising sea levels in the city of Semarang. This chapter will address two main problems: (a) how climate change, especially sea level rise, impacts women in Semarang City, and (b) the role of women in climate change adaptation and mitigation in Semarang City.

Findings

Small-Scale and Medium Enterprises

Fish Smoking

> The capital for the smoking business comes from borrowing from fish suppliers from Jakarta, Indramayu, and Tegal. The results of the smoking are then sold to hotels, restaurants, and regular buyers.

– Mrs Khadijah, fish smoker, Interview, 25 February 2025, Tambaklorok, Tanjung Mas Village, Semarang City.

In response to climate and environmental changes, the women of Tambaklorok, Tanjung Mas Village, and Mangkang Wetan Village, Semarang City, have emerged as central figures in strengthening economic resilience. The fishing community in this area has long been vulnerable to the fluctuating availability of fish, extreme weather events, and the changing marine ecosystem. In response to these challenges, women in the community have found innovative ways to secure their families' economic stability. One such initiative, the fish-smoking business, has offered an alternative income stream and demonstrated women's significant role in driving local economic development.

The Mina Asri Group, formed in 1998, was initially composed of 40 members. Today, the group is a well-established collective of 18 women, with its official registration with the Fisheries Department adding credibility to its operations. The group's primary activities include fish smoking, producing salted fish, and marketing its products. Over the years, the women of Mina Asri have received a variety of governmental and institutional support, including ice storage containers for preserving fish, a freezer from the Ministry of Home Affairs, and an oven from oil and natural gas corporation PT Pertamina to support the fish-smoking process. The skill of smoking fish in the Tambaklorok area is generally passed down from generation to generation; Mrs Khadijah, one of the senior smokers and leader of the Mina Asri Group, acquired her fish-smoking skills from her mother-in-law. The popular smoked fish dish is 'Manyung fish', priced at IDR 30,000/kilogram for the fish head and IDR 70,000/kilogram for the fish body. However, workers in this sector face challenges including capital, access to preservatives such as freezers, and product cleanliness.

Ismaroh, a female fish smoker and the head of the Semarang City Milkfish Processor, lives in Mangunharjo Village, Tugu District, Semarang City. Smoked fish and milkfish are obtained from the sea in the Indramayu Regency area, West Java, not from the coastal regions of Semarang City. Ismaroh uses two marketing strategies: participating in exhibitions or bazaars where she distributes business cards/brochures, and posting product statuses on her WhatsApp account. This promotional method has been successful and her business has survived for more than ten years, with many loyal buyers. Resilience through fish smoking and milkfish processing, while not the family's primary income source (her husband is an office worker), has increased the family's budget. Figures 14.1 and 14.2 illustrate the products.

Fig. 14.1 Smoked fish, milkfish, and crackers that have not been packaged. (*Source: Ade Angelia Marbun, BRIN Documentation*)

Fig. 14.2 Ready-to-sell products that have been packaged and granted permits from the Semarang City Government. (*Source: Ade Angelia Marbun, BRIN Documentation*)

Shrimp Paste

> Since I was born in Tambaklorok, the sea was far away, so I could look for firewood after school. Now, many villages have disappeared. If the sea wave season only relies on work from people's orders - if rely on the results of fishermen, the economy could die.

– Mrs Suratih, drying fish for shrimp paste with 13 children, Interview, 24 January 2024, Tambakrejo, Semarang City

Semarang City, a coastal and marine area, has a shrimp paste industry including Tambakrejo Village, Gayamsari District. This centre is managed by multiple generations of the same family. Their shrimp paste products are widely distributed in large markets in Semarang City, such as Johar Market, and in other areas with regular customers, especially in Central Java Province. Four women were interviewed on 25 January 2024 in Tambakrejo regarding their resilience in becoming fish dryers and daily workers in the home industry making shrimp paste. All four are relatives working to help their families due to their fishermen husbands facing increasing difficulty with catches (fish, shellfish, crabs) over the past five years. The lives of fishermen are also very dependent on the weather and seasons, meaning that there are frequently lean times with no income. Consequently, the women of these families must work odd jobs, becoming daily labourers in home industries making shrimp paste and drying fish and shrimp for around US$6–10/per day.

In the interviews, the women stated that they have no alternative jobs other than being daily labourers and are very dependent on the owners of shrimp paste-making businesses, as almost all of them have only an elementary school education. When the rainy season arrives, and there are no shrimp products, such as raw materials for shrimp paste, these women lose their jobs; they are then re-employed once the shrimp paste business starts producing again. Thus, they have no protection regarding job security and decent wages, and even when working they are exposed to sunlight for almost 10 hours a day. Protection for their health is through the Social Security Administering Body (BPJS), through which the government issues regular citizen contributions, and the women can seek treatment at an affordable community health centre if they are sick.

Mangrove Products

Musdalifah is a mangrove batik craftsman from Mangunharjo Village, Tugu District, Semarang City. The sea level has been rising here since 1998; in the past, the Mangunharjo area was prosperous because it relied on shrimp ponds, but rising sea levels have turned the area into a 2-kilometre-long swamp and major flooding occurred in 2023, which entered people's homes. As a result, residents took the initiative to begin planting mangroves, which not only reduce tidal flooding but provide the raw materials for making mangrove batik.

The batik artisans group in Mangunharjo initially comprised 20 people, but overtime this fell to ten, and now five female artisans are left. The product name is Srikandi Pantura, which is then transformed into 'Mas Bamat' and 'Mba Bamat' brands. Making mangrove batik requires perseverance as the mangrove fruit must be collected, cooked, and dried and then the dye extracted. The distinctive colours and patterns created using natural mangrove dyes have garnered significant interest and appreciation from the public, including students and visitors. This positive reception has motivated the group to innovate further and expand their craft. The motifs employ around 20 models and are widely liked by the wider community, including those from as far as Papua. The marketing model utilised by the batik group includes participating in exhibitions, coaching by the department, including

Fig. 14.3 Mangrove Batik Centre in Mangunharjo Village, Semarang City. (*Source: Laely Nurhidayah, BRIN Documentation*)

strategically building product image, identifying market segments, managing customer relations, and selling to visiting tourists and students. However, in recent times enthusiasts of mangrove batik have decreased and it is hoped that the government will help market the products of these coastal women. Figure 14.3 depicts the mangrove batik centre.

Making coffee from mangrove fruit is another form of resilience based on efforts to maintain the economic resources of residents, especially around Mangunharjo, Mangkang Wetan, and Mangkang Kulon villages, Semarang City. Some residents had difficulty migrating to new jobs after their land was affected by tidal waves and rising sea levels, and was sold to the private sector because the government was unwilling to buy the area for conservation. Residents began planting mangroves around 2000 to protect residential areas from tidal waves, and since 2010 mangrove fruit has been used as an ingredient in coffee. The process is complex; the fruit must be extracted, cleaned, cut, soaked, and roasted, which takes about seven days. The initial trial of making mangrove coffee in 2012 gave rise to a feeling of 'fly' and the process continued to be developed until the finished product was fit for consumption. To ensure safety, health, and hygiene standards, training was provided by the Semarang City Government, and in 2019 Indonesia Power (an electricity company), as part of its corporate social responsibility (CSR) program, assisted in the process of food and home industry licensing (PIRT), halal certification, design, and copyright. Laboratory tests to ensure coffee safety were carried out at the University of 17 August 1945, Semarang City. Coffee artisans adhere to the principle of limited mangrove coffee production to avoid excessive exploitation of mangroves.

Labour

Garment

> It is better to work in a factory, but textile factories are evacuated due to flooding – companies are moving because they cannot get orders.

– Interview on January 6, 2024, Tambakrejo Village, Semarang City

Wiwi is a coastal resident of Tambakrejo Village and a textile factory worker at Lucky Textiles Semarang. The factory is in the Tanjung Emas Port area, Semarang City, which is prone to tidal flooding. She has been a factory worker since 2005 after graduating from high school; most factory workers are women. Working hours are from 07.30 a.m. to 5.00 p.m. five days a week, with wages paid according to the provisions of the Semarang City Government. Working as a textile factory worker significantly supports Wiwi's family budget. Currently, however, the company is planning to relocate outside Semarang City because the factory location is prone to tidal flooding. For Wiwi, the implications of moving to the factory location are significant given that she has a family; moreover, even if she does moves to the new factory location, her wages will decrease.

Likewise, Mrs Ida was laid off from the Lucky Textile Semarang company after more than 11 years and, now approaching 40 years old, finds it difficult to compete with the new workforce or to move to another city. Due to the low income of her family, her husband works odd jobs and Mrs Ida opened a small shop selling coffee and iced drinks. She supplements this with piecework using her sewing skills, but work from small companies earns only IDR 60,000 a day, or is paid by order.

Green Mussel Peeler

Shellfish peeling is another form of work for women living in coastal areas, providing an additional source of family income and a buffer against economic vulnerability. Many fishermen's wives use their time peeling green mussels while their husbands are out at sea. This work is often done independently or with the help of their children, who help increase productivity. The women we interviewed shared that wages ranging from IDR 4000 to IDR 5000 per kilogram of peeled mussels, which can significantly contribute to meeting their families' daily needs.

This job is considered lighter and more flexible than formal employment, making it appealing to women, especially those who also need to manage household responsibilities or lack access to other job opportunities. With a work rhythm not bound by specific hours, they can balance their time between earning income and fulfilling domestic duties, such as caring for children and managing the household. Moreover, this work allows them to operate from home or within their communities, creating a more comfortable and supportive work environment. In addition to providing economic benefits, this activity also opens opportunities for local economic development. Many green mussels are processed into value-added products, such as ready-to-cook peeled mussels or other mussel-based food products, which are then sold in local or regional markets. Through their impactful yet straightforward skills, women in the coastal areas of Semarang have demonstrated their ability to actively contribute to improving their families' welfare while simultaneously strengthening the local economy.

Women and Conservation

Urban Farming

Community adaptation to the impact of climate change through urban farming represents an innovative solution to address environmental challenges. Urban agriculture helps enhance local food security and contributes positively to ecosystems by greening urban areas. By utilising small plots of land around residential areas, communities can grow various plants, including vegetables, fruits, and medicinal plants. The practice of urban farming has been implemented by the community in Citizens Association 15 Tambaklorok, Semarang. The head of Citizens Association (RW) 15, Slamet Rianto, stated that residents work collaboratively to create aesthetically pleasing and functional green spaces. A flower garden concept was designed to provide a recreational space for children and residents while beautifying the area. The available land is divided into 10 Neighbourhood Associations (RTs), with each RT responsible for managing a 7 × 15-metre plot of land. Each RT plants different crops, including strawberries, butterfly pea flowers, medicinal plants, spinach, and others. This division fosters diversity and uniqueness in the yields across different areas. Figure 14.4 depicts an urban farm.

The agricultural products are marketed in fresh form or processed into value-added products. The community also utilises available land for livestock farming. The Women Farmers Group Cantik Bahari manages urban farming activities in Citizens Association 15. Sri Wahyuni, the leader of Cantik Bahari, stated that a schedule has been developed to ensure the maintenance of urban farming practices by all group members. Residents routinely engage in communal work to care for the land and livestock, whether for cleanliness competitions, hosting guests, or ensuring the sustainability of agriculture and livestock farming. The results of urban farming are also utilised to support the Jumat Berkah ('Blessed Friday') Program, where the Women Farmers Group provides nutritious food and milk for children experiencing stunting. Sri Wahyuni noted that the number of stunted children, initially 39, has been successfully reduced to just nine. Figure 14.5 depicts the distribution of nutrition-rich food to the community.

The community, particularly the women involved in Women Farmers Group Cantik Bahari, is not only responsible for agricultural success but is also actively engaged in reducing stunting rates and improving the environment. Initially, this program was purely community-driven. However, its positive impact has drawn the attention of both the government and private sectors. This flagship program has earned the Citizens Association various national and local awards, including recognition as a Climate Village from the Ministry of Environment of Indonesia.

Fig. 14.4 Urban farming in the coastal area of Tambaklorok, Tanjung Mas Village, Semarang City, is managed by a women's community. (*Source: Agus Suntoro, BRIN Documentation*)

Fig. 14.5 Distribution of vegetables every Friday to stunted children from urban farming in Semarang City. (*Source: Ade Angelia Marbun, BRIN Documentation*)

Mangrove Community

Recognising the importance of restoring mangrove ecosystems, local initiatives have been spearheaded by community leaders such as Nurhayati and her husband Sururi, recipient of the Kalpataru Award from the Ministry of Environment, an award for his services in preserving the environment through mangrove restoration. Their efforts aimed to rehabilitate mangrove forests to restore their ecological functions and improve community livelihoods. The initiative was supported by various stakeholders, including NGOs, academics, government bodies, and private sector CSR programs. Although initially met with scepticism, continuous education and outreach programs successfully demonstrated the vital role of mangroves in protecting coastal areas and supporting climate change adaptation. Since 1998 more than one million mangroves have been planted by the community led by Mr Sururi and Mrs Nurhayati.

Academic contributions have also been integral to the success of the mangrove restoration efforts. Educational research has provided insights into the optimal soil and water conditions required for mangrove growth and effective planting techniques. This evidence-based approach has ensured that rehabilitation efforts are informed by scientific principles, enhancing their sustainability and long-term impact. The mangrove planting initiative in Semarang exemplifies how community-driven efforts, supported by multi-stakeholder collaboration, can address climate change adaptation while promoting local economic resilience. Integrating scientific knowledge, community participation, and institutional support has resulted in significant environmental and socio-economic benefits. However, ensuring the long-term sustainability of these efforts requires ongoing commitment, supportive policies, and adaptive management strategies. These initiatives highlight the potential of mangrove restoration as a model for holistic and sustainable climate adaptation strategies. Figure 14.6 depicts mangrove conservation.

Discussion

Challenges Faced by Micro, Small, and Medium Enterprises (MSMEs)

Business Licensing and Regulatory Compliance

Business licensing is a critical legal mechanism to ensure the legitimacy and accountability of economic activities, particularly within the micro, small, and medium enterprises (MSMEs) sector involved in fish-based food processing. In the Indonesian context, licensing requirements extend beyond formal legal compliance; they are designed to ensure that products adhere to government-mandated standards of quality and food safety. This is especially significant considering the high risk of

Fig. 14.6 Mangrove conservation in Mangunharjo, Semarang City, initiated by residents and women's communities. (*Source: Ade Angelia Marbun, BRIN Documentation*)

microbiological contamination in processed fish products when hygiene and handling practices are inadequate.

A substantial proportion of business actors in this sector are women entrepreneurs operating at the household level. As these women often assume dual responsibilities in both domestic and economic spheres, formalising their business activities is essential—not only for safeguarding public health but also for promoting the socio-economic empowerment of women.

The key types of business licenses relevant to women-led MSMEs in the fish-based food sector include:

- Business Identification Number (Nomor Induk Berusaha, NIB). This foundational document establishes the legal status of the enterprise and is issued through the Online Single Submission (OSS) system administered by the Ministry of Investment/BKPM. The NIB integrates other essential business registrations, including the Company Registration Certificate (TDP), Importer Identification Number (API), and access to labour-related social security programs.
- Home Industry Food Production Certificate (Sertifikat Produksi Pangan Industri Rumah Tangga, SPP-IRT). This certificate, issued by local health authorities, is required for women entrepreneurs producing food and beverage products at the household scale. It involves compliance with specific technical standards, including completion of food safety training, inspection of production facilities, and evaluation of product labelling.
- Distribution Permit from the National Agency for Drug and Food Control (BPOM). This permit is mandatory for processed fish products with longer shelf lives, wider distribution reach, or advanced packaging forms such as frozen items.

- Halal certification. Issued by the Indonesian Ulema Council (MUI) through the Halal Product Assurance Organising Agency (BPJPH), this certification is vital for women entrepreneurs targeting Muslim consumer markets, where compliance with halal standards is a growing concern.

In practice, the SPP-IRT remains the most obtained licence among women-led MSMEs in the food processing sector. Applications are submitted through the OSS system and must include a declaration of commitment, which outlines the entrepreneur's readiness to participate in food safety training, adopt Good Processed Food Production Practices (CPPB-IRT), and comply with documentation and labelling regulations. The SPP-IRT is valid for five years and renewable, subject to periodic supervision by local health officials. Nevertheless, many women entrepreneurs perceive the licensing process as complex and burdensome due to limited access to relevant information, technical skills, and financial resources.

This gap illustrates the disconnect between formal regulatory frameworks and grassroots business practices. While government regulations are intended to protect consumers from unsafe food products, many women-led enterprises continue to operate informally due to financial constraints and limited incentives to formalise. Field observations in Semarang revealed that many women engaged in fish-based food production continue to use household kitchens with inadequate separation between domestic and production areas, substandard sanitation, and limited awareness of the importance of production documentation and informative product labelling. These conditions create significant barriers during facility inspections and often result in application rejection.

According to the Semarang Health Department, inadequate production infrastructure is the primary cause of SPP-IRT rejections. Ensuring proper hygiene and sanitation remains a cornerstone of food safety assurance. Many women entrepreneurs lack access to dedicated production spaces, sufficient lighting and ventilation, or compliant storage systems for raw materials. However, acquiring the SPP-IRT holds strategic value, enhances consumer trust, expands product access to modern retail channels, and enables partnerships with public and private sector actors.

Licensing-related challenges also underscore structural disparities among women entrepreneurs. Those with higher levels of education and stronger social networks are generally more capable of meeting administrative and technical requirements. Consequently, there is a critical need for collaborative support mechanisms involving local governments, educational institutions, and civil society organisations. These efforts should move beyond punitive regulatory approaches and emphasise capacity-building, mentorship, and hands-on support to foster greater regulatory compliance and long-term sustainability among women-led MSMEs.

Marketing

Marketing is a crucial aspect that determines the sustainability and growth of MSMEs, particularly in Semarang's fish-based food processing sector. Marketing activities are not limited to product promotion efforts but also encompass the entire strategic process of building product image, identifying market segments, choosing distribution channels, setting appropriate prices, and managing long-term customer relationships. In fisheries-based MSMEs, marketing poses challenges due to limited resources, small-scale production, and restricted access to modern markets and technology.

The Semarang City Government, through the Fisheries Department, has implemented various initiatives to support the marketing capabilities of women entrepreneurs in MSMEs. One of these efforts is periodic outreach and mentoring, particularly targeting women MSME actors in coastal areas such as Kelurahan Mangunharjo, Tanjung Mas, and Tambakrejo. This outreach includes education on hygienic fish processing, application of household food quality standards, attractive and informative product packaging techniques, and training using social media and e-commerce platforms for product promotion.

In practice, many women entrepreneurs in Semarang have begun adapting to information and communication technology developments. They use social media such as Facebook, Instagram, and WhatsApp as one of the main channels to sell their products. These platforms are considered adequate due to their ease of use, low cost, and ability to reach consumers directly and personally. For instance, WhatsApp status is used to display the latest products to contacts who are loyal customers or part of the entrepreneur's social network. Meanwhile, Facebook and Instagram are utilised to reach a broader market, including consumers outside the city.

On the other hand, the use of e-commerce is also growing. Some women entrepreneurs have started selling their products through Tokopedia and Shopee. These digital channels enable fish-based processed products such as shrimp crackers, fish balls, presto fish, and mangrove coffee to reach regional and national markets without needing a physical store. This is particularly helpful for women-led MSMEs in coastal areas with limited physical market access.

Despite significant progress, women entrepreneurs in MSMEs face several serious marketing challenges. First, there is intense market competition from other MSMEs and large-scale industrial products. Many similar products are offered at lower prices and in more professional packaging, making it difficult for MSMEs to compete. Second, limited production capacity and human resources make it challenging to fulfil large orders or maintain consistent supply. Third, a lack of knowledge in digital marketing and product branding means that many women entrepreneurs have yet to optimise their product identity in the digital market.

External factors such as infrastructure development also influence marketing patterns. For example, the construction of toll roads passing through key areas in Semarang has closed several souvenir shops, previously serving as primary distribution channels for fish-based products. As a result, women MSME entrepreneurs have lost one of their important sales outlets. Moreover, some souvenir shops set selling prices too high, making the products unsellable and eventually returning them to the producers, creating financial losses for the women entrepreneurs.

The Fisheries Department of Semarang City, represented by Ms Dumas, stated that one of the most effective approaches to MSME development is strengthening partnership models with large-scale businesses. Large companies can act as mentors in such collaborations by providing training, expanding distribution access, and supporting product development and promotion. These partnerships not only extend the market reach of local products but also foster the creation of an inclusive and sustainable business ecosystem. For example, large restaurants or hotels can incorporate MSME products into their menus, while large cooperatives can serve as distribution intermediaries for MSME products in their regions.

Capital Access

Capital access is one of the fundamental aspects supporting the existence and development of MSMEs, particularly in the fish-based food processing sector, which requires consistent operational, processing, packaging, and distribution funding. In the city of Semarang, challenges related to capital accessibility and sufficiency remain key obstacles for women entrepreneurs in the MSME sector. The capital owned by most women entrepreneurs tends to be limited and is generally only sufficient to support daily business activities such as raw material procurement, product processing, and basic marketing.

These limitations hinder business expansion, including increasing production capacity, diversifying products, and developing broader marketing networks. Most women-led MSMEs operate using personal funds, family savings, or informal assistance that lacks long-term sustainability. Access to formal financial institutions, such as banks, remains relatively low, partly due to concerns about loan risks and the uncertain nature of business income.

Moreover, administrative factors represent a significant barrier. Many women entrepreneurs lack the necessary documentation to apply for business credit, such as Business Identification Numbers, financial statements, cash flow records, and systematic business plans. This administrative unpreparedness often disqualifies them from formal financing schemes such as the People's Business Credit (KUR) or other government loan programs.

Capital-related challenges also affect the low participation rate in product legalisation and certification. The process of obtaining certifications—such as Home Industry Food Production Certificate (SPP-IRT), halal certification, or distribution permits from the Food and Drug Monitoring Agency (BPOM)—incurs high administrative costs, technical training, and laboratory testing expenses, which are burdensome when self-funded by women entrepreneurs. Inadequate legal compliance restricts these products from entering modern markets, limits consumer reach, and negatively impacts consumer trust in the safety and quality of the products.

To address these issues, the Semarang City Government, through relevant agencies such as the Department of Fisheries and the Department of Cooperatives and

MSMEs, has made efforts to enhance the capacity of women-led MSMEs by offering financial management training, business planning assistance, and facilitation of equipment grants and access to financing sources. Another strategy is encouraging women entrepreneurs to join collective business groups such as Poklahsar (Processing and Marketing Groups) to strengthen the bargaining position of women MSME actors in accessing government aid programs and financial institutions.

Support also comes from the central government through the Ministry of Marine Affairs and Fisheries, notably via the Marine and Fisheries Business Capital Management Institution (LPMUKP). This institution provides low-interest financing and business mentoring, including business planning, funding proposal development, and financial reporting. This approach helps women-led MSMEs gain access to capital and improve managerial capacity and business competitiveness.

However, the effectiveness of these programs remains highly dependent on the internal readiness of the women entrepreneurs. Available capital schemes cannot be optimally utilised without regular financial recordkeeping, awareness of the importance of business legality, and willingness to participate in training and mentoring programs. Therefore, a more integrative approach is required—one that combines financial support with institutional capacity building, improved financial literacy, and continuous end-to-end mentoring.

In the long term, creating a sustainable and healthy entrepreneurial ecosystem requires policies that foster collaboration between women-led MSMEs and various stakeholders, including government, financial institutions, training providers, and large-scale industry players. Local governments can enhance their role as business facilitators and promote alternative community-based financing systems, such as cooperatives or revolving funds, which are more flexible and accessible. Thus, capital challenges can be addressed through technical solutions and by establishing a strong foundation for women-led MSMEs to grow and compete sustainably in an increasingly competitive economic environment.

Labour

Relocation

The Central Java Trade Service confirmed that in the Central Java province there are 671 large-scale textile companies—and 41 of these are located in Semarang City. The textile industry includes preparation and spinning, textile fibres and weaving, textile and apparel finishing, and manufacturing of non-clothing textile goods such as bed sheets, tablecloths, curtains, blankets, rugs, ropes, etc. (CJV, 2024). However, in recent years, there has been a trend of companies deciding to relocate their businesses from Semarang City. Relocation is a strategic decision that has a long-term impact, with the motivation of minimising costs to maximise profits. The two main factors in relocation are lower production costs and more significant market opportunities (Rahayu, 2023). However, for Semarang City, the relocation of companies to other areas is also motivated by the impact of tidal floods and tidal waves, which are detrimental to company operations (Sufan, 2023). Deputy Chairman of the Indonesian Chamber of Commerce and Industry (Kadin), Benny Soetrisno said that production activities stop when there is tidal flooding, as happened in May 2022, causing significant losses for factories in the northern area of Semarang City, especially the industrial centre at Tanjung Emas Port (Sandi, 2022).

Layoffs of female workers due to textile factories closing due to tidal flooding, rising sea levels, and declining export demand occurred in several industrial areas of Semarang City; examples include the Fuji Metec company in 2023 with 500 workers, the Sai Apparel Industries company in 2023 with 1482 workers, the Grand Best Indonesia company with 300 workers, and the Lucky Textile company with 100 workers and the relocation of factories to Demak Regency to reduce workers' wages (Syafali, 2025). Achmad Nur Hidayat, an economist and public policy expert with UPN Veteran Jakarta, emphasised that the position of women in the textile industry means that if there is a termination of employment (PHK) against them, the problems that arise will impact the family economy and create social issues (Ivani, 2024).

Female workers who face being laid off due to the relocation of the textile industry outside the city of Semarang face a problematic reality, in that relocating to keep their job can mean (a) leaving their families, namely husbands and children, to work in other cities; (b) income will decrease because the minimum wage in the largest city in Central Java Province is in Semarang City; (c) transportation or finding a new place to live incurs extra costs; and (d) many are over 35 years old, which is the average benchmark for new worker acceptance in Indonesian industry.

Guarantees of Workers' Rights

Work is a source of livelihood that serves to guarantee human life. Guarantees of the right to work have been regulated in various international instruments such as the Universal Declaration of Human Rights, especially Article 23, and the Covenant on Economic, Social and Cultural Rights in Articles 6, 7, and 8 and specifically in the Committee on Economic, Social and Cultural Rights' General Comment No. 18 on the Right to Work. The Committee states that work is an individual right that everyone has. At the same time, it is also a collective right. This right includes all forms of work, whether it is self-employment (self-employment) or paid work. The right to work should not be understood as the absolute and unconditional right to obtain work (Buhaerah, 2015), but labour rights include the right to freely chosen employment, fair and decent working conditions, wage protection, freedom of association, freedom from layoffs, and the right to strike (Mantouvalou, 2012). This respect for the individual and their dignity is reflected through the individual's freedom of choice in work while at the same time emphasising the importance of work for personal development and social and economic inclusion. The International Labour Organization (ILO) Convention No. 122 on Employment Policy (1964) states that 'regular, productive and freely chosen employment' links the State's obligation to create conditions favourable to the elimination of unemployment with the commitment to ensure the elimination of forced labour (UN Human Rights Commission, 2019).

In the context of workers' rights protection, it turns out that women in the coastal areas of Semarang City are very vulnerable to violations, both as formal workers and informal workers. For example, for formal women workers, especially those working in textile companies, the tendency is currently for 'piecework' meaning that their status is not that of permanent employees, and their labour rights are not protected. Second, they are vulnerable to layoffs, because these women are very dependent on company policies, and their work is primarily based on orders. Additionally, the disparity in employment guarantees between companies, in policies regarding menstrual leave, breastfeeding, and wages, differ from those for men in equivalent job positions (Pratami, 2018). For women workers in the informal sector, such as green mussel peelers, shrimp paste makers, fish dryers, and seafood processing workers, there is certainly no protection under employment guarantees. Almost all these jobs have no fixed wages, are below the city wage standard, offer no health and life insurance, and are not touched by various policies or interventions from the Manpower Office. In an interview, Danang Ardiyanto, from the Manpower Office of Central Java Province, emphasised that the wage standards for informal workers refer to the Job Creation Law No. 11 of 2020, which is based on an agreement between the employer and the worker. Thus, the relationship is not that of employer–employee but rather that of—highly unequal—partners.

Conservation

Despite the initiatives of the local community, especially women, in urban farming and mangrove restoration and conservation, there are many challenges to implementing this ground-level initiative of nature-based solutions. The most notable challenge is that the degree of participation of women in every Neighbourhood Units (RT) is different. Some RTs are more active in urban farming, others less so, which results in different results. Every RT has been allocated space for urban farming, but those with active women's groups have seen more diverse trees and vegetables planted and grown. In addition, active women's groups produce more goods and vegetables that can be sold or shared—for example, to help with stunting cases in their areas. Women's leadership is the key to practical urban farming in Tambaklorok. In addition, social capital plays an important role, including engagement, resource mobilisation, and access to information. Training for women's groups in urban farming is needed to enhance their participation.

In addition, in mangrove restoration, it is notable that women's participation is limited to specific groups of women. For example, Mrs Khadijah is actively conducting mangrove restoration as the wife of Mr Sururi, who received the Kalpataru award. She is thus familiar with the networks and funding available through CSR, which can assist with mangrove restoration. Despite her efforts, however, she feels that the local government does not support her activities in restoring mangroves, as the lack of central government funding makes it difficult for the local government to support all restoration projects in all areas of Central Java. The government needs to increase incentives for local communities to support mangrove restoration activities and enhance the communities through empowerment, training, and infrastructure support. Many local communities are keen to establish mangrove tourism but have not received any support from the government due to difficulties with land ownership, as discussed earlier.

Empowerment

One of the challenges to realising women's empowerment, including coastal women, is through the development planning deliberation mechanism or 'Musrenbang'. Women's Musrenbang is an essential forum for identifying women's needs and advancing their aspirations regarding development in their region, especially coastal areas. The legal basis for the Semarang City Government policy is Article 24(a) of Semarang Mayor Regulation No. 7 of 2003 concerning the Protection of Women and Children. An annual budget of IDR 40 million per subdistrict supports the commitment to implementing women's empowerment programs. Sugiarti, Head of Mangkang Kulon Village, which is a coastal area, gave examples of seven programs resulting from the 2025 Musrenbang, namely (a) training for women to increase resilience in climate-prone villages; (b) recycling used cooking oil; (c) sex education (preventing early marriage); (d) disaster response training; (e) making milkfish crackers; (f) maggot cultivation training; and (g) forming a Women's Farmer Group. Although the Women and Children's Service is given the authority to empower by regulation, its scope is limited to the process of formulating programs and implementing small start-up businesses. The main problem is the government's budget efficiency policy, which impacts the program's continuity. Therefore, the Women and Children's Office encourages the development of coastal women to be managed by the Cooperatives and Small and Medium Enterprises Office.

However, during an interview with the Semarang City Women and Children Empowerment Service in February 2025 it was noted that the issue of women's and children's empowerment is hampered by the prevalence of underage marriages. The rate of early marriages has risen sharply since the COVID-19 pandemic; in 2017, there were 57 cases; in 2018, this increased to 64 cases; in 2019, there were 109 early marriages recorded; and in 2020, it jumped sharply to 217 cases of early marriage in Semarang City (Azinar & Sholahuddin, 2022). The Mayor of Semarang has noted the importance of socialising the adverse impact of child marriage to ensure the community understands that child marriage creates a range of problems, including health risks such as cervical cancer and stunted children (Laeis, 2023).

Conclusion

Climate change has impacted coastal communities, especially fishermen and local communities. Women play a significant role in climate change adaptation in coastal areas through economic and ecological resilience, such as urban farming and mangrove conservation. Fishermen depend on their livelihood on fishing, which is increasingly impacted by climate change, and women thus need to find alternative sources of income in the informal sector to support their families. It can be seen from the case study of Semarang that most women work in the informal sector, such as in small-scale and medium enterprises or as labourers in garment factories. However, these sectors are also vulnerable to weather patterns: SMSE production is stopped if, for example, there are no raw materials for shrimp paste during the rainy season, while many factories near Tambaklorok are affected by climate-change-related flooding, and some are moving to safer locations. This has caused many women in these areas to lose these jobs.

There are difficulties in enhancing the market access of SMSEs in this area; for example, many cannot meet the requirement of food health certificates, such as PIRT and BPOM, which are essential to expanding their market. The local government needs to assist with micro-credit facilities, hygiene training, and awareness among SMSEs. The government has initiated training for local communities through the local Women Empowerment and Protection of Children agency; however, due to funding cuts from the central government, this training has been reduced from previous years.

Musrenbang Perempuan, the Development Planning Meeting for Women, is one of the best places to provide women with a voice for deciding the priority to assist them with diversifying their livelihood, sustainable management of natural resources, strengthening social networks, and utilising information technology. The Musrenbang can facilitate training to help women find alternative livelihoods—for example, teaching massage skills or other skills that can provide women with economic resilience. Women champions—for example, those who have successfully established their businesses—could help other women by asking them to join their supply chain or empowering them by sharing tips for building or running small-scale businesses.

The government and private sectors could help develop more appropriate mitigation and adaptation programs that reflect the characteristics of local communities. For example, through CSR, women could be trained for alternative livelihoods, assisted with market access, and given support for mangrove restoration and urban farming in this area.

References

Agency, Semarang City Regional Planning. (2019). Regional action plan for climate change adaptation in Semarang City. *Semarang*. https://ikupi.org/rencana-aksi-daerah-adaptasi-perubahan-iklim-rad-api-kota-semarang-2019/

Azinar, M., & Sholahuddin, I. (2022). Perception of early marriage in high school students in Gunungpati District, Semarang City. *Higeia Journal of Public Health Research and Development, 6*(2), 206–217.

Brian Walker, C. S., Holling, S. R., Carpenter, A. P., & Kinzig. (2004). Resilience, adaptability and transformability in social-ecological systems. *Ecology and Society, 9*(2), 5. https://doi.org/10.5751/ES-00650-090205

Buhaerah, P. (2015). *Measuring Ecosoc Rights*. Komnas HAM.

CJV. (2024). Distribution of the number of textile companies in Central Java. *Central Java Provincial Trade Office*. 2024. https://disperindag.jatengprov.go.id/v3/publik/berita_baca/MDIxZDE1MmY4MjBkZjEyYjg3MWQwOGYyZmUyMzQ2NDFlNmViYzNkNDY3NTljZjllZmU3YjM4NDc0ZTE0ZTMz

Hallegatte, S. (2014). Economic resilience: Definition and measurement. *Policy Research Working Paper*. https://documents1.worldbank.org/curated/en/350411468149663792/pdf/WPS6852.pdf

IDLO. (2022). Climate justice for women and girls: A rule of law approach to feminist. *Idlo*. Rome, Italy. https://www.idlo.int/publications/climate-justice-women-and-girls-rule-law-approach-feminist-climate-action

Ivani, O. (2024). Sritex declared bankrupt, economist: Mass layoffs in textile industry trigger social crisis. *Tempo.Co*. 2024. https://www.tempo.co/ekonomi/sritex-dinyatakan-pailit-ekonom-phk-massal-industri-tekstil-picu-krisis-sosial-1160250

Kaczan, D., Nurhabni, F., Cheung, W., Frölicher, T., Kuswardani, A., Lam, V., Muawanah, U., et al. (2022). *Hot water rising: The impact of climate change on Indonesia's fisheries and coastal communities*. Washington, D.C.. https://doi.org/10.1596/40564

Kesemat. (2024). Highlighting the negative impacts of climate change on women and children, KeSEMaT Attends Griffith University, BRIN, and UNDIP Research Collaboration. *Kesemat.or.Id*. 2024. https://kesemat.or.id/2024/05/27/soroti-dampak-buruk-perubahan-iklim-terhadap-perempuan-dan-anak-anak-kesemat-hadiri-kolaborasi-riset-griffith-university-brin-dan-undip/

Laeis, Z. (2023). Mayor of Semarang: Intensify socialization of the impact of child marriage. *Antaranews.Com*. 2023. https://www.antaranews.com/berita/3529695/wali-kota-semarang-intensifkan-sosialisasi-dampak-pernikahan-anak

Li, B., Li, X., & Liu, C. (2023). Dynamic Evolution of the Ecological Resilience and Response under the Context of Carbon Neutrality. *Ecosystem Health and Sustainability. 9*. https://doi.org/10.34133/ehs.0130

LIPI. (2017). *Data analysis and interpretation of social science research results*. Center for Basic Research Education and Training.

Mantouvalou, V. (2012). Are labour rights human rights? *European Labour Law Journal, 3*(2), 151–172. https://doi.org/10.1177/201395251200300204

Mayesa Alam, et al. (2015). *Women and climate change: Impact and agency in human rights, security, and economic development*. The Georgetown Institute for Women, Peace and Security. https://giwps.georgetown.edu/wp-content/uploads/2017/09/Women-and-Climate-Change.pdf

Moghim, S., & Garna, R. K. (2019). Countries' classification by environmental resilience. *Journal of Environmental Management, 230*(September 2018), 345–354. https://doi.org/10.1016/j.jenvman.2018.09.090

OHCHR. (2020). Climate change: Protecting women's rights. *Ohchr.Org*. 2020. https://www.ohchr.org/sites/default/files/Documents/Issues/ClimateChange/materials/2PGenderLight.pdf

Osman-Elasha, B. (2009). Women...In The Shadow of Climate Change. Https://Www.Un.Org/. 2009. https://www.un.org/en/chronicle/article/womenin-shadow-climate-change

PEMSEA. (2019). State of oceans and coasts 2018: Blue economy growth in the East Asian Seas Region. *PEMSEA*. https://pemsea.org/sites/default/files/2023-12/Regional_SOC_20190611.pdf

D. R. Pratami. 2018. Women's labor protection policy analysis: Study of fulfillment of women's work rights in the garment industry in Semarang Regency. *Journal of Politic and Government Studies* 8 (1): 1–11. file:///C:/Users/DELL/Downloads/22624-45803-1-SM.pdf

Rahayu, H. G. (2014). Semarang Mayor's Work Accountability Report 2014. *Semarang*. https://semarangkota.go.id/content/image/files/1. BAB I ATA 23-03-2015.pdf

Rahayu, Y. S.. (2023). Industrial Relocation 'Questioning Industrial Relocation to the Job Creation Law Polemic.' *LBH Semarang*. 2023. https://lbhsemarang.id/relokasi-industri-menyoal-relokasi-industri-hingga-polemik-uu-cipta-kerja/

Resurrección, B. P. (2013). Persistent women and environment linkages in climate change and sustainable development agendas. *Women's Studies International Forum, 40*, 33–43. 10.1016/j.wsif.2013.03.011.

Rusmadi, et al. (2017). Gendering the climate change policy: A study of gender analysis on Semarang's Integrated City Climate Strategy. *Advanced Science Letters, 23*(3), 2556–2558. https://doi.org/10.1166/asl.2017.8697

Sandi, F. (2022). Chaos! Semarang Floods Submerge Factories, Disrupt Textile Exports. *CNBC Indonesia*. 2022. https://www.cnbcindonesia.com/news/20220524150122-4-341514/kacau-banjir-semarang-rendam-pabrik-ganggu-ekspor-tekstil

Simmie, J. (2009). The economic resilience of regions: Towards an evolutionary approach. *Cambridge Journal of Regions, Economy and Society, 3*, 27–43. https://doi.org/10.1093/cjres/rsp029

Sufan, A. (2023). Tidal floods in Semarang cause many companies to move from Pantura. *Media Indonesia*. 2023. https://mediaindonesia.com/nusantara/577978/rob-di-semarang-sebabkan-banyak-perusahaan-pindah-dari-pantura

Syafali, M. (2025). Double track exploitation: Ecological damage and robbery of workers' rights in Pantura Semarang. *Semarang Legal Aid*. 2025. https://lbhsemarang.id/jalur-ganda-eksploitasi-kerusakan-ekologis-dan-perampokan-hak-buruh-di-pantura-semarang/

UN Human Rights Commission. (2019). General Comment 18 Right to Work. In *Komentar Umum Kovenan Hak Ekonomi, Sosial Dan Budaya*, 205–18. Komnas HAM.

UN Women. (2009). Women, gender equality and climate change. http://www.un.org/womenwatch/feature/climate_change/

Wigley, T. M. L. (2021). The link between climate change and sexual and reproductive health and rights. *Women Deliver*. https://womendeliver.org/wp-content/uploads/2021/02/Climate-Change-Report.pdf

Young, A., et al. (2024). Forced labour and climate change: Keeping a focus on women and children. *Griffith.Edu*. 2024. https://www.griffith.edu.au/research/disrupting-violence/forced-labour-and-climate-change-keeping-a-focus-on-women-and-children

Part III
Migration

Chapter 15
Shifting Sands: Immobility and Adaptation in Rising Seas

Inayah Hidayati and Jianqiang Liang

Introduction

Global sea level rise (SLR) presents a significant threat to coastal regions. Throughout the twentieth century, sea levels rose at an average rate of 1.7 mm per year; however, since 1993, this rate has escalated to approximately 3.2 mm per year (Schedel & Schedel, 2018). This trend is anticipated to persist, potentially rendering low-lying areas susceptible to more severe flooding and erosion, jeopardising community stability and undermining livelihoods. While migration is often regarded as a viable adaptive response to these challenges, many communities remain immobilised due to complex interactions between economic constraints, cultural entanglements and a lack of feasible alternatives (Adams, 2015; Benveniste et al., 2022). Understanding the dynamics of immobility and its implications for adaptation is therefore critical to the discourse on climate change. Immobility concerning sea level rise heightens vulnerability, especially in areas where economic deprivation, social connections and cultural identity significantly influence residents' decisions to stay (Farbotko & McMichael, 2019). Coastal ecosystems, which serve as vital sources of livelihoods, are increasingly degraded by salinisation, flooding and habitat loss, further straining the socioeconomic fabric of affected communities (Akib et al., 2012). For those unable or unwilling to migrate, adaptation becomes essential. Strategies such as constructing physical barriers, implementing community-driven disaster planning and integrating indigenous knowledge into

I. Hidayati (✉)
Directorate of Human Development, Demography, and Culture Policy, National Research and Innovation Agency, Jakarta, Indonesia
e-mail: inay001@brin.go.id

J. Liang
Griffith University, Brisbane, QLD, Australia

L. Nurhidayah et al. (eds.), *Climate Change, Labour and Migration in Indonesia*, Springer Climate, https://doi.org/10.1007/978-3-032-15334-0_15

resilience-building efforts present potential pathways for in situ adaptation (Ghani & Hidayati, 2024; Handayani et al., 2025; Hidayati et al., 2021; Seftiani et al., 2021). However, the effectiveness and sustainability of these strategies rely on the availability of resources, community engagement and a supportive policy framework (Bustamante, 2023).

Several studies have shown that the sustainability of coastal communities is at risk due to rising sea levels, which also affects the phenomenon of immobility. This is a result of the loss of investment opportunities, declining property values and financial losses, which limit the ability of communities to migrate under capital and resource constraints (Amin et al., 2019; McMichael et al., 2021; Nicholls & Tol, 2006; Wu et al., 2015). This trend is further supported by government assistance in the form of adaptive infrastructure development and inclusive adaptation policies; communities remain in high-risk areas due to the lack of reasonable relocation options (Völz et al., 2024). Alongside external factors, people's cultural attachment to their birthplace and reliance on traditional livelihoods are also significant internal influences. Traditional values, rituals and family ties reinforce the decision to stay, despite the increasing severity of environmental hazards (Benveniste et al., 2022; Ghani & Hidayati, 2024).

The interaction between sea level rises and the immobility of certain demographic segments presents a challenging problem. While migration is often viewed as a viable adaptive option to the impacts of climate change, many individuals and groups are either unable or unwilling to relocate (Ali et al., 2023). This inability to move can heighten vulnerability, as those in high-risk areas cannot adapt (Molinsky & Forsyth, 2023). Understanding the underlying reasons for the population's reluctance to relocate is crucial, followed by identifying adaptive strategies that enable people to remain safe in their current locations.

This chapter explores the interactions between sea level rise, immobility and adaptation, focusing on the northern coastal areas of Java Island, including Jakarta, Semarang, Demak and Pekalongan. These regions face significant challenges from tidal flooding and land subsidence, which exacerbate the vulnerability of communities unable to migrate. Jakarta and Semarang exhibit similar conditions in their responses to the impacts of climate change, while Demak and Pekalongan also indicate comparable conditions in that regard. By examining the factors that contribute to immobility and evaluating the feasibility of in situ adaptation strategies, this research aims to provide insights into how communities can enhance resilience in the face of sea level rise. Furthermore, the research emphasises the necessity for policymakers to address socioeconomic barriers and support local adaptation initiatives to ensure that vulnerable groups are not left behind in climate resilience planning.

Immobility and Adaptation: Trapped Population in Coastal Areas

Sea level rises has dramatically disrupted Indonesia's coastal ecosystems. Coastal flooding, particularly during high tides, exacerbates erosion and facilitates saltwater intrusion into freshwater aquifers, threatening critical habitats such as mangroves and coral reefs. These ecosystems play a vital role in supporting coastal biodiversity, providing protection against natural disasters, sustaining fisheries and bolstering local livelihoods (Garner et al., 2015; Marfai, 2014). The degradation of these ecosystems significantly diminishes ecological resilience, leaving coastal populations more vulnerable to climate-related risks (El-Malky et al., 2018).

Indonesia's coastal populations are among the most vulnerable to the effects of climate change due to frequent flooding, land subsidence and degradation of coastal ecosystems caused by SLR (Dwirahmadi et al., 2019; Marfai, 2014). Recent events, such as devastating floods in Java's northern coastal regions, particularly in Jakarta and Semarang, have brought attention to the pressing need for action (Aprijanto et al., 2025; Muzakki et al., 2022; Pramono, 2021). Coastal communities are more vulnerable to poverty due to the economic impact of tidal flooding and the recent COVID-19 pandemic (Seftiani et al., 2021). The socioeconomic consequences are most acutely felt by coastal residents who depend on agriculture and fisheries, two sectors that are increasingly susceptible to changing environmental conditions. These communities struggle to relocate because of cultural ties, financial limitations and a lack of viable relocation options. Immobility, defined as the unwillingness or inability to leave one's location, is crucial to understanding these dynamics. Voluntary immobility arises from personal preferences, such as deep cultural connections, attachment to one's home, proximity to work, or perceived benefits of remaining (Ofori et al., 2023). In contrast, involuntary immobility results from external constraints like economic hardship, limited opportunities or information, physical disabilities or health-related challenges (Ayeb-Karlsson et al., 2020).

The coexistence of voluntary and involuntary immobility is particularly evident among 'trapped populations' in Indonesia, especially in the face of rising sea levels. Significant barriers such as poverty and limited access to relocation resources prevent migration. These dynamics demonstrate how voluntary and involuntary immobility coexist in vulnerable communities. Environmental degradation exacerbates this issue, as dwindling natural resources further restrict mobility, often leaving individuals feeling trapped and helpless (Ayeb-Karlsson et al., 2020). This highlights the psychological aspect of immobility, which is frequently overlooked in traditional migration narratives (Ayeb-Karlsson et al., 2020). In Semarang, a city heavily affected by tidal flooding, strong attachment to place and family ties are key factors that lead residents to remain in disaster-prone areas. Social and employment networks, alongside economic dependence, further limit relocation, emphasising the complex interplay between environmental risks and social dynamics (Amin et al., 2021).

A comprehensive understanding of immobility in climate-affected regions, such as Indonesia's coastal areas, is essential for developing effective policies and interventions. While analysing the efficacy of adaptive strategies employed by populations in Jakarta, Semarang, Demak and Pekalongan, this study aims to identify the primary causes of immobility, including poverty, inadequate infrastructure and a lack of relocation assistance. Solutions must address the underlying causes of immobility, such as poverty, poor infrastructure and insufficient relocation support, while promoting resilience through ecological restoration and sustainable livelihoods. This multifaceted approach can help alleviate the constraints that trap populations, enabling them to adapt to the increasing threats posed by climate change.

Methodology

The research discussed in this chapter was conducted from 2024 to early 2025. The study employed a qualitative research design, which is well-suited to exploring the complex and nuanced experiences of residents and communities facing climate change challenges in Jakarta, Semarang, Demak and Pekalongan. The qualitative approach provides an in-depth understanding of the social, cultural and economic factors that influence community adaptation strategies. The framework for this research includes comprehensive face-to-face interviews, focus group discussions (FGDs) with community members and leaders, field observations and participatory action research. Each method has a distinct yet complementary role: FGDs encourage collective reflection and articulation of shared experiences, while individual interviews facilitate a deeper exploration of personal narratives, motivations and constraints. Participatory action research actively involves community members in identifying challenges and collaboratively creating adaptive solutions, which fosters local ownership and agency. Ethical considerations are integral to the research design, with strict adherence to the principles of informed consent, voluntary participation, confidentiality and the respectful treatment of all participants at various stages of data collection and engagement. This multifaceted approach ensures a thorough understanding of community dynamics and adaptation measures to address climate change.

Our research involved in-depth interviews with participants, including community leaders and members, to gather detailed narratives of their experiences, perceptions and responses to climate change. The community leaders interviewed in this study were adult men and women who played important roles in their communities, such as village heads, community group administrators, religious figures and individuals recognised by residents for their social influence, experience or contribution to communal welfare. These semi-structured interviews provided flexibility to explore emerging topics, focusing on understanding mobility barriers, perceived risks related to climate change and community adaptation strategies.

In addition to the interviews, we also facilitated FGDs to encourage community members to share insights and collaborate in identifying common challenges and solutions. Community members participating in the FGDs were adult men and women who had lived in the local community for at least the past year. The selection of participants emphasised inclusivity based on age, gender, occupation and socio-economic status to capture a broad spectrum of experiences and perspectives related to immobility and adaptation to sea level rise. These guided discussions ensured that all voices were heard and conversations remained focused on the research objectives. We conducted field observations to gain contextual insights into the community environment and physical manifestations of climate change. Our observations focused on infrastructure, housing conditions and people's interactions with their environment, providing a richer understanding of community members' lived experiences and their practical challenges. We also highlighted significant characteristics such as stilt houses, makeshift flood defences and signs of chronic flooding. Through participatory action research, we collaborated with residents to map vulnerable areas.

Findings and Discussion

Factors Contributing to Immobility

The phenomenon of immobility in the face of rising sea levels and climate change is not unique to Indonesia; it aligns with the global trend of 'trapped populations', particularly in vulnerable areas. This section examines how various interrelated factors contribute to immobility in four coastal regions of the north coast of Java, grouped as Jakarta–Semarang and Demak–Pekalongan, by emphasising economic, social and cultural dimensions.

Jakarta and Semarang

The Livelihood and Place Attachment

The coastal populations of Jakarta and Semarang face increasingly severe environmental challenges, including tidal flooding, land subsidence and rising sea levels. However, instead of opting to relocate to safer areas, many residents are choosing to stay. This decision is influenced by economic constraints and structural, social and cultural factors unique to urban areas like Jakarta and Semarang. As urban regions, the coastal areas of Jakarta and Semarang provide more livelihood opportunities, particularly in the informal sector. Many coastal residents in Jakarta work as labourers at the port, processing fishing catches, or as small traders who rely on local fishing and trading activities. The economy in this region is continually evolving, making it easier for residents to meet their daily needs than if they were to move to distant locations with similar job prospects. 'It's easy to find work here. Even if it floods, we can still work at the port or sell seafood,' one Muara Baru—Jakarta resident noted during an in-depth interview.

Coastal communities in the Semarang coastal areas, such as Tambaklorok, Bandarharjo, Kemijen and Gayamsari, remain reluctant to relocate despite frequent tidal flooding. For many individuals, a place of residence is not merely a physical dwelling but also a symbol of identity, social relationships and cultural ties that are difficult to replace. They view it as an integral part of their lives, the place where they grew up, nurturing social bonds with family and the community. This situation fosters a strong sense of belonging that cannot be replaced.

Many residents depend on the coastal area for their livelihoods, both in the fishing sector and other sectors, making the decision to move or stay even more complex. Dependence on the place where they live and earn money strengthens place attachment. The economic context and the proximity to the site that affect individuals' choices to remain must be considered while formulating adaptation strategies. This attachment makes the coastal areas of Jakarta and Semarang more appealing, despite their susceptibility to disasters.

Recognition and Political Processes

Jakarta's coastal population tends to adapt and remain stay due to access to various forms of recognition and assistance from the local government. Jakarta has social assistance programs for vulnerable individuals, such as the elderly, disabled and children—which include food subsidies, direct cash payments and infrastructure repair facilities—that are frequently provided based on official residence documents (Gubernur DKI Jakarta, 2022). For legal residents who possess such documents, remaining in coastal areas becomes a strategy to continue receiving these benefits. One of our research participants noted, 'We often receive assistance from the government, from necessities to house renovations during floods. If we move, we may not receive the same support.' Additionally, urban coastal areas like Jakarta frequently serve as grounds for practical politics, particularly during elections, with residents holding legal status often gaining advantages such as assistance prior to elections.

Similar conditions also exist in Semarang, where residents continue to inhabit land owned by a state-owned enterprise (BUMN). The land remains registered in the name of BUMN, meaning that it cannot be traded. This situation restricts residents, preventing them from selling their land to migrate elsewhere. Furthermore, the community believes that relocating would mean losing an integral part of their lives that cannot be replaced. The presence of corporate social responsibility (CSR) programs also significantly influences community decisions to remain in place. CSR initiatives from various companies and organisations have made notable contributions to enhancing the quality of life for Semarang's coastal communities by providing different forms of support (Warsilah, 2023). These include public facilities that promote urban farming development, which not only generates additional income but also strengthens food security in an area often subjected to tidal flooding. In the health and education sectors, the assistance provided alleviates the burden on economically challenged communities, particularly in delivering essential medical and educational services. Additionally, food security programmes, such as food seedling assistance and mangrove planting, offer ecological and economic benefits to the community. Mangrove planting also plays a crucial role in environmental protection by providing natural barriers against coastal erosion and mitigating the impacts of tidal flooding. This creates a reliance on local government and CSR, further entrenching the community's decisions to remain. The politicisation of aid further reinforces residents' inclination to stay, as they feel more economically and socially secure in their current location.

Government relocation efforts, supported by temporary infrastructure, often encounter resistance due to the distance of new residential locations from coastal workplaces. Additionally, the costs associated with relocation, including transportation and lost income during the transition, present a significant obstacle. 'The flats offered are decent, but too far from the port where I work. If I move, my income will decrease, so it's better to stay here despite the flooding,' said a dockworker, one of our research participants in Jakarta. Similarly, many young individuals in Semarang are employed as factory labourers, earning a limited regional minimum wage that is frequently insufficient to cover their basic needs. These limitations create a complex cycle of poverty. As environmental conditions worsen, the community's economic vulnerability intensifies. Their land and homes are often submerged by tidal floods, but due to limited options and resources they continue to remain in place. Moreover, the cost of relocating to a better area is well beyond their means. Land and property prices in other regions are higher and require savings they are unable to accumulate under challenging economic conditions. This exacerbates people's inability to access resources that could assist them in escaping poverty.

The decision to stay in coastal Jakarta and Semarang reflects a combination of economic necessity, reliance on government support and urban social dynamics. Although these areas remain exposed to significant environmental risks, residents prefer to adapt locally rather than confront the economic and social uncertainties that relocation might entail. This decision also illustrates the importance of understanding the economic, political and cultural context when designing more inclusive adaptation policies for urban coastal communities.

Family and Community Attachment

Most populations have resided in this part of coastal Jakarta and Semarang for generations, fostering deep-seated social ties that render relocation a complex emotional decision. Residents often convey that despite persistent flooding, they prefer to stay close to their family and friends, who offer essential mutual support during crises. For instance, during a particularly severe tidal flood in January 2020, many families in one of Jakarta's coastal villages banded together to construct temporary embankments from sandbags to safeguard their homes. One resident articulated, 'We know this is not a permanent solution, but we support one another because that's what community signifies to us.' This collective action exemplifies the community's resilience and underscores its reliance on temporary fixes rather than sustainable solutions. A local fisherman shared his view: 'I can't simply move. My boat and nets are here, and my friends are here. If I relocate, I lose everything.' This sentiment mirrors a broader trend among coastal residents of prioritising established social networks and cultural ties over the uncertainty of relocation.

The presence of family and community in coastal areas also plays an important role in influencing the decision to remain and the aspect of immobility. Strong kinship ties often support various facets of life, from sharing resources to offering assistance during disasters (Aldrich & Meyer, 2015). Moreover, traditions passed down through generations, such as participation in the fisheries sector, further deepen individuals' attachment to the coastal environment. The decision to remain is not solely a rational choice based on economic factors but also a form of respect for familial and community values. In this context, a profound connection to family and community is one of the key aspects that must be considered in formulating adaptation policies. Any approach that overlooks these social and emotional dimensions risks failing to grasp the complexity of the needs of coastal communities. Therefore, more inclusive adaptation policies must explore ways to bolster local social networks while providing adaptation options that do not undermine cultural and community ties.

Demak and Pekalongan

Place and Community Attachment

In Bedono Village, Demak, the residents are reluctant to leave their homes even though they have been submerged and partially separated from the land. This attitude reflects a strong attachment to the area. They regard their homes and land as a family legacy that must be preserved. Additionally, one of the submerged locations contains an ancestral grave that is sacred to the local community: Dukuh Tambaksari, Bedono Village, is the site of the tomb of Sheikh Abdullah Mudzakkir, also known as Mbah Mudzakkir, a great scholar who played an important role in the spread of Islam in Demak. The presence of this tomb is one of the reasons residents choose to remain in place to safeguard their land and the burial sites of their ancestors.

Comparable conditions exist in Pekalongan. One of the main reasons for not moving is residents' strong emotional attachment to their homes and communities. For them, this place is not just a physical location but also an integral part of their social identity and life. The phrase 'live and die here' illustrates the depth of their bond with their neighbourhood. A 50-year-old male fisherman on the Pekalongan coast told the research team, 'I have been here all my life, from birth until now. My family and neighbours are also here. We live side by side with the sea, and even though the tidal flooding worsens, I choose to stay here.' Despite the increasing severity of tidal flooding, they have built strong relationships with nature and adapted, such as through fishing activities, to survive. These attachments make them feel comfortable and reluctant to consider moving. Relating to a safer place is perceived as daunting and impractical, as it involves major changes in social and economic life and losing access to the social networks they have known for years.

The concept of *place attachment* refers to the emotional bond between an individual and their environment, which involves a deep connection to place that is physical but also social and symbolic (Gustafson, 2001). This bond can influence a person's decisions and behaviour, especially when faced with significant environmental challenges or changes. Place attachment often arises from personal experiences, collective memories and cultural values associated with a place, creating a strong sense of belonging and identity. For local populations, their houses are not merely places to live but also symbols of family heritage, tradition and a spiritual connection to their ancestral lands. Although the houses are now separated from the land and often surrounded by water, the residents' profound attachment to this environment illustrates the crucial role place plays in shaping their social and cultural identity.

In response to extreme environmental changes, the villagers continue to adapt to maintain their presence on the land. One measure they have implemented is the construction of an embankment to prevent seawater from inundating ancestral tombs. These tombs hold spiritual value and serve as the centre of the community's social and cultural activities. Additionally, houses in the village have been raised repeatedly to anticipate rising water levels, although this requires considerable financial resources and labour. The access road to the settlement, now surrounded by water, has also been creatively modified. Residents use floating bamboo as a makeshift bridge, demonstrating their capacity for innovation in the face of physical challenges. This kind of adaptation highlights the villagers' resilience and affirms their commitment to remaining and sustaining their lives in a place of profound emotional and cultural significance to them.

Social Capital

Demak is a significant study area, particularly in Sayung Sub-district, where residents face severe challenges due to tidal flooding. Rising sea levels and land subsidence, worsened by erosion, create increasingly critical environmental conditions, leading to the gradual submersion of the area. Many residents, especially fishermen and farmers, encounter substantial economic constraints that limit their ability to relocate to safer regions. With low incomes and dependence on local resources such as fish farming, numerous families cannot afford the costs associated with migration. The barriers preventing residents from moving are illustrated in the following interview excerpt with a resident of Timbusloko Village:

> Yes, that's right. My house is submerged when the sea rises, and it seems to be getting higher every year. But I have lived here all my life, from childhood until now. Where else would I move to? I am a fisherman; our source of livelihood is here. Moving is not easy. We need quite a bit of money, and in the new place, I would have to find another job, even though my expertise is solely in the sea. (PRT—fisherman, male, 42 years old)

Similarly, an interview with a coastal resident of Pekalongan highlights their profound attachment to their land and community. 'We have lived here for generations; this land belongs to our parents. Even though tidal floods occur every year, we remain here. We assist each other in repairing our houses. Relocation is not a solution for us; it is better to stay and find a way to live.' This statement illustrates residents' deep connection with their environment, where land represents not just a physical space but also a symbol of their heritage, identity and cultural sustainability.

Moreover, deep-rooted social and cultural ties to land and heritage compel individuals to remain in their homes despite the risks posed by climate change. These emotional attachments often outweigh the practical necessity to relocate, rendering them immobile (Chen & Lee, 2021). One of our research participants in Demak stated in an interview, '[T]his village has become part of our family history. All our stories, traditions and customs are here. If we move, all of that will be lost.' The decision to move or stay hinges on these cultural and social factors, resulting in immobility or an inability to relocate, even though they are aware of the practical need to leave vulnerable areas.

In response to worsening conditions, communities in Timbusloko–Demak have constructed basic infrastructures, such as bamboo walkways linking their frequently flooded homes. While this infrastructure facilitates access and mobility within the village, it is a temporary solution that does not sufficiently address the ongoing threat of flooding. Residents have also built houses on stilts to avoid rising water levels; however, these measures are again inadequate for long-term protection. The decision to remain and build temporary infrastructure rather than relocate implies that residents prioritise adaptation over displacement, influenced by emotional attachment and economic constraints that compel them to stay in this location. They not only preserve their physical home but also their psychological and cultural sense of belonging, leading them to choose to remain despite the existing infrastructure's insufficient protection. In other words, these circumstances establish a pattern of adaptation that is defensive and unsustainable, only capable of tackling temporary challenges, which fosters immobility due to a combination of economic and social factors, along with limited long-term adaptation options.

The reliance on social capital leads to a form of immobility in which the emotional, social and cultural costs of relocation surpass the perceived benefits of moving to safer areas. Residents' reluctance to move reflects a broader trend in communities facing environmental threats: the fear of losing not only possessions but also the intangible resources of social cohesion, cultural identity and a sense of belonging. Despite the increasing risk of flooding and environmental degradation, people's attachment to their land and social networks keeps them rooted in place even as their physical environment becomes increasingly inhospitable.

Immobility

Economic constraints stemming from low income are among the primary factors affecting individual decisions regarding immobility in areas impacted by tidal flooding due to climate change (Benveniste et al., 2022). Strong links to social, cultural and community attachment are also significant considerations in decision-making (Adams, 2015; Ayeb-Karlsson, 2020). It is important to emphasise that government policies can also contribute to immobility, with some overlooking environmental sustainability (Thiede, 2023; Thornton et al., 2023), while others promoting adaptation informed by local knowledge (Ayeb-Karlsson et al., 2018; Farbotko & McMichael, 2019).

According to the Risk Index for Climate Displacement (IOM, 2024), Jakarta, Semarang, Demak and Pekalongan are experiencing a slow-onset disaster manifested through land subsidence and tidal flooding, which have severe consequences due to the population's low adaptive capacity. Temporary adaptations, such as stilt houses and bamboo pathways, illustrate the limited resources available for permanent relocation (Benveniste et al., 2022; Heslin et al., 2019). Alongside economic factors, risk perceptions and fatalistic attitudes reinforce the decision to remain (McMichael et al., 2021; Chen & Lee, 2021). Community resilience, including support from local institutions to enhance livelihoods, further bolsters this choice and demonstrates that immobility stems from the interplay between local limitations and capacities (Thornton et al., 2023).

Adaptation Strategies Employed

In the face of climate change, governments and communities in Demak, Semarang, Pekalongan and Jakarta have implemented various adaptation strategies, including physical adaptation, livelihood adaptation and community-based initiatives. Physical adaptation, such as constructing sea walls, elevating houses and improving drainage systems, plays a crucial role in protecting infrastructure and communities from the direct impacts of flooding and sea level rise (Anfuso et al., 2021; Duvat et al., 2022; Dwirahmadi et al., 2019; Marfai, 2014; Piggott-McKellar et al., 2019; Takagi et al., 2017). Conversely, livelihood adaptation is undertaken through economic diversification, such as developing ecotourism or adopting sustainable agriculture and fisheries techniques, which help mitigate the economic vulnerability of coastal communities (Piggott-McKellar et al., 2019; Wutzler et al., 2022).

Furthermore, community-based initiatives promote collective action and local governance by encouraging active community participation in adaptation planning, integrating traditional knowledge and fostering social cohesion (Bronen & Chapin, 2013; Bronen, 2015; Palermo & Hernandez, 2020; Marfai, 2014). Notable successes have been observed in the construction of sea walls, drainage systems and community collaborations across various cities (Duvat et al., 2022; Hinkel & Bisaro,

2016). However, limited funding, unequal access to resources and political-economic instability continue to impede long-term environmental sustainability (Agustina, 2023; Ferris & Weerasinghe, 2020; Bustamante, 2023). Strengthening adaptive capacity is crucial for enhancing community participation and ensuring an equitable distribution of resources.

Evaluating the Adaptation Measures to Climate Change

In coastal regions such as Demak, Semarang, Pekalongan and Jakarta—which are particularly vulnerable to the effects of climate change—various metrics are used to evaluate adaptation strategies. These metrics include both short- and long-term sustainability, community resilience and overall quality of life. Infrastructure developments, including embankments and improved drainage systems, have provided short-term relief from flooding in Jakarta (Silver, 2014). However, environmental degradation and rapid urbanisation have hindered efforts to address the impacts of climate change. Long-term sustainable environmental action needs to be developed to support the coastal community. One such initiative is the Jakarta Smart City project, which integrates technology and urban planning for a more adaptive future (de Roo & Miller, 2017; Damar et al., 2020).

The government should encourage public participation in climate action to enhance resilience and quality of life. Village improvement programmes have made significant progress in fostering environmental awareness among residents, leading to a reduction in flood risks (Arrozaaq & Firmansyah, 2023; Simanjuntak et al., 2022). However, broader issues remain. Deficiencies in public health infrastructure and declining air quality continue to undermine living conditions, particularly as climate change exacerbates these pre-existing vulnerabilities (Syuhada et al., 2023).

The situation becomes more severe for residents in informal settlements as they face multiple stressors that reduce their capacity to adapt. In Muara Angke and Cilincing, Jakarta, marginal coastal communities frequently affected by tidal flooding require external support to survive and adapt (Dyah & Iskandaria, 2019). These groups endure the brunt of environmental injustices. Consequently, climate adaptation must prioritise equity, which entails ensuring access to sanitation, healthcare and safer housing. Active community participation in planning and execution will enhance outcomes and foster stronger, more resilient communities.

Policy Implications

To effectively support immobilised populations, policymakers should adopt a multidimensional approach. Recommendations include integrating local knowledge and preferences into broader climate adaptation plans through participatory planning mechanisms such as village assemblies (*musyawarah desa*), community mapping, or local adaptation forums to ensure that policies are culturally sensitive and contextually relevant (Gemenne & Blocher, 2017; Farbotko & McMichael, 2019). Policymakers should also prioritise creating financial mechanisms that provide resources for adaptation projects, empowering communities to implement and sustain their strategies (Benveniste et al., 2022) and supporting local adaptation initiatives undertaken by communities (Bustamante, 2023). This can take the form of grants for climate adaptation of rural communities through a competitive or needs-based mechanism. Additionally, encouraging collaboration between local governments, NGOs, community members and local businesses can enhance the effectiveness of adaptation initiatives and ensure that the voices of immobilised populations are heard in decision-making processes. Government should form multi-stakeholder climate adaptation task forces at the district or sub-district level, involving NGOs, the private sector, community representatives and technical experts, including university and research agencies.

Governments play a crucial role in facilitating migration for residents who wish or need to move. Relocation should be planned with the economic, social and cultural needs of residents in mind. For example, the location of replacement residences should be close to economic centres or workplaces, thereby minimising disruptions to livelihoods. The government can offer incentives in the form of financial support, training in new skills, or access to jobs in the new location. Such incentives can alleviate the economic and social burdens often faced by migrating communities. The relocation process should also consider the emotional and psychological aspects of the displaced population. The government can provide social and psychological assistance programs to help residents adapt to their new environment. This is important to ensure that the migration process does not trigger trauma or diminish their quality of life. Relocation will be more successful if accompanied by the strengthening of infrastructure, such as housing, transportation, education and health facilities in the new location. Additionally, building social networks in the new area can help residents feel welcome and adapt more swiftly.

In this context, government support must not disproportionately favour urban areas. Existing aid and adaptation programmes have often been concentrated in cities, while rural and remote communities, many of which are more vulnerable and have fewer resources, remain under-supported. The government must adopt a rural–urban equity lens in allocating adaptation resources, ensuring that assistance is equitably distributed regardless of whether a population resides in urban or rural areas. Prioritising inclusivity in climate resilience funding and services will help bridge existing development gaps and ensure no community is left behind.

Future research should prioritise longitudinal studies to track long-term adaptation outcomes among both immobile and mobile populations, focusing on differences across age groups. Recent studies can offer valuable insights into the effectiveness of various adaptation strategies over time and help identify best practices (Koubi et al., 2022). Exploring new technologies and adaptation strategies is essential, particularly under rapidly changing environmental conditions. Research should aim at innovative solutions that can enhance community resilience and reduce vulnerability to climate impacts (Farbotko, 2019). Lastly, understanding the subjective dimensions of immobility, such as the emotional and psychological factors influencing individuals' decisions to remain, can provide a more comprehensive understanding of the challenges faced by these communities (Bergmann, 2023).

Conclusion

Global sea level rise poses significant and complex challenges, particularly for coastal populations. These communities face the dual threats of environmental change and immobility. This study highlights the intricate interactions between economic, social and cultural factors that contribute to immobility. The study emphasises that relocation is not always feasible or desirable for many individuals in these communities. Instead, they have developed and implemented various adaptive strategies, including physical modifications and community-based initiatives, to cope with the effects of SLR while remaining in their original locations.

However, the effectiveness and sustainability of these in situ adaptations vary widely and they are often constrained by inadequate resources, insufficient support and external factors such as political and economic instability. This research underscores the importance of inclusive and equitable adaptation planning that recognises and integrates local knowledge, cultural sensitivities and the specific needs of immobile populations. Policymakers should adopt a comprehensive approach that provides necessary financial and technical support and encourages collaboration among all stakeholders to ensure that adaptation strategies remain effective and sustainable in the long term.

Future research should continue to explore the long-term effects of adaptation strategies among immobile populations, concentrating on innovative solutions that enhance resilience. Moreover, understanding the emotional and psychological aspects of immobility will be crucial in developing more holistic and effective responses to the challenges presented by climate change. By addressing these factors, we can better support the efforts of vulnerable coastal communities to adapt to a changing environment while protecting their livelihoods and cultural heritage.

References

Adams, H. (2015). Why populations persist: Mobility, place attachment, and climate change. *Population and Environment, 37*(4), 429–448. https://doi.org/10.1007/s11111-015-0246-3

Agustina, N. (2023). Rob flood as an impact of sea level rise around Kenjeran Beach tourism Surabaya. *IOP Conference Series: Earth and Environmental Science, 1273*(1), 012084. https://doi.org/10.1088/1755-1315/1273/1/012084

Akib, S., Jahangirzadeh, A., Kamali, B., & Mamat, N. (2012). Sea level rise and implication on coastal process: A review. https://doi.org/10.1115/omae2012-84205

Aldrich, D. P., & Meyer, M. A. (2015). Social capital and community resilience. *American Behavioral Scientist, 59*(2), 254–269. https://doi.org/10.1177/0002764214550299

Ali, A., Khan, M. Z., Khan, B., & Ali, G. (2023). Migration, remittances and climate resilience: Do financial literacy and disaster risk reduction orientation help to improve adaptive capacity in Pakistan? *GeoJournal, 88*(1), 595–611. https://doi.org/10.1007/s10708-022-10631-6

Amin, C., Sukamdi, S., & Rijanta, R. (2019). Modeling (Im) mobility: The decision to stay in disaster prone area amongst fishermen community in Semarang. In *E3S Web of Conferences* (Vol. 76, p. 03012). EDP Sciences. https://doi.org/10.1051/e3sconf/20197603012

Amin, C., Sukamdi, S., & Rijanta, R. (2021). Exploring migration hold factors in climate change hazard-prone area using grounded theory study: Evidence from coastal Semarang, Indonesia. *Sustainability, 13*(8), 4335. https://doi.org/10.3390/su13084335

Anfuso, G., Postacchini, M., Luccio, D., & Benassai, G. (2021). Coastal sensitivity/vulnerability characterization and adaptation strategies: A review. *Journal of Marine Science and Engineering, 9*(1), 72. https://doi.org/10.3390/jmse9010072

Aprijanto, P. T., Wibawa, B., Fithor, A., Santoso, M. A., Irfani, M., Cholishoh, E., Murtiaji, C., Apriyanto, D., & Sukmana, C. I. (2025). Innovative strategies for tidal flood protection: A systematic literature review on spatial management in coastal city (case study: Semarang City, Central Java, Indonesia). *Marine Systems & Ocean Technology, 20*(1), 9. https://doi.org/10.1007/s40868-024-00158-5

Arrozaaq, D. L. C., & Firmansyah, M. I. (2023). *Adaptive governance or Status Quo? Unpacking the City of Jakarta's Approach to the Air Pollution Control Program Adaptive Governance atau Status Quo?* Membongkar Pendekatan Kota Jakarta dalam Program Pengendalian Polusi Udara.

Ayeb-Karlsson, S. (2020). 'When we were children we had dreams, then we came to Dhaka to survive': Urban stories connecting loss of wellbeing, displacement and (im)mobility. *Climate and Development, 13*(4), 348–359. https://doi.org/10.1080/17565529.2020.1777078

Ayeb-Karlsson, S., Kniveton, D., & Cannon, T. (2020). Trapped in the prison of the mind: Notions of climate-induced (im)mobility decision-making and wellbeing from an urban informal settlement in Bangladesh. *Palgrave Communications, 6*(1). https://doi.org/10.1057/s41599-020-0443-2

Ayeb-Karlsson, S., Smith, C., & Kniveton, D. (2018). A discursive review of the textual use of 'trapped' in environmental migration studies: The conceptual birth and troubled teenage years of trapped populations. *Ambio, 47*(5), 557–573. https://doi.org/10.1007/s13280-017-1007-6

Benveniste, H., Oppenheimer, M., & Fleurbaey, M. (2022). Climate change increases resource-constrained international immobility. *Nature Climate Change, 12*(7), 634–641. https://doi.org/10.1038/s41558-022-01401-w

Bergmann, J. (2023). Introduction. In *Climate change and forced migration* (pp. 3–16). https://doi.org/10.1007/978-3-658-42298-1_1

Bronen, R. (2015). Climate-induced community relocations: Using integrated social-ecological assessments to foster adaptation and resilience. *Ecology and Society, 20*(3). https://doi.org/10.5751/es-07801-200336

Bronen, R., & Chapin, F. (2013). Adaptive governance and institutional strategies for climate-induced community relocations in Alaska. *Proceedings of the National Academy of Sciences, 110*(23), 9320–9325. https://doi.org/10.1073/pnas.1210508110

Bustamante, M. (2023). Ten new insights in climate science 2023. *Global Sustainability, 7*. https://doi.org/10.1017/sus.2023.25

Chen, T., & Lee, B. (2021). Income-based inequality in post-disaster migration is lower in high resilience areas: Evidence from U.S. internal migration. https://doi.org/10.31235/osf.io/yvfqz

Damar, A., Colijn, F., Hesse, K. J., Adrianto, L., Yonvitner, Fahrudin, A., et al. (2020). Phytoplankton biomass dynamics in tropical coastal waters of Jakarta Bay, Indonesia, in the period between 2001 and 2019. *Journal of Marine Science and Engineering, 8*(9), 674. https://doi.org/10.3390/jmse8090674

de Roo, G., & Miller, D. (2017). *Integrating city planning and environmental improvement: Practicable strategies for sustainable urban development*. Routledge. https://doi.org/10.4324/9781315252162

Duvat, V., Magnan, A., Goeldner-Gianella, L., Grancher, D., Costa, S., Maquaire, O., et al. (2022). Internal relocation as a relevant and feasible adaptation strategy in Rangiroa Atoll, French Polynesia. *Scientific Reports, 12*(1). https://doi.org/10.1038/s41598-022-18109-8

Dwirahmadi, F., Rutherford, S., Phung, D., & Chu, C. (2019). Understanding the operational concept of a flood-resilient urban community in Jakarta, Indonesia, from the perspectives of disaster risk reduction, climate change adaptation and development agencies. *International Journal of Environmental Research and Public Health, 16*(20), 3993. https://doi.org/10.3390/ijerph16203993

Dyah, A., & Iskandaria, H. (2019, November). Community empowerment towards green settlement in RT. 03/RW. 01 Petukangan Utara Village, Pesanggrahan Sub-district, South Jakarta. *In ICCD* (Vol. 2, No. 1, pp. 381–386). https://doi.org/10.33068/iccd.Vol2.Iss1.115

El-Malky, M., Al-Ansi, M., & Al-Marri, A. (2018). Assessing the coastal risks in Doha municipality spatial development plans through strategic environmental assessment (SEA). *Journal of Environmental Science, 42*(2), 1–31. https://doi.org/10.21608/jes.2018.21858

Farbotko, C. (2019). Climate change displacement: Towards ontological security. In *Rethinking displacement: Asia Pacific Perspectives* (pp. 251–267). https://doi.org/10.17875/gup2019-1219

Farbotko, C., & McMichael, C. (2019). Voluntary immobility and existential security in a changing climate in the Pacific. *Asia Pacific Viewpoint, 60*(2), 148–162. https://doi.org/10.1111/apv.12231

Ferris, E., & Weerasinghe, S. (2020). Promoting human security: Planned relocation as a protection tool in a time of climate change. *Journal on Migration and Human Security, 8*(2), 134–149. https://doi.org/10.1177/2331502420909305

Garner, K., Chang, M., Fulda, M., Berlin, J., Freed, R., Soo-Hoo, M., et al. (2015). Impacts of sea level rise and climate change on coastal plant species in the central California coast. *PeerJ, 3*, e958. https://doi.org/10.7717/peerj.958

Gemenne, F., & Blocher, J. (2017). How can migration serve adaptation to climate change? Challenges to fleshing out a policy ideal. *Geographical Journal, 183*(4), 336–347. https://doi.org/10.1111/geoj.12205

Ghani, M. W., & Hidayati, I. (2024, March). Adapting to sea level rise: Migration decisions and disaster mitigation strategies in small-scale fishing communities. In *IOP Conference Series: Earth and Environmental Science* (Vol. 1314, No. 1, p. 012013). IOP Publishing. https://doi.org/10.1088/1755-1315/1314/1/012013

Gubernur DKI Jakarta. (2022). Peraturan Gubernur DKI Jakarta Nomor 44 Tahun 2022 tentang Pemberian Bantuan Sosial dalam Rangka Pelindungan Sosial. Jakarta. https://jdih.jakarta.go.id/dokumen/detail/4946/peraturan-gubernur-nomor-44-tahun-2022-tentang-pemberian-bantuan-sosial-dalam-rangka-pelindungan-sosial

Gustafson, P. (2001). Roots and routes: Exploring the relationship between place attachment and mobility. *Environment and Behavior, 33*(5), 667–686. https://doi.org/10.1177/00139160121973188

Handayani, W., Tatius, C., Hillmann, F., Young, A., Esariti, L., & Nurhidayah, L. (2025). (Im) mobility and environmental change in the coastal sinking cities of Java, Indonesia. *Cosmopolitan Civil Societies: An Interdisciplinary Journal, 17*(1), 64–80.

Heslin, A., Deckard, N. D., Oakes, R., & Montero-Colbert, A. (2019). Displacement and resettlement: Understanding the role of climate change in contemporary migration. In *Loss and damage from climate change: Concepts, methods and policy options* (pp. 237–258). https://doi.org/10.1007/978-3-319-72026-5_10/

Hidayati, I., Putri, I. A. P., Ghani, M. W., Situmorang, A., & Widayatun. (2021, April). Small-scale fishing families and their daily multiple-stressor on climate change and COVID-19: Preliminary findings. In *IOP Conference Series: Earth and Environmental Science* (Vol. 739, No. 1, p. 012047). IOP Publishing. https://doi.org/10.1088/1755-1315/739/1/012047

Hinkel, J., & Bisaro, S. (2016). Methodological choices in solution-oriented adaptation research: A diagnostic framework. *Regional Environmental Change, 16*(1), 7–20. https://doi.org/10.1007/s10113-014-0682-0

IOM. (2024). Risk index for climate displacement. *Climate Mobility Innovation Lab Asia and the Pacific*. https://roasiapacific.iom.int/sites/g/files/tmzbdl671/files/documents/2024-12/ricd-ap-factsheet.pdf

Koubi, V., Schaffer, L., Spilker, G., & Böhmelt, T. (2022). Climate events and the role of adaptive capacity for (im-) mobility. *Population and Environment*, 1–26. https://doi.org/10.1007/s11111-021-00395-5

Marfai, M. A. (2014). Impact of sea level rise to coastal ecology: A case study on the northern part of Java Island, Indonesia. *Quaestiones Geographicae, 33*(1), 107–114. https://doi.org/10.2478/quageo-2014-0008

McMichael, C., Farbotko, C., Piggott-McKellar, A., Powell, T., & Kitara, M. (2021). Rising seas, immobilities, and translocality in small island states: Case studies from Fiji and Tuvalu. *Population and Environment, 43*, 82–107. https://doi.org/10.1007/s11111-021-00378-6

Molinsky, J., & Forsyth, A. (2023). Climate change, aging, and well-being: How residential setting matters. *Housing Policy Debate, 33*(5), 1029–1054. https://doi.org/10.1080/10511482.2022.2109711

Muzakki, A., Rizky, A. C., & Nafisah, A. S. (2022, December). Analisis banjir rob dan penanggulangannya di Kota Semarang. In *Indonesian Conference of Maritime* (Vol. 1, No. 1, pp. 13–25). http://proceedings.upi.edu/index.php/INCOMA/article/view/2691

Nicholls, R., & Tol, R. (2006). Impacts and responses to sea-level rise: A global analysis of the SRES scenarios over the twenty-first century. *Philosophical Transactions of the Royal Society a Mathematical Physical and Engineering Sciences, 364*(1841), 1073–1095. https://doi.org/10.1098/rsta.2006.1754

Ofori, D. O., Bandauko, E., Kutor, S. K., Odoi, A., Asare, A. B., Akyea, T., & Arku, G. (2023). A systematic review of international and internal climate-induced migration in Africa. *Sustainability, 15*(22), 16105. https://doi.org/10.3390/su152216105

Palermo, V., & Hernandez, Y. (2020). Group discussions on how to implement a participatory process in climate adaptation planning: A case study in Malaysia. *Ecological Economics, 177*, 106791. https://doi.org/10.1016/j.ecolecon.2020.106791

Piggott-McKellar, A. E., McNamara, K. E., Nunn, P. D., & Watson, J. E. (2019). What are the barriers to successful community-based climate change adaptation? A review of grey literature. *Local Environment, 24*(4), 374–390. https://doi.org/10.1080/13549839.2019.1580688

Pramono, I. B. (2021, October). Nature-based solutions for integrating flood and land subsidence: A case study in Jakarta and Semarang. In *IOP Conference Series: Earth and Environmental Science* (Vol. 874, No. 1, p. 012001). IOP Publishing. https://doi.org/10.1088/1755-1315/874/1/012001

Schedel, J. R., Jr., & Schedel, A. L. (2018). Analysis of variance of flood events on the US east coast: The impact of sea-level rise on flood event severity and frequency. *Journal of Coastal Research, 34*(1), 50–57. https://doi.org/10.2112/JCOASTRES-D-16-00205.1

Seftiani, S., Putri, I. A. P., Hidayati, I., Katherina, L. K., Ningrum, V., & Vibriyanti, D. (2021, July). Climate change and COVID-19: The double whammy for coastal community in Demak, Indonesia. In *IOP Conference Series: Earth and Environmental Science* (Vol. 824, No. 1, p. 012101). IOP Publishing. https://doi.org/10.1088/1755-1315/824/1/012101

Silver, C. (2014). Spatial planning for sustainable development: An action planning approach for Jakarta. *Jurnal Perencanaan Wilayah dan Kota, 25*(2), 115–125.

Simanjuntak, C., Gaiser, T., Ahrends, H. E., & Srivastava, A. K. (2022). Spatial and temporal patterns of agrometeorological indicators in maize producing provinces of South Africa. *Scientific Reports, 12*(1), 12072. https://doi.org/10.1038/s41598-022-15847-7

Syuhada, G., Akbar, A., Hardiawan, D., Pun, V., Darmawan, A., Heryati, S. H. A., et al. (2023). Impacts of air pollution on health and cost of illness in Jakarta, Indonesia. *International Journal of Environmental Research and Public Health, 20*(4), 2916. https://doi.org/10.3390/ijerph20042916

Takagi, H., Fujii, D., Esteban, M., & Yi, X. (2017). Effectiveness and limitation of coastal dykes in Jakarta: The need for prioritizing actions against land subsidence. *Sustainability, 9*(4), 619. https://doi.org/10.3390/su9040619

Thiede, B. C. (2023). *Climate variability and internal migration to urban and rural areas: Evidence from global microdata* (No. 55). KNOMAD working paper. https://knomad-ois.assyst.in/sites/default/files/publication-doc/knomad_paper_55_climate_migration_urban_brain_c_thiede_november_2023.pdf

Thornton, F., Serraglio, D. A., & Thornton, A. (2023). Trapped or staying put: Governing immobility in the context of climate change. *Frontiers in Climate, 5*, 1092264. https://doi.org/10.3389/fclim.2023.1092264

Völz, V., Hinkel, J., Kupfer, S., MacPherson, L. R., & Wulff Norrby, C. J. (2024). Learning about sea level rise uncertainty improves coastal adaptation decisions. *Earth's Future, 12*(10), e2024EF004704. https://doi.org/10.1029/2024EF004704

Warsilah, H. (2023). Building resilient city in coastal urban areas: Case study of community adaptation and response toward climate change and tidal floods in Semarang, Indonesia. In *Climate change, community response and resilience* (pp. 75–92). Elsevier. https://doi.org/10.1016/B978-0-443-18707-0.00004-7

Wu, T., Hou, X., & Chen, Q. (2015). Coastal economic vulnerability to sea level rise of Bohai Rim in China. *Natural Hazards, 80*(2), 1231–1241. https://doi.org/10.1007/s11069-015-2020-3

Wutzler, B., Hudson, P., & Thieken, A. H. (2022). Adaptation strategies of flood-damaged businesses in Germany. *Frontiers in Water, 4*, 932061. https://doi.org/10.3389/frwa.2022.932061

Chapter 16
Mobility in the Context of Climate Change: Addressing Relocation Challenges and Potential Human Rights Issues in Demak, from a Legal Perspective

Adis Nur Hayati, Aditya Rahmadhony, Inayah Hidayati, and Ana Borges Jelinic

Introduction

The issue of population mobility due to rising sea levels from climate change remains a significant topic of discussion. As is known, many studies show that climate change impacts or implies population mobility. Climate change can prompt migration by altering environmental conditions, such as sea level rise, shifts in weather patterns, and extreme weather events (Floranza & Asio, 2022). It makes certain areas of the earth uninhabitable, thereby encouraging new population movement patterns (McAdam & Ferris, 2015). For example, climate change causes rising sea levels and increases the frequency of tidal flooding, making coastal and lowland areas uninhabitable (Martin, 2017). This then forces people to leave their homes for safer locations, either on their own initiative or at the government's behest. It is projected that by 2100, approximately 88 million to 1.4 billion individuals worldwide will be compelled to relocate due to rising sea levels (Seeteram et al., 2023).

Planned relocation is a form of adaptation strategy in dealing with these mobility issues due to climate change. Referring to the 2010 Conference of the Parties (COP) to the UN Framework Convention on Climate Change (UNFCCC), planned relocation, displacement, and migration are forms of adaptation strategies in dealing with

A. N. Hayati (✉) · A. Rahmadhony
Research Center for Law, National Research and Innovation Agency (BRIN), Jakarta, Indonesia
e-mail: adis011@brin.go.id

I. Hidayati
Directorate of Human Development, Demography, and Culture Policy, National Research and Innovation Agency, Jakarta, Indonesia

A. B. Jelinic
Disrupting Violence Beacon, Griffith University, Brisbane, Australia

L. Nurhidayah et al. (eds.), *Climate Change, Labour and Migration in Indonesia*, Springer Climate, https://doi.org/10.1007/978-3-032-15334-0_16

population mobility issues due to climate change. When executed meticulously and thoroughly, planned relocation is deemed capable of yielding beneficial results by allowing a government to proactively mitigate potential risks associated with the displacement of individuals from their residences (Ferris & Weerasinghe, 2020; Garimella, 2022). However, relocation programs will create new human rights and security issues if their implementation is not planned thoroughly.

The Demak Regency area is one of the regions in Indonesia experiencing flooding problems due to rising sea levels and land subsidence (Prasetyo et al., 2019; Susanto et al., 2010). In low-lying coastal areas like Demak, rapid sea level rise may force entire communities to move, exacerbating current vulnerabilities and pressures (Maldonado, 2014). Flood events due to rising sea levels began to occur in Demak in the 1980s. There are at least four subdistricts badly affected by frequent tidal flooding: Sayung, Bonang, Karang Tengah, and Wedung (Kusuma, 2016). If not addressed urgently, it is predicted that tidal flooding due to sea level rise and land subsidence will submerge villages in the Demak coastal area in the next 10–20 years (Sushmita, 2021). As previously noted, constant tidal flooding has already forced people in the coastal area of Demak to move to more suitable locations.

In response to this problem, the Demak district government has conducted several relocation programs for a number of the village areas worst affected by tidal flooding. However, several problems arose in implementing these relocation programs, sparking concerns about potential violations of the human rights of relocated communities. Our investigations in Demak, especially in Bedono Village, revealed that several problems arose due to practical complexities and the lack of standards for implementing relocation in the field. This chapter explains how fieldwork validated those original observations. Community roundtables were conducted at which it was discovered that, due to the sub-optimal implementation of the relocation programs, most people have become reluctant or refuse to participate in future programs, indicating that relocation processes often negatively impact people's livelihoods, community relations, and land rights (McCaughey et al., 2018). Individuals also remain in disaster-prone regions due to their emotional attachment to the locale, familial ties, social networks, and employment opportunities (Amin et al., 2021).

This chapter aims to identify the issues in the relocation process that occurred in Demak and analyse the gaps in the Indonesian legal framework to guarantee people's safety and human rights. Our research reveals that the implementation of planned relocation in Demak faces significant challenges, including violations of human rights such as the right to legal certainty, adequate housing, and secure livelihoods for relocated communities. Considering Indonesia is an archipelagic country with more than 17,000 islands, and that two thirds of its territory is ocean (Setiawan, 2023), this research suggests that the Indonesian government must consider establishing comprehensive regulation or guidelines regarding standards for handling planned relocation, particularly in the context of migration due to climate change. To explain this argument, this chapter is divided into three parts. First, it explains the relationship between climate change, mobility, and human rights. Second, the chapter examines the problems and potential human rights issues related to the

Demak government's mobility or planned relocation program. Finally, it analyses the gaps in and way forward for Indonesia's current legal frameworks in addressing the issue of planned relocation.

Method

This study combined qualitative approaches employing field observations, in-depth interviews, a community roundtable (Goodson & Phillimore, 2012), and an analysis of statutory regulations. The results presented here are a subset of a larger research effort in the region. The research was conducted in the Demak Regency area over nearly 20 days in 2024. In the first stage, observations were conducted in areas of Demak that were heavily impacted by sea level rise, as well as in the relocation destinations for the affected community. They were followed by in-depth interviews with residents and community leaders affected by sea level rise and subsequently relocated by the government. These interviews aimed to explore their experiences, feelings, and responses to the government's planned relocation programs.

In addition to engaging with the community, in-depth interviews were held with government institutions, such as the Demak Department of Housing and Settlement, to gather further insights into these planned relocation programs. A community roundtable in Bedono Village was then conducted to gain a more comprehensive understanding of the community's perceptions of climate change and the planned relocation program in general (Goodson & Phillimore, 2012). Finally, in addition to collecting primary field data, legislative and regulatory materials were analysed.

The Relationship Between Climate Change, Mobility, and Human Rights

Climate change, population mobility, and human rights are interrelated with each other. As previously mentioned, several studies have shown that climate change can drive migration and overall population mobility by making the environment in certain areas unsuitable for habitation (McAdam & Ferris, 2015). In such cases, if the migration or relocation process is not planned and implemented carefully and comprehensively, it has the potential to result in human rights violations.

The complex interplay among climate change, population mobility, and human rights necessitates a comprehensive understanding of their interrelations. The scholarship highlighting the significance of incorporating human rights principles into migration discourse has been influential in policy proposals in this area (Bettini et al., 2017; Knox, 2016; Quirico, 2017). Climate change can significantly impact human rights as outlined in international law, leading to state intervention to uphold these rights and prevent their violation (Knox, 2016). Furthermore, acknowledging climate change events as direct human rights violations highlights the necessity of taking a human rights-based approach to addressing climate change (Mukherjee & Mustafa, 2019).

Many population mobility patterns have been identified in the context of climate change, including migration, displacement, and planned relocations. The UNFCCC has officially acknowledged these population mobility strategies as effective adaptation measures in the context of global climate action (McAdam & Ferris, 2015). However, this chapter will only focus on planned relocation as one of the adaptation strategies implemented in Indonesia. Planned relocation—also referred to as climate-induced resettlement, community relocation, or climate-induced relocation—involves the organised transfer of people and assets from areas that have become uninhabitable or are at risk due to the impacts of climate change (McMichael et al., 2019; Piggott-McKellar et al., 2019). This type of mobility is seen as a response to climate-related shocks and gradual changes (Piggott-McKellar & Vella, 2023). More specifically, according to the *Guidance on Protecting People from Disasters and Environmental Change through Planned Relocation* published by the United Nations High Commissioner for Refugees (UNHCR) in collaboration with the Brookings Institution and Georgetown University, planned relocation is defined as:

> A planned process in which persons or groups of persons move or are assisted to move away from their homes or places of temporary residence, are settled in a new location, and provided with the conditions for rebuilding their lives. Planned Relocation is carried out under the authority of the State, takes place within national borders, and is undertaken to protect people from risks and impacts related to disasters and environmental change, including the effects of climate change. Such Planned Relocation may be carried out at the individual, household, and/or community level. (Brookings Institution et al., 2015, p. 5)

The complex issue of climate-induced displacement and relocation is closely intertwined with justice and human rights matters. Scholars have emphasised the significance of addressing these concerns within the context of climate mobility (Farbotko et al., 2022). The European Court of Human Rights has established a direct connection between climate change and human rights, elucidating governments' obligations to uphold human rights standards amid climate challenges (Mladenov et al., 2022). Moreover, academic research has explored the interconnection between human rights and legal responses to climate change, highlighting the intertwined nature of these domains (McDave & Dagadu, 2023).

The process of planned relocation in response to climate change should prioritise minimising the exposure of affected communities to the adverse impacts of climate change. It is essential to ensure that these communities can rebuild their livelihoods successfully in new locations (Piggott-McKellar et al., 2019). In formulating strategies for climate-induced relocation, addressing multiple dimensions of social justice, encompassing distributive, procedural, and restorative justice, is essential for empowering relocated and host communities' agency and ensuring the dignity, and security of those migrating (Peters & Van Den Hoek, 2021). Furthermore, effective governance and institutional strategies are essential for successfully managing climate-induced community relocations, highlighting the significance of expertise and contributions in this domain (Bronen & Chapin, 2013).

The Problems and Potential Human Rights Issues of the Demak Government's Planned Relocation Program

The Demak Regency government has conducted several relocation programs in response to climate-change-related tidal floods and the looming threat of rising sea levels and land subsidence. There have been relocation initiatives for various hamlets within Bedono Village; according to one of the participants in this research, these include the transfer of Tambaksari Hamlet to Purwosari Hamlet in 1998, the relocation of Rejosari Hamlet to Gemulak and Sidogema hamlets in 2006, and the movement of Mondoliko Hamlet to Dombo Hamlet in 2022. This article examines the specific relocation processes in Rejosari Hamlet and Mondoliko Hamlet between 2006 and 2022.

The implementation of the relocation programs carried out by the Demak government shows a good effort by the government to help resolve the problems of communities affected by tidal flooding. However, this endeavour has sparked concerns regarding protecting human rights and the practical complexities of its execution. Our observations and interviews in the field found several problems or challenges in implementing the government's relocation program due to the lack of standards and regulations. These problems include the absence of adequate relocation land for settlements, inadequate housing infrastructure, resistance from those in the destination area, burdensome relocation costs, and issues concerning finding paid work. These situations give rise to potential human rights violations, such as of the right to guaranteed legal certainty, adequate housing and a good, healthy living environment, and the right to a decent livelihood for relocated communities.

Issues in the Process of Relocating Rejosari Hamlet to Gemulak Village

The relocation of Rejosari Hamlet to Gemulak Hamlet and Sidogema Hamlet occurred in 2006. As Mbaye and Okara (2023) explain, climate change and natural disasters reduce people's living space, causing migration and mass movement of people from their original places of residence to other places considered better. A comparable situation occurred in Rejosari Hamlet, where residents were compelled to opt for relocation due to the deteriorating state of their homes and land, and diminishing accessibility, all adversely impacted by escalating tidal flooding. One of the research participants said that

> we were forced to move because the previous place had been continuously affected by [erosion] (and tidal flooding) since the 1990s. If the road to our hamlet was good and passable, if we could buy mountain or landfill land, maybe the residents could still raise their houses. However, because the road was small, trucks could not pass at that time.

The government did not direct the implementation of the Rejosari Hamlet relocation program itself at that time; instead, the initiative came from the residents. The process was reportedly arduous; the residents persistently submitted proposals for relocation, yet the government exhibited no sense or urgency in resolving the community's issues. The community ultimately protested at the Sayung Demak District Office in an attempt to highlight their requirements, and the government subsequently acquiesced, according to a secretary of Gemulak Village.

The government's response to the problems of communities affected by tidal flooding in Rejosari Hamlet was realised in 2006. Most Rejosari Hamlet residents were relocated to the river border region in Gemulak Village. The Demak government allocated a budget of IDR 5 million per house for 98 families, each given a land area of 6×9 m^2. In addition to relocation, the government offered affected communities the opportunity to migrate to Sumatra and Kalimantan. However, most people wanted to move to areas close to their source of livelihood. This ultimately prompted certain residents who had relocated to Kalimantan to return to Java. A participant from Gemulak Village told researchers, '[t]he residents believe it is challenging to attain a superior quality of life akin to that in Java despite being assigned roles as fishermen in Kalimantan based on their skills'.

Our interviews and observations in the Gemulak Village region, the designated site for the Rejosari Hamlet relocation, revealed at least three issues encountered by the community during the relocation program: (1) the absence of adequate relocation land for settlements, (2) inadequate infrastructure for the relocation of houses, and (3) resistance towards the relocated community from the original residents of the relocation site. These are explored further below.

Lack of adequate relocation land for settlements. Our research found that the relocation area occupied by Rejosari Hamlet is vulnerable to being hit by future disasters. Furthermore, besides being disaster-prone, the legality or ownership status of the land occupied by the relocated community also remains uncertain. As mentioned previously, the people of Rejosari Hamlet were relocated to the river border (irrigation) area in the Gemulak Village area. The river border area itself is an area at high risk of flooding and tidal inundation, especially during the rainy season. Since 2010, Gemulak Village has been experiencing flooding and erosion, and the conditions worsened between 2016 and 2017. The former Rejosari Hamlet inhabitants who relocated to evade flooding and erosion have been impacted again, according to a participant in our research. According to Indonesian Disaster Data, hydrometeorological disasters influenced by weather and surface flows dominate disaster events in Indonesia, namely 99.35% of all geological disasters (Rosyida et al., 2024). Figure 16.1 below shows a view of the rear of the relocation houses in Gemulak Village, illustrating the problem.

Furthermore, in addition to being disaster-prone, the legal ownership or status of the land occupied by the relocated community also remains uncertain. Our research

Fig. 16.1 Rear of relocation houses in Gemulak Village. (Source: Authors' documentation, 2024)

found that the relocated communities do not have legal property rights in the form of ownership title over the land occupied. The land to which the community was relocated is a river border (irrigation) area with a residence permit granted by the Public Works, Water Resources and Spatial Planning Department of Central Java Province. The relocated community was not granted clear land ownership status, only given a permit to occupy the land for a certain period (specifically, until 2016). According to one participant, '[a]t this moment, the permit to occupy the land has expired, yet the community is unable to extend it and lacks the knowledge on how to do so'. One of the sources we interviewed said that every year people pay land tax on the location of the house they currently occupy. However, because the occupancy permit has expired, people cannot pay the tax anymore. This leaves the relocated community feeling as though they are living without legal certainty. The absence of clear legality means that relocated communities can be evicted at any time. According to Article 26 of the Regulation of the Minister of Public Works and Public Housing No. 28/PRT/M/2015 concerning the Implementation of River Boundary Lines and Lake Boundary Lines, this situation arises because:

> Buildings within river borders and lake borders which were erected based on permits obtained based on correct procedures are declared as status quo and are gradually being put to restore the function of river borders and lake borders.

Notably, aside from land rights issues, the process for issuing occupancy permits to relocated communities frequently reveals numerous discrepancies. Some of these include (1) the recurrent transaction of occupancy permits by the relocated community to third parties, leading to a deviation from the intended implementation, and (2) the occupancy permits issued by the government are exclusively for semi-permanent residences, yet many individuals have constructed permanent dwellings in this region.

Inadequate housing infrastructure. As previously explained, the relocated community received cash assistance of IDR 5 million per household and land

Fig. 16.2 A typical relocation house in Gemulak Village. (Source: Authors' documentation, 2024)

measuring 6 × 9 m^2 to construct semi-permanent houses, as described by one participant who is also a government worker. This arrangement is certainly less than ideal for families with many residents or family members. Semi-permanent housing is vulnerable to floods and theft, and is thus unsuitable for long-term living. In response to environmental changes, people are compelled to elevate their residences, permanently altering the house's structure. Figure 16.2 depicts relocation houses in Gemulak Village.

Tensions between original residents and relocated individuals. Highlighting the resistance of established communities to accepting displaced populations, one of our research participants reported that there was resistance or rejection in the early days of the community being relocated to Gemulak Village. Even now, this difference in attitude is still felt by the head of the community (*Ketua RT*) which was relocated:

> Every time I apply for assistance in this village, it gets rejected. For example, many alleyways get flooded and submerged. Those alleyways are completely underwater during high tides. However, thankfully, I pushed the residents to pool money to buy soil for this one alley, which is now passable through our self-help efforts. However, I asked for assistance for just one alley … and they responded that it was impossible because that is irrigation land owned by the irrigation department. The village budget cannot be used for that. However, it turns out that the village budget could be used to raise the alleyways in Sidogemah Village [another village with similar conditions]. We are like the marginalised in this village…

In this case, the resistance of original residents in relocation destination villages towards newcomers can result in social inequality, discrepancies in treatment, and issues related to governance and decision-making (Krishnadas, 2007). In Demak, social disparities are evident as relocated individuals often receive unequal treatment or services from village authorities compared to original residents. These multifaceted factors underscore the need for more equitable and inclusive relocation planning and implementation approaches to effectively manage climate-induced relocations.

Issues in the Process of Relocating Mondoliko Hamlet to Dombo Village

The relocation of Mondoliko Hamlet to Dombo Village took place in 2022. Like Rejosari Hamlet, the implementation of the Mondoliko Hamlet relocation program at that time resulted from an initiative or request from the community. In contrast to the relocation process in Rejosari Hamlet, this relocation required that the community independently acquire and possess land for their resettlement. The process of requesting the relocation of Bedono Village began with an audience with the leadership of the Regional People's Representative Council (DPRD) of Demak Regency in 2022 (DPRD, 2021). The Regency government responded to the results of the public hearing by launching a new housing program for residents affected by flooding or disasters, namely the *tuku lemah oleh omah* ('buy land, get a house') program, through changes to the 2022 Regional Revenue and Expenditure Budget (APBD). The Demak Regency government provides housing assistance worth IDR 50 million per unit through this program. However, it should be noted that this assistance is entirely in the form of materials and excludes construction services and land purchases (Demak Regency, 2023). As stated by a participant from the Demak Department of Housing and Settlement:

> We assisted in the form of IDR 50 million needed to purchase house materials. From the regional budget (APBD) to the community, it is in the form of money. However, the money is directly used to purchase house materials. So, they do not handle the money themselves. It goes into their bank accounts, and the funds are directly transferred from those accounts. The purchases are made based on the material needs and are directly sent to the building supply stores.

Based on the interviews and observations we conducted in the Dombo Village area, the relocation destination, there are at least two main issues faced by the relocated community: (1) the costly relocation process and (2) the relocated residents losing or changing jobs. These are explored further below.

Cost of relocation. As previously mentioned, under the *tuku lemah oleh omah* program the government assisted with house construction materials worth IDR 50 million per unit. In this case, residents felt that the government assistance was insufficient to build new houses. Additional expenses remain—including remuneration for construction labour, acquiring missing building materials, and procuring land for the new housing—and must be met by the community. Labour, in particular, constitutes a significant supplementary expense as people cannot build their own houses and need professional workers. These labour costs are unavoidable and are often one of the largest costs when building a house. Apart from labour costs, while the material assistance provided by the government meets most basic needs, there are often deficiencies in the type and quantity of materials provided. Communities must therefore purchase materials not covered by aid, such as additional sand, bricks, cement, and wood, among other materials. The ever-increasing price of building materials is another challenge for residents in managing their budgets, an issue raised by a community leader during interviews.

Apart from labour and material costs, another cost that must be considered is the cost of purchasing a plot of land. Our research revealed that the head of Bedono Village has initiated a program to facilitate the granting of land valued at IDR 25 million to residents under a loan scheme with lifetime instalment payments. This assistance initiative in the form of 'social plots' was carried out by the village head using personal funds, to help affected people to a safer place and to build a new life. The lending process was designed to provide access to residents who did not have the financial capacity to purchase land on their own.

> The district government provides material assistance in the form of house materials, but residents must have the requirements to own land. Because they could not afford land, I helped them prepare the land. With IDR25 million per plot. So, it is a social plot. (Head of Bedono Village, personal communication, January 2024)

However, while this mechanism provides land ownership assistance, filling the gap left by the government program, the relocated community still bears the burden of paying off the land purchase loan.

Loss or change of work. As noted by McDermott and Gibbons (2016), many relocated individuals are expected to transition to new vocations, such as fishpond farming or other non-coastal-related occupations. However, most lack the requisite skills and training, leading to substantial challenges during this transitional phase. In the village of Dombo, we found that residents forced to move from coastal regions to inland areas often need help to adapt to new non-fishing jobs. Despite the challenges, the coastal residents are resilient and strive for a better future. Below is an excerpt from an interview with the wife of a fisherman who relocated to a non-fishing area:

> Yeah, I typically just go along with my husband. He is a fisherman and sets off early in the morning, returning in the afternoon. We have to factor in extra expenses for gasoline to reach the coast, so we rely on a motorbike to reach the boat. It can be challenging for me sometimes, especially if we only catch a small fish. I used to assist at sea, but now I mainly stay home. I have not found a steady job yet; I aspire to open a small shop but lack the funds. For now, the most I can do is odd jobs, such as assembling food boxes, but the pay is quite low. (Participant in Dombo Village)

The interview excerpt underscores a multifaceted issue that goes beyond economic hardship and job insecurity, extending to the emotional impact of transitioning from a fishing-based livelihood to a new area with limited employment prospects. The relocation to a non-fishing area has resulted in increased financial burdens, including increased travel costs, while also intensifying the emotional and social difficulties associated with the transition. The wife's struggle to contribute to fishing or secure stable employment may lead to frustration and helplessness. This situation underscores the need for comprehensive support programs that address relocated individuals' financial and emotional needs. This could involve customised training for alternative employment and financial aid to overcome initial obstacles.

Significant challenges arise from the insufficient training and support provided during the transition. A notable number of residents have expressed concerns about the lack of tailored training programs that align with the job market in their new area. Consequently, even though they invest considerable effort and resources in continuing with fishing or similar activities, some residents opt to stay in these roles due to the absence of viable alternatives. This underscores the pressing need for enhanced training and support initiatives to facilitate a smoother transition for these individuals.

The current situation presents a challenging scenario, with many individuals needing greater clarity and preparation to start a new life chapter. For instance, a 50-year-old fisherman who has dedicated most of his life to the sea may now be required to transition into working in a factory or as a farmer in a new area, as noted by one participant who was also a former community head of Dombo Village. While this transition was necessary, the individual often lacked the necessary skills and support for this job adjustment, thus creating significant stress and apprehension about the future. Despite the impending dangers brought on by climate change, including increased sea levels and the devastation of coastal ecosystems, economic insecurity and a shortage of new skills often persuade these individuals to stay in their familiar environment. This situation emphasises the complex nature of voluntary relocation amid rapid environmental transformations. It also underscores the need for improved training programs and more robust support for affected coastal residents. Government involvement in this process is crucial, as it ensures a smoother transition and enhances their readiness for life in the relocation area.

The process of relocating coastal residents affected by climate change presents significant challenges. Our research identified that many coastal inhabitants, whose primary source of income is fishing, encounter difficulties when transitioning to new employment post-relocation. Despite these challenges, the residents of Dombo have shown remarkable resilience. They strive to adapt to non-fishing occupations when compelled to leave coastal areas for inland locations. They face numerous obstacles, including inadequate training and support during the transition. Such training as is provided frequently does not align with the residents' current skills or meet the job market demands in the new area. However, their determination is evident as many individuals continue to work in fishing-related roles despite the increased effort and expenses associated with the profession.

Human Rights Issues of the Demak Government's Planned Relocation Program

As mentioned earlier, the issues within the government's planned relocation create the potential for new human rights violations against the relocated community. The relocation process in Bedono Village may infringe upon various human rights of the displaced communities, including the right to legal certainty, adequate housing, a healthy living environment, and a dignified livelihood. The description is as follows:

First, there are potential violations of the right to legal certainty. Referring to the provisions of Article 28D paragraph (1) of the 1945 Constitution, Article 3 paragraph (2) of Law No. 39 of 1999 concerning Human Rights (Law No. 39 of 1999) states that '[e]veryone has the right to recognition, guarantees, protection, and fair legal certainty, as well as an equal treatment before the law'. Nonetheless, as previously stated, the ambiguity or uncertainty experienced by relocated communities concerning their residency status or the legality of their land rights is unequivocally inconsistent with the right to legal certainty as mandated by these laws and regulations. The current policies or laws lack a clear mandate for addressing migration issues, as people are only given land on riverbanks by the government for temporary residence (Nurhidayah & McIlgorm, 2019).

Second, there are potential violations of the right to residence and a good and healthy living environment. The provisions of Article 28H paragraph (1) of the 1945 Constitution state that '[e]veryone has the right to live in physical and spiritual prosperity, to have a place to live, and to have a good and healthy living environment and the right to obtain health services'. This guarantee of the right to housing and a good environment was later reaffirmed in Law No. 39 of 1999, Article 9, paragraph (3): 'Everyone has the right to a good and healthy living environment'. Furthermore, Article 40 of Law No. 39 of 1999 also stipulates that '[e]veryone has the right to live and live a decent life'. However, as explained above, there are numerous problems—relocation areas that are still vulnerable to disasters, inadequate infrastructure for relocation houses, relocation costs that are not necessarily affordable for the entire community, and resistance from original residents of the relocation area—that clearly demonstrate that the fulfilment of these rights has not been achieved in the implementation of the relocation program carried out by the government.

Third, there are potential violations of the right to a decent life. The provisions of Article 27, paragraph (2) of the 1945 Constitution state that '[e]very citizen has the right to work and a living worthy of humanity'. Furthermore, as previously mentioned, Article 40 of Law No. 39 of 1999 also reiterates that everyone has the right to a decent life. However, the existence of issues regarding relocated communities who have lost or had to change their livelihoods as a result of being relocated certainly has the potential to violate this right. Without prior skills training, residents expected to transition to new jobs may struggle to sustain a living, compromising their right to work and a decent livelihood. It is crucial to recognise that relocation without adequate training can violate these individuals' human rights, including the right to work and a decent standard of living. This highlights the importance of prioritising the rights of individuals when executing relocation programs.

Moreover, ignoring the skills and contributions of the affected residents can lead to resistance and reluctance to participate in the relocation process. In some instances, individuals opt not to relocate because the available livelihood options do not match their skills and preferences (McDermott & Gibbons, 2016). This creates a situation wherein many individuals feel unprepared and uncertain about starting a new life. Lacking the necessary skills and support for job transition, these individuals often report stress and anxiety about their future. Despite the threats of climate change, such as rising sea levels and harm to coastal ecosystems, economic insecurity and the absence of new skills often compel them to remain in their original homes.

In this case, it should be noted that while relocation aims to protect residents from flooding, it is crucial for the process and its requirements to prioritise residents' long-term welfare and safety. Relocation without adequate support can heighten their vulnerabilities and has the potential to raise new issues of human rights violations. The Demak Regency government's initiative to address migration underscores the intricate challenges associated with forced relocations resulting from the impacts of climate change. Governments must prioritise protecting individuals' rights, provide comprehensive support encompassing financial aid and psychological assistance, and offer robust training for transitioning livelihoods. This commitment to comprehensive support and training should reassure the communities of the government's dedication to the well-being of affected individuals. Furthermore, meaningful engagement of local communities in decision-making processes is vital to ensure that relocation practices are sustainable and uphold the rights of affected individuals.

It should be noted that while the implementation of the relocation program in Demak raises several problems threatening the human rights of relocated communities, many of our research participants still expressed gratitude for the implementation of the relocation program. Despite the problems in the relocation process, the interviewees felt that at least they now had a new house or place to live that—so far—has not been subject to flooding. For example, several people who relocated to Gemulak Village in 2006 have adapted to their new livelihood and feel at home in their current residence. The relocated individuals appear to lack awareness of their rights and therefore merely accept, and express gratitude for the assistance rendered by the government. This situation illustrates the injustice affecting relocated communities. In this case, the government must acknowledge that unresolved issues in executing the relocation will diminish public confidence in future governmental initiatives. This was evident during one of the community roundtables we conducted, at which numerous individuals expressed reluctance or declined to engage in the forthcoming relocation program.

Gaps and Way Forward of Indonesia's Current Legal Frameworks in Addressing the Issue of Climate-Induced Mobility or Planned Relocation

One of the issues complicating the assurance of human rights during relocation in Indonesia is the lack of comprehensive and adequate regulations and/or guidelines for managing relocation programs. Even though such programs are complex, Indonesia does not currently have national-level legislation or guidance that specifically regulates the implementation of relocation programs due to climate change. Currently, the main rules regarding climate change adaptation in Indonesia are contained in Law No. 16 of 2016 concerning the Ratification of the Paris Agreement to the United Nations Framework Convention on Climate Change (Law No. 16 of

2016) and Law No. 32 of 2009 concerning the Environment, as amended by Law No. 6 of 2023 concerning the Ratification of Regulations instead of Law No. 2 of 2022 concerning Job Creation (Law No. 32 of 2009). These two regulations explain the government's obligations and responsibilities for implementing climate change adaptation but do not specifically mention or regulate relocation programs.

At the statutory level, one of the legal bases for implementing relocation programs due to disasters is regulated in Law No. 24 of 2007 concerning Disaster Management, which stipulates that the government is responsible for implementing disaster management, including 'ensuring the fulfilment of the rights of communities and refugees affected by disasters fairly and by minimum service standards'. According to Article 1 of this law, a refugee is 'a person or group of people who are forced or compelled to leave their place of residence for an uncertain period due to the negative impacts of a disaster'. Article 54 further explains: 'Handling communities and refugees affected by disasters is carried out through activities including data collection, placement in safe locations, and meeting basic needs'. Although Law No. 24 of 2007 does not explicitly mention migration, displacement, or planned relocation policies in disaster management efforts, 'placement in safe locations', Article 54 can be interpreted as a basis for implementing relocation due to disasters.

Several regulations explicitly govern relocation programs, but these regulations tend to be more technical and less comprehensive than legal regulations. An example is the National Disaster Management Agency Regulation No. 03 of 2018 concerning the Handling of Refugees in Disaster Emergency Situations (Perka BNPB 3). Article 20 of Perka BNPB 3 stipulates that refugee relocation activities include: (a) a study of the feasibility of relocation sites; (b) socialisation of relocation; (c) preparation of land and processing of land permits for relocation; (d) establishment of residential premises for relocation; and (e) preparation of transportation and accommodation facilities. However, these regulations do not further guide or explain the appropriateness of standards for relocation sites, socialisation mechanisms, or the processing of relocation permits. Another relevant instrument is the Minister of Public Works and Public Housing Regulation No. 29/PRT/M/2018 concerning Technical Standards for Minimum Service Standards for Public Works and Public Housing, which presents guidelines or minimum service standards for providing and rehabilitating habitable houses for disaster victims. This regulation addresses infrastructure concerns in delivering habitable housing for disaster victims, including methodologies for assessing property damage, quality benchmarks for goods and services offered to aid recipients and other pertinent issues. However, other factors, such as economic, legal, social, cultural, and other issues in the implementation of relocation, are not addressed.

A similar situation occurs with laws and regulations at the regional government level. The absence of regulations or guidelines for implementing relocation programs as a comprehensive form of climate change adaptation at the national level means that standards or rules for implementing relocation programs in regions depend on the policy initiatives of each region. Despite frequent disasters, significant erosion, and tidal flooding in the Demak Regency area, there are currently no specific regulations governing disaster relocation programs. One of the legal bases for efforts to implement a disaster-related relocation program in Demak is contained in Regional Regulation No. 9 of 2016 concerning the Implementation of Disaster Management in Demak Regency. Article 36 of this regulation stipulates that one form of handling communities and refugees affected by disasters is 'placement in safe locations' and fulfilment of basic needs, which includes 'assistance in providing shelter or a place of residence'. The technical provisions should be regulated more specifically through a Regent's regulation; however, the mandate from regional regulations to create such a Regent's regulation has not yet been issued.

In addition to regulations relating to disaster management, Demak also has regulations that specifically address efforts to deal with tidal flooding, namely Demak Regency Regulation No. 14 of 2022 concerning Collaborative Strategies for Infrastructure Development with an Incremental Planning Approach in Coastal Inundation-Affected Areas. However, these Regent's regulations only focus on infrastructure development to mitigate the impact of tidal floods and do not provide guidance on relocation programs. This situation contrasts with the policy in the Kaur Regency area in Bengkulu Province. The Kaur Regency government has specific regulations that provide guidelines for relocation due to disasters, namely Kaur Regent Regulation No. 62 of 2023 concerning Guidelines for Residential Relocation for Disaster Victims. However, despite having specific guidelines, these regulations, as with the previously mentioned technical regulations, are not comprehensive and focus on relocation infrastructure issues such as building area, house type, and related aspects.

Furthermore, apart from problems related to the absence of comprehensive guidelines or regulations in implementing relocation programs at the national level, problems in implementing climate-change-related relocation programs arise due to differences in interpretation regarding the terminology of 'disaster' in government agencies. The legal basis for implementing relocation due to climate change is generally based on legal regulations relating to disaster management. This is because the definition of 'disaster' in these regulations is broad, relating not only to rapid-onset disasters but also to slow-onset disasters due to climate change. In this case, in Law No. 24 of 2007 'disaster' is defined as '[a]n event or series of events that threatens and disrupts people's lives and livelihoods caused by natural and/or non-natural factors and human factors, resulting in human casualties, environmental damage, property loss, and psychological impacts'. Natural disasters are defined as '[d]isasters caused by events or a series of events caused by nature include earthquakes, tsunamis, mountains, eruptions, floods, droughts, hurricanes, and landslides'.

Such a definition should include slow-onset disasters such as tidal floods due to climate change, as intended in the Law and the technical regulations below. More technically, in the Guidelines for the Implementation of Disaster Event and Impact Data Standards No. 07 of 2023 issued by the head of the National Disaster Management Agency (BNPB), the categorisation of tidal floods as a type of disaster is explicitly stated. However, in implementation in the field, government agencies often do not consider that tidal floods are included in the category of 'disasters'. One reason tidal floods are not considered a 'disaster' is that their slow-onset nature is not considered to give rise to an element of urgency. Additionally, the broader interpretation has the potential to burden local government budgets. For example, if tidal floods are considered a 'disaster' then the government is required to assist in meeting basic needs such as food, clean water, etc. for all affected communities. Tidal floods occur almost every year and cover very wide areas. This issue of interpretation is a serious problem because if the government does not consider tidal floods to be included in the 'disaster' category, then the assistance and guarantees of rights outlined in the previous regulations cannot be accepted by the community.

Relocation programs are complex and require careful preparation. It is therefore necessary to establish special guidelines for its implementation; it is not sufficient to rely on the initiative and policies of each regional government. It is also necessary to understand that the problems in relocating people are not confined to the process or infrastructure of the house being relocated but also include other aspects, relating to legal, social, and health matters. There are at least six aspects or issues that need to be considered and may arise in the implementation of the relocation program: (1) political issues—for example, regarding public distrust of government intervention in the relocation process which arise due to political factors; (2) legal and governance issues—for example, related to the lack of a legal framework, institutional issues, and policy gaps in relocation plans; (3) technical and planning issues—for example, related to relocation implementation issues such as planning, land, funding, and community participation; (4) human rights issues—for example, social and economic rights issues, and politics of relocated communities; (5) social and cultural issues—for example, issues regarding social and cultural challenges, and the identity of relocated communities; and, finally, (6) health issues—for example, related to the physical and mental health of the community during the relocation process (Garimella, 2022). Therefore, special and comprehensive regulations or guidance are needed to implement this program.

At present, there are no binding legal regulations at the international level that specifically govern standards for implementing relocation due to climate change. Numerous international legal instruments have underscored the significance and immediacy of executing planned relocation. Several organisations such as the UNHCR, Georgetown University, and the Brookings Institution, as well as the International Organization for Migration (IOM), UN University, and the World Bank, have at different times collaborated to report on or provide guidance regarding planned relocation practices due to climate change (Ferris & Weerasinghe, 2020). Furthermore, several countries, such as Fiji and the Solomon Islands, also have special guidance regarding implementing relocation programs due to climate change.

The first international legal instrument to explicitly address the issue of planned relocation was the Cancun Adaptation Framework, which resulted from a resolution from the 2010 COP (McAdam & Ferris, 2015). In this case, the provisions of the Adaptation Framework categorise planned relocation as a form of adaptation to the adverse effects of climate change. Article 14(f) of the Cancun Adaptation Framework

> [i]nvites all Parties to enhance action on adaptation under the Cancun Adaptation Framework ... by undertaking ... (f) Measures to enhance understanding, coordination, and cooperation about climate change induced displacement, migration and planned relocation, where appropriate, at the national, regional and international levels.

Another legal instrument is the Nansen Initiative Agenda for the Protection of Cross-Border Displaced Persons in the Context of Disasters and Climate Change (2015). This instrument explicitly encourages 'improving the use of planned relocation as a preventative or responsive measure to disaster risk and displacement'. Furthermore, the 2015 Paris Agreement associated with the COP instrument requires the formation of a Task Force to consider planned relocation as a last resort in the context of the Warsaw International Mechanism on Loss and Damage (UNFCCC, 2018). The Sendai Framework for Disaster Risk Reduction 2015–2030 also explains the importance of 'formulating public policies, where applicable, aimed at addressing the issues of prevention or relocation, where possible, of human settlements in disaster risk zones, subject to national law and legal systems' (Ferris & Weerasinghe, 2020; McMichael, 2023).

Although these regulations acknowledge the significance of planned relocation, they remain general and lack specific provisions for their regulation. Without regulations governing specific relocation protocols, various international organisations have partnered to develop literature or guidelines for executing planned relocation practices in response to climate change (Ferris & Weerasinghe, 2020). The Brookings Institution, Georgetown University, and UNHCR have issued a document on *Guidance on Protecting People through Planned Relocations from Disasters and Environmental Change*. Furthermore, in 2017, Georgetown University, IOM, and UNHCR also published the *Toolbox: Planning Relocations to Protect People from Disasters and Environmental Change*. This guidance offers a toolbox containing five essential factors that need to be considered in decision-making in implementing relocation: (1) establishing and complying with an appropriate legal framework, (2) understanding and addressing the needs and impacts of planned relocations on affected populations, (3) providing information to, undertaking consultation with, and ensuring the participation of, affected populations, (4) understanding and addressing complexities related to land issues, and (5) undertaking monitoring and evaluation, and ensuring accountability (Ferris, 2019 p. 441; Ferris & Weerasinghe, 2020, p. 139).

Moreover, while there are no binding international legal regulations governing planned relocation, several nations, including Fiji and the Solomon Islands, have developed specific guidelines for executing relocation programs in response to climate change. In 2018, Fiji published *Planned Relocation Guidelines—A Framework to Undertake Climate Change-related Relocation*. This makes Fiji the first country to create a guideline or national framework regarding the implementation of its relocation program. The guideline contains basic principles for implementing planned relocation, stages of planned relocation, and stakeholders involved (Republic of Fiji, 2018). At a more granular level, in 2023 Fiji also published *Standard Operating Procedures for Planned Relocation in the Republic of Fiji*. In 2022, both Fiji and the Solomon Islands published Planned Relocation Guidelines for their countries. These documents guide the relocation process and includes guidance for the period before relocation, such as identification, planning, and the process after the relocation.

Other countries' instruments, guidelines, and practices should serve as a reference for the Indonesian government in developing legislation or directives that specifically govern the implementation of relocation programs necessitated by climate change. Although planned relocation is the ultimate recourse in climate change adaptation strategies, developing legal instruments remains essential. In certain regions, such as Demak, relocation is occasionally the sole recourse for communities impacted by tidal flooding resulting from climate change. A comprehensive relocation guidance would be fundamental to prevent potential human rights violations for individuals involved in the relocation program, with relocation thus more likely to garner community support.

In addition to formulating comprehensive planned relocation guidelines, governments must implement adaptive governance strategies to mitigate the effects of climate change. This includes developing comprehensive climate change adaptation plans, establishing dedicated working groups for climate change adaptation, and prioritising programs that reduce communities' vulnerability to climate-related risks (Rusnaedy et al., 2021). Additionally, engaging communities in climate change adaptation initiatives, such as the Climate Village Program, is crucial as it enhances public awareness of the impacts of climate change, with communities better informed and potentially more involved in the process, facilitating adaptive measures (Safrina et al., 2022).

Conclusion

The mobility challenge resulting from climate change is a global phenomenon, and Demak is one of several regions of Indonesia that are confronting this issue. The Demak government has repeatedly executed relocation initiatives to mitigate the issue of tidal flooding resulting from elevated sea levels and land subsidence impacting its communities. However, the relocation process has encountered several problems due to the complexities involved and the lack of standardised procedures for implementation. These problems include the lack of adequate relocation land for settlements, inadequate housing infrastructure, resistance from within the relocation destination areas, burdensome relocation costs, and livelihood issues. These issues also lead to potential human rights violations for relocated communities, including violations of the right to guarantee legal certainty, the right to adequate housing and a healthy living environment, and the right to a decent livelihood. Given these concerns, the Indonesian government should consider establishing specific rules and guidelines for managing planned relocations, particularly in relation to environmental migration due to climate change.

References

Amin, C., Sukamdi, S., & Rijanta, R. (2021). Exploring migration hold factors in climate change hazard-prone area using grounded theory study: Evidence from Coastal Semarang, Indonesia. *Sustainability, 13*(8), 4335. https://doi.org/10.3390/su13084335

Bettini, G., Nash, S. L., & Gioli, G. (2017). One step forward, two steps back? The fading contours of (in)justice in competing discourses on climate migration. *The Geographical Journal, 183*(4), 348–358. https://doi.org/10.1111/geoj.12192

BNPB. (2018). *Regulation Number 03 of 2018 on the handling of refugees in disaster emergency situations*. BNPB.

BNPB. (2023). *Guidelines for the implementation of disaster event and impact data standards number 07 of 2023*. BNPB.

Bronen, R., & Chapin, F. S. (2013). Adaptive governance and institutional strategies for climate-induced community relocations in Alaska. *Proceedings of the National Academy of Sciences, 110*(23), 9320–9325. https://doi.org/10.1073/pnas.1210508110

Brookings Institution, Georgetown University, & UNHCR. (2015). *Guidance on Protecting People from Disasters and Environmental Change through Planned Relocation*. https://environmentalmigration.iom.int/sites/g/files/tmzbdl1411/files/Guidance%20on%20Planned%20Relocations%20-%20Split%20PDF.pdf

Demak Regency. (2016). *Regional Regulation Number 9 of 2016 on the implementation of disaster management in Demak Regency*. Demak Regency Government.

Demak Regency. (2022). *Regent's Regulation Number 14 of 2022 concerning collaborative strategies for infrastructure development with an incremental planning approach in in Rob-Affected Areas*. Demak Regency Government.

Demak Regency. (2023, 7 November). *Pemkab Demak Salurkan Bantuan Jaring Sosial Bagi Warga Terdampak Rob* Konten ini telah terbit di. https://demakkab.go.id/news/pemkab-demak-salurkan-bantuan-jaring-sosial-bagi-warga-terdampak-rob

DPRD. (2021, 1st February). *Audiensi Warga Dukuh Mondoliko Desa Bedono Kabupaten Demak.* https://dprd.demakkab.go.id/berita/AUDIENSI-WARGA-DUKUH-MONDOLIKO-DESA-BEDONO-KABUPATEN-DEMAK

Farbotko, C., Thornton, F., Mayrhofer, M., & Hermann, E. (2022). Climate mobilities, rights and justice: Complexities and particularities. *Frontiers in Climate, 4*, 1026486. https://doi.org/10.3389/fclim.2022.1026486

Ferris, E. (2019). Climate change, migration, law and global governance. *North Carolina Journal of International Law, 44*(3), 425–455.

Ferris, E., & Weerasinghe, S. (2020). Promoting human security: Planned relocation as a protection tool in a time of climate change. *Journal on Migration and Human Security, 8*(2), 134–149. https://doi.org/10.1177/2331502420909305

Floranza, J. M., & Asio, F. J. M. R. (2022). The impact of disasters and climate change on migration and displacement. *Technoarete Transactions on Climate Change and Disaster Management Research, 1*(2). https://doi.org/10.36647/TTCCDMR/01.02.A004

Garimella, P. P. (2022). Planned relocation: An unusual case for developed countries. *Current Research in Environmental Sustainability, 4*, 100177. https://doi.org/10.1016/j.crsust.2022.100177

Goodson, L., & Phillimore, J. (Eds.). (2012). *Community research for participation: From theory to method* (1st ed.). Bristol University Press. https://doi.org/10.2307/j.ctt1r33pzb

Kaur Regency. (2023). *Regent Regulation Number 62 of 2023 concerning guidelines for residential relocation for disaster victims*. Kaur Regency Government.

Knox, J. H. (2016). Human rights principles and climate change. In K. R. Gray, R. Tarasofsky, & C. P. Carlarne (Eds.), *The Oxford handbook of international climate change law* (1st ed., pp. 213–236). Oxford University Press. https://doi.org/10.1093/law/9780199684601.003.0011

Krishnadas, J. (2007). Relocating the master's domain: Social and legal locations of gender from post-disaster to everyday life. *Social & Legal Studies, 16*(1), 131–147. https://doi.org/10.1177/096466390707377

Kusuma, M. A. (2016). *Dampak Rob terhadap Perubahan Sosial Masyarakat di Kawasan Rob Desa Bedono Kecamatan Sayung Kabupaten Demak.*

Maldonado, J. (2014). A multiple knowledge approach for adaptation to environmental change: Lessons learned from coastal Louisiana's tribal communities. *Journal of Political Ecology, 21*(1). https://doi.org/10.2458/v21i1.21125

Martin, S. F. (2017). Environmental change and human mobility: Trends, law and policy. *Comparative Population Studies, 42*. https://doi.org/10.12765/CPoS-2017-13

McAdam, J., & Ferris, E. (2015). Planned relocations in the context of climate change: Unpacking the legal and conceptual issues. *Cambridge Journal of International and Comparative Law, 4*(1), 137–166. https://doi.org/10.7574/cjicl.04.01.137

McCaughey, J. W., Daly, P., Mundir, I., Mahdi, S., & Patt, A. (2018). Socio-economic consequences of post-disaster reconstruction in hazard-exposed areas. *Nature Sustainability, 1*(1), 38–43. https://doi.org/10.1038/s41893-017-0002-z

McDave, K. E., & Dagadu, P. P. (2023). Forging the nexus of climate change and human rights. *European Journal of Law and Political Science, 2*(1), 1–5. https://doi.org/10.24018/ejpolitics.2023.2.1.54

McDermott, R., & Gibbons, P. (2016). Human rights and proactive displacement: Determining the appropriate balance between the duty to protect and the right to remain. *Disasters, 41*(3), 587–605. https://doi.org/10.1111/disa.12217

McMichael, C. (2023). Climatic and environmental change, migration, and health. *Annual Review of Public Health, 44*(1), 171–191. https://doi.org/10.1146/annurev-publhealth-071421-045148

McMichael, C., Katonivualiku, M., & Powell, T. (2019). Planned relocation and everyday agency in low-lying coastal villages in Fiji. *The Geographical Journal, 185*(3), 325–337. https://doi.org/10.1111/geoj.12312

Ministry of Public Works and Public Housing. (2018). *Regulation Number 29/PRT/M/2018 concerning technical standards for minimum service standards for public works and public housing (Peraturan Menteri Pekerjaan Umum dan Perumahan Rakyat Nomor 29/PRT/M/2018 tentang Standar Teknis Standar Pelayanan Minimal Pekerjaan Umum dan Perumahan Rakyat).* Ministry of Public Works and Public Housing.

Mladenov, M., Prelević-Plavšić, S., & Staparski, T. (2022). Climate change from the perspective of the European Court of Human Rights. *International Review, 3–4*, 118–122. https://doi.org/10.5937/intrev2204125M

Mukherjee, V., & Mustafa, F. (2019). Climate change and right to development. *Management and Economics Research Journal, 5*, 1. https://doi.org/10.18639/MERJ.2019.735041

Nurhidayah, L., & McIlgorm, A. (2019). Coastal adaptation laws and the social justice of policies to address sea level rise: An Indonesian insight. *Ocean & Coastal Management, 171*, 11–18. https://doi.org/10.1016/j.ocecoaman.2019.01.011

Okara, A., & Mbaye, L. M. (2023). Climate change, natural disasters, and migration. *IZA World of Labor, 346v2*, 1–10. https://doi.org/10.15185/izawol.346v2

Peters, L. E. R., & Van Den Hoek, J. (2021). Charting a justice-based approach to planned climate relocation for the world's refugees. In I. J. Ajibade & A. R. Siders (Eds.), *Global views on climate relocation and social justice* (1st ed., pp. 48–61). Routledge. https://doi.org/10.4324/9781003141457-4

Piggott-McKellar, A., McNamara, K., Nunn, P., & Sekinini, S. (2019). Moving people in a changing climate: Lessons from two case studies in Fiji. *Social Sciences, 8*(5), 133. https://doi.org/10.3390/socsci8050133

Piggott-McKellar, A., & Vella, K. (2023). Lessons learned and policy implications from climate-related planned relocation in Fiji and Australia. *Frontiers in Climate, 5*, 1032547. https://doi.org/10.3389/fclim.2023.1032547

Prasetyo, Y., Bashit, N., Sasmito, B., & Setianingsih, W. (2019). Impact of land subsidence and sea level rise influence shoreline change in the coastal area of Demak. *IOP Conference Series: Earth and Environmental Science, 280*(1), 012006. https://doi.org/10.1088/1755-1315/280/1/012006

Quirico, O. (2017). Systemic integration between climate change and human rights in international law? *Netherlands Quarterly of Human Rights, 35*(1), 31–50. https://doi.org/10.1177/0924051917695210

Republic of Fiji. (2018). *Planned Relocation Guidelines – A framework to undaertake climate change related relocation.*

Rosyida, A., et al. (2024). *Data Bencana Indonesia 2023.* Jakarta: Badan Nasional Penanggulangan Bencana (BNPB).

Rusnaedy, Z., Haris, A., Congge, U., & Prianto, A. L. (2021). Adaptive climate change governance in Makassar, Indonesia. *Journal of Governance, 6*(2). https://doi.org/10.31506/jog.v6i2.12384

Safrina, S., Roesa, N., & Rosemary, R. (2022). Community participation for adaptation and mitigation of climate change: Case study the implementation of Program Kampung Iklim (Proklim). *Batulis Civil Law Review, 3*(2), 137. https://doi.org/10.47268/ballrev.v3i2.945

Seeteram, N. A., Ash, K., Sanders, B. F., Schubert, J. E., & Mach, K. J. (2023). Modes of climate mobility under sea-level rise. *Environmental Research Letters, 18*(11), 114015. https://doi.org/10.1088/1748-9326/acfe22

Setiawan, A. (2023, July). *Menyiapkan Tata Kelola Laut Berkelanjutan.* https://indonesia.go.id/kategori/ragam-ais-forum-2023/7295/menyiapkan-tata-kelola-laut-berkelanjutan?lang=1

Susanto, K. E., Marfai, M. A., & Mardiatno, D. (2010). Proyeksi Kenaikan Permukaan Laut dan Dampaknya Terhadap Banjir Genangan Kawasan Pesisir. *Majalah Geografi Indonesia, 24*(2).

Sushmita, C. I. (2021, May 26). *Demak Terancam Tenggelam dalam 20 Tahun!* [News]. https://jateng.solopos.com/demak-terancam-tenggelam-dalam-20-tahun-1127657

Legislation

Law Number 16 of 2016 on the ratification of the Paris Agreement to the United Nations Framework Convention on Climate Change. Jakarta: Indonesia.

Law Number 24 of 2007 on disaster management. Jakarta: Indonesia.

Law Number 32 of 2009 on environmental protection and management. Jakarta: Indonesia.

Law Number 39 of 1999 on human rights. Jakarta: Indonesia.

Law Number 6 of 2023 on the Stipulation of Government Regulation in Lieu of Law Number 2 of 2022 on Job Creation as Law. Jakarta: Indonesia.

The 1945 Constitution. Jakarta: Indonesia.

Chapter 17
Climate Migration of Fishermen in Coastal North Jakarta: Impact and Policy Interventions

Laely Nurhidayah, Intan Hapsari Surya Putri, Agus Suntoro, Ahmad Fathony, and Nurrahman Aji Utomo

Introduction

Climate change impacts fishermen around the world through fish migration, reduced fish stock and more frequent extreme weather events, leading to economic losses and migration. These phenomena are also observed on the northern coastal of Java, with fishermen and their families from Indramayu, Tegal and Cirebon relocating to Jakarta. Migration is a strategy adopted by these fishermen as a way out of poverty. However, the large-scale movement of fishermen to Jakarta has created social and legal challenges, including illegal settlements, lack of clean water and sanitation, lack of access to health, education and other public services, and conflicts between fishermen. Despite these problems, however, many of these migrants are determined to remain in Jakarta as it provides greater economic opportunities than their places of origin. Many of these migrant fishermen reside in Muara Angke in the north of Jakarta. Muara Angke is a fishing port and melting pot of migrant fishermen from surrounding regions, as well as their wives and other migrant women who work peeling green mussels and salting fish.

This migration is triggered by both push and pull factors. The push factors include decreasing fish stocks in their home regions as a result of overfishing and climate change, as well as weather conditions forcing them to sail further afield, increasing costs. In contrast, fish are more plentiful in the seas surrounding Jakarta,

L. Nurhidayah (✉) · A. Suntoro · A. Fathony · N. A. Utomo
Research Center for Law, National Research and Innovation Agency (BRIN), Jakarta, Indonesia
e-mail: lael003@brin.go.id

I. H. S. Putri
Center for Urban and Regional Resilience Research (CURE) – Faculty of Engineering, Diponegoro University, Semarang, Indonesia

L. Nurhidayah et al. (eds.), *Climate Change, Labour and Migration in Indonesia*, Springer Climate, https://doi.org/10.1007/978-3-032-15334-0_17

which include the 'thousand islands', many of which are national parks with protected areas. Another pull factor is that Jakarta, as the capital of Indonesia, offers far greater economic opportunities, providing these fishermen with better market access and higher prices and profit margins.

However, the scale of this migration and the resulting urbanisation have created complex social and legal problems in Jakarta. Most of these fishermen settle illegally on the shores or banks of coastal areas and rivers and erect temporary housing, leading to the creation of slum areas without proper sanitation or clean water. The local government has in the past simply evicted these people and sent back to their region of origin. However, many resist such relocation, and many not only but also invite other members of their households to stay with them in Jakarta to find better livelihoods, despite these areas being prone to coastal flooding. The government is finally beginning to recognise and accommodate this population, with moves in recent years to build them stilt and floating housing and provide them with Jakarta ID cards, making them eligible for social security, health and education.

This chapter will examine the impact of climate change on the migration of fishermen from Indramayu, Cirebon and Tegal to Jakarta, as well as the impact of, and policy interventions in response to, this migration and urbanisation. The research involved literature reviews and interviews with and field observation of fishermen in Muara Angke, north Jakarta, as well as government officials, NGOs and fishermen in Indramayu, Cirebon and Tegal. The chapter aims to provide a better understanding of the impact of climate change on fishermen migration, its social and legal implications and the complex policy interventions employed to solve the problem.

Climate Change and Migration of Fishermen: Push and Pull Factors

Climate change has increasingly become a dominant factor influencing migration patterns, particularly among fishing communities who rely heavily on coastal ecosystems for their livelihoods. The effects of climate change are multifaceted, driving migration through a combination of push and pull factors that span environmental, economic and social domains. Understanding these factors is critical to addressing the challenges faced by fishermen, especially as they adapt to climate change impacts.

One of the most pressing push factors for fishermen is environmental degradation. Rising sea levels and extreme weather events such as tropical cyclones and flooding have become more frequent due to global warming. These environmental shifts threaten the stability of coastal communities, forcing fishermen to abandon their traditional settlements in search of safer environments, typically in urban areas or higher ground (Ahmed et al., 2024; Purnomo et al., 2024; Rahman et al., 2023). Furthermore, changes in fish populations—driven by rising ocean temperatures, ocean acidification and overfishing—are impacting the availability of fish, leading to declining catches. As a result, fishermen often face reduced income and must seek new livelihoods or relocate (Jakariya et al., 2020; Mbaye et al., 2023; N'Souvi et al., 2024). For example, in Bangladesh coastal fishing communities are highly vulnerable to climate change, with marine fisheries significantly affected by natural disasters, saltwater intrusion and shifting coastal morphologies (Ahmed et al., 2024; Jakariya et al., 2020). The uncertainty in fish populations pushes these communities to find more stable environments.

Economic hardships also act as significant push factors. The economic sustainability of fishing has been severely impacted by climate change. The cost of fishing operations, particularly fuel for boats and investment in advanced equipment to adapt to changing fishing conditions, has risen. This financial strain often compels fishermen to move to urban centres in search of alternative employment opportunities (Le Bras et al., 2024; Shaffril et al., 2017a). In regions like the Gulf of Alaska, the decline in fish stocks due to warmer waters has caused a sharp decrease in fishing revenues, forcing fishermen to consider other sources of income (Szymkowiak et al., 2023).

Health and safety risks, exacerbated by climate change, further contribute to the migration of fishermen. As coastal communities experience the spread of diseases such as malaria, respiratory issues and skin conditions linked to changing climates, the health risks associated with continued fishing become untenable. Additionally, the increasing frequency of extreme weather events, including storms and floods, poses significant safety risks to fishermen at sea. This has led many to relocate to safer environments, particularly inland or in areas less vulnerable to the harsh impacts of climate change (Mbaye et al., 2023; Rahman et al., 2023; Samah et al., 2019).

On the other hand, there are also several pull factors that attract fishermen to new areas. Economic opportunities in urban centres or more resilient coastal areas offer a strong incentive for migration. As fishing becomes less viable in some regions, alternative livelihoods in industries like construction, trade and services provide better prospects. In some cases, migration also allows fishermen to access better financial resources and aquaculture opportunities, which are more sustainable and less dependent on fluctuating marine resources (Jakariya et al., 2020). Social networks also play an important role in facilitating migration. Fishermen often rely on established communities in destination areas, where they can receive practical and emotional support. These communities can help new arrivals integrate into the workforce and adapt to their new environment (Shaffril et al., 2017b).

The availability of adaptation strategies also plays a role in pulling fishermen to new regions. Certain coastal areas are investing in infrastructure improvements such as mangrove restoration, the construction of breakwaters and the development of sustainable aquaculture. These efforts not only help protect fishing resources but also provide safer and more sustainable environments for fishermen to continue their livelihoods (N'Souvi et al., 2024; Szymkowiak et al., 2023). Additionally, policy support can create favourable conditions for migration. Government policies that address climate change and provide support for fishermen's adaptation strategies can make relocation a viable option, allowing fishermen to continue their work under better conditions.

The migration of fishermen due to climate change is shaped by a complex interplay of push and pull factors. Environmental degradation, economic pressures and health risks push fishermen to leave their traditional homes, while economic opportunities, social networks and effective adaptation strategies pull them to new regions. A deeper understanding of these factors is essential for developing policies that not only address the immediate impacts of climate change on fishing communities but also ensure the long-term resilience and well-being of fishermen (Jakariya et al., 2020; Le Bras et al., 2024; Mohamed Shaffril et al., 2019; Samah et al., 2019).

Findings

Patterns of Migration

There are four types of migration—permanent, temporary, voluntary and forced. Based on our findings, all four types are evident among the migrants in Muara Angke. For example, in an interview, one of the fishermen leaders in Muara Angke commented that he had migrated permanently to Jakarta long ago, settling on the bank of a river with others and eventually forming an illegal settlement. Other fishermen commented that they had migrated permanently after initially renting homes in Muara Angke for years; in many cases these were relatives or neighbours of fishermen who had relocated to Jakarta. Temporary, and particularly seasonal, migration is also common. According to an interview with fishermen in Cirebon and Indramayu, they temporarily base themselves in Jakarta to catch certain fish, such as squid, shrimp, mackerel, etc. in seasons when these are abundant in the Jakarta Bay area. This sometimes leads to conflict between fishermen over fishing grounds, but our interviewees stated that in such cases they will simply find other fishing ground nearby to resolve the conflict. The crew of larger ships are also often temporary migrants, usually from Cirebon and Indramayu. Usually they have three-month contracts, and they travel to Jakarta by shuttle bus or public transport.

The above are all also examples of voluntary migration. Forced migration in Jakarta is generally the result of local government actions such as relocating communities due to flooding risks or as part of reclamation or infrastructure projects, or returning migrant workers to their place of origin. In the latter case many of those forced to migrate will usually return to Jakarta as it provides them with opportunities that are simply not available in their place of origin.

Regardless of whether their presence in Jakarta is temporary or permanent, or whether their migration was forced or voluntary, most of these fishermen have limited education and no skills other than as fishermen. Most of them are small-scale fishers, who under the Fisheries Law are permitted to fish everywhere in Indonesia's waters, while a smaller proportion are hired crew on larger vessels fishing in other areas of Indonesia.

Push and Pull Factors of Migration in Jakarta Bay

Human migration is intrinsically linked to societal change and development, including climate change impacts. The International Organization for Migration (IOM) identifies that people leave their places of origin for various reasons, including better economic or educational opportunities elsewhere, family concerns or to flee conflict or persecution (de Bruin et al., 2024). There are several phenomena related to temporary migration, including 'missing men'—those who have migrated from rural areas to urban centres for a specific, often lengthy, period, but not permanently (Roshania et al., 2025). Interestingly, the results of a study by the University of Indonesia indicate that the migration pattern of fishermen in the north coast of Java, such as Indramayu and Cirebon to Jakarta, is not solely based on fishing areas but was based on the advantages offered by urban areas and the opportunity to improve the standard of living of fishermen and their families (Akmalia, 2016).

In the context of migration of fishermen (including family members) to Jakarta from various regions such as Cirebon, Indramayu, Tegal Regency and other coastal areas to Jakarta, there are several factors. The first is geographical advantages and natural conditions. Jakarta Bay is in the Jakarta Province on the north coast of Java; it has 72 kilometres of coastline, a shallow sea area with a depth of 1–24 metres and more suitable weather for fishermen (Kuswanto, 2024). Fishermen can operate small boats and fish without having to operate in groups. They also benefit from the many small islands around Jakarta Bay, which provide resting places and shelter during bad weather, storms or winds. The push factor for temporary migration of fishermen in the Indramayu area to Jakarta due to natural conditions was confirmed by Kadbari, head of Karangsong Village, Indramayu Regency, in an interview in July 2024, who noted that the primary consideration is that the sea in the northern coastal areas of Indramayu and Cirebon is deep water, and thus requires a larger ship operated by more than one person, with no small islands for shelter. This increases fuel costs—due to the need to operate large ships and go further out to sea—as well as the safety of fishermen (Fathoni, 2024).

The ability to operate smaller vessels also eases the licensing process. The Ministry of Maritime Affairs and Fisheries Circular Letter Number 7 of 2014, effectively provides an exemption from licensing for vessels under 10 gross tons (GT), making it easier for fishermen to operate in shallow seas. The Circular Letter is a corrective to the former Regulation of the Minister of Maritime Affairs and Fisheries Number 30 of 2012 concerning Capture Fisheries Business in Fisheries Management Areas, and the Regulation of the Minister of Fisheries Number 11 of 2016 concerning Minimum Service Standards for Re-Measurement Fishing Vessel Licensing Outlets, which required licensing even for vessels under 10 GT and imposed additional operational regulations for the catchment area (Hanafi, 2024).

However, the main pull factor for the temporary migration of regional fishermen to Jakarta is economic. Jakarta, as the largest centre of trade in Indonesia, provides greater market opportunities and better prices for the sale of fish and seafood. Good sales results and high demand from consumers in Jakarta increase the income of fishermen and cover ship operations. This is not the case for fishermen who remain in their hometowns, which have lower prices due to need to sell catches to middlemen who will then mark up the product and sell it in Jakarta. The Muara Angke Port Management Unit confirmed that the catch from Jakarta-based fishermen meets only 20% of daily consumption in Jakarta; the rest relies on shipments from various regions to the fishing port or directly to markets, restaurants, hotels or *tengkulak* (intermediaries) as distributors (Fathony, 2024). Particularly during the fishing season in Jakarta, fishermen from the regions will temporarily migrate to Jakarta because they believe all the catch will be absorbed by the market (Barid, 2025).

The head of Fisheries and Marine Affairs, North Jakarta KPKP Sub-department, added that a second pull factor is the availability of human resources, in this case, the ship's crew. Many ships with a capacity of up to 10 GT or larger need additional crew, and this need is met by the arrival of fishermen from Indramayu, Tegal and Cirebon.

Discussion

This section discusses the impact of climate migration on key sectors affecting coastal communities in North Jakarta. The discussion begins with the challenges in informal settlements, clean water and sanitation, and health risks and pollution that worsen living conditions. It then transitions to policy interventions, focusing on the importance of recognising housing rights, alongside access to education and healthcare, as essential elements for enhancing the resilience and well-being of climate migrants in the area.

Impacts

Informal Settlement

Muara Angke area is a coastal neighbourhood located in Penjaringan District, North Jakarta. It has long served as one of the city's main fishing ports and is home to traditional fishing communities. Since being officially designated as a Nusantara Fishing Port in 1977 by then-Governor Ali Sadikin, the area has grown into a hub of fishery-related activities and densely populated informal housing settlements. Over time, Muara Angke attracted waves of migrants, primarily fishermen from coastal areas such as Indramayu, Cirebon, Pekalongan and Jepara, who moved to Jakarta seeking better livelihoods. This influx has contributed to the organic growth of the settlement, but also presented complex challenges, especially regarding land tenure, service provision and social integration.

Many houses in Muara Angke were built informally on reclaimed lands, riverbanks or mangrove conservation zones, places not officially zoned for residential

Fig. 17.1 Informal settlement in Muara Angke, Jakarta. (Source: Researcher documentation, 2024)

use, see Fig. 17.1. As a result, much of the settlement is considered illegal or unplanned. For example, in the Muara Angke area about 3000 households live in informal housing, and only 800 residents have Jakarta ID cards. This legal ambiguity has led to conflicts between residents and authorities, particularly concerning relocation plans and the recognition of residency rights. Some communities have eventually received administrative recognition, forming local neighbourhood associations (RW), but others continue to live under the threat of eviction or lack access to basic services.

Muara Angke reflects the broader challenges of rapid urbanisation and coastal development in Jakarta. It is a community shaped by migration, informality and resilience—but also strained by policy, environmental vulnerability and spatial marginalisation. Addressing its challenges requires inclusive, community-based approaches that balance urban governance, environmental protection and the rights of coastal communities.

Clean Water and Sanitation

Muara Angke is lagging in infrastructure development and access to public services, including clean water. The communities must buy clean water from the kiosk or mobile water sellers. The residents spend an average of IDR 13,000 per day per family for drinking and cooking water, while, for washing clothes and bathing they use around 200 litres of groundwater per family per day and spend IDR 25,000 (Tempo, 2022). The communities were not provided with their own toilet facilities and have built emergency public toilets near the coastal areas. They usually bathe in the river or use rainwater for bathing.

Health and Pollution

Water pollution is becoming a major issue, especially in Jakarta Bay, with pollution from land-based sources, waste from communities, and industrial and marine plastics from the river. Many residents have reported that they suffered from itchy skin because of polluted water. The problem of pollution in Jakarta Bay has long been a concern for fishermen in Muara Angke and other surrounding areas. This pollution means fish are not available near the shore, forcing the fishermen to sail further for their catch. In addition, based on a study by Institut Pertanian Bogor (IPB), green mussels and other seafood from Jakarta Bay is contaminated with pollution and micro-plastics and is not considered safe for consumption. Despite this, fishermen still catch and sell green mussels as they have no alternative.

Policy Intervention

Political Recognition and Housing

The Jakarta government has attempted several policy interventions over the years, such as offering relocation to subsidised rental flats (*rusunawa*) and enforcing coastal normalisation projects. In 2004 for example, the government introduced a normalisation project with the result that the community in Muara Angke needed to be moved. They were relocated to apartment housing elsewhere in Muara Angke; however, as this required them to pay rent and utilities fees, many have subsequently returned. Some live in the Fisheries Port Management Unit (UPPP) areas owned by the Ministry of Marine Affairs and Fisheries. However, relocation programs have frequently met with resistance from local residents, who fear losing their livelihoods and social networks. Critics argue that many of these policies are top-down and fail to include community voices in the decision-making process.

During Anies Baswedan's term as governor of Jakarta (2017–2022), there was increased recognition of this informal settlement following a political commitment during the gubernatorial election campaign. The Jakarta government subsequently recognised them as a new RW in Muara Angke, which in turn resulted in administrative recognition of population by the Jakarta Provincial Government. With administrative recognition, they became entitled to Jakarta Resident Identity Cards, allowing them to access public services (Secretary of Pluit Village, 2024) including free healthcare, free education until elementary school and social security. Since the 2024 elections, new President Prabowo has established stilt housing for 200 households in Muara Angka (see Fig. 17.2) in an initiative to provide access to better housing and infrastructure and mitigate flooding impacts.

Education

In Jakarta, government programs like the Jakarta Smart Card (KJP) and Jakarta Superior Student Card (KJS) have made significant strides in increasing access to education by providing subsidies for school fees and educational materials. However, these programs are limited to residents with a Jakarta ID card, which excludes many migrants, particularly those from rural or coastal areas. As a result,

Fig. 17.2 Stilt house built by government. (Source: Laely Nurhidayah documentation)

migrants who move to Jakarta seeking better economic opportunities often cannot access these educational programs, leaving them without the support they need to reduce the financial burden of school fees and supplies.

This issue is particularly pronounced in Muara Angke Jakarta coastal communities, where many children from fishing families are unable to attend school regularly. They are often required to accompany their parents on fishing trips to help support the family income. In many cases, there is a belief that formal education holds less value compared to earning through fishing activities. As a result, children miss school, leading to high dropout rates and low school attendance in these communities. Migrant women also face challenges in accessing education and training. They often have fewer opportunities than men, and their migration to Jakarta makes it even harder for them to access education and participate in community decision-making processes, further contributing to gender inequality in education.

Despite the government's efforts through the KJP and KJS programs, the exclusion of migrants limits their effectiveness. Furthermore, these programs often fail to reach more remote communities, where schools are frequently overcrowded and under-resourced. To address these challenges, new policies should ensure that children from fishing communities, regardless of migration status, have better access to education. Government programs should be expanded to include non-residents, promoting inclusivity and reducing the educational gap. Additionally, community awareness campaigns can help parents in coastal areas understand the importance of education and its long-term benefits. Flexible education options, such as vocational training or online courses, would also benefit children with parents who work irregular hours.

While the government's efforts have made progress, there is a need for more inclusive and targeted policies to address the specific challenges faced by migrant children and communities. By expanding access to education and providing flexible learning options, the government can foster a more equitable education system for all children.

Health

Poverty, water pollution, inadequate urban sanitation and lack of access to drinking water disproportionately impacted the health of local people in Muara Angke, especially women and children. The community are vulnerable to a variety of health issues, such as water-borne diseases, malnutrition and negative maternal and child health indicators.

Women and children in Muara Angke, Jakarta, suffer from the 'double burden', whereby not only must they face the physical impacts of climate change, such as floods, extremes of weather and sea level rise, but they also experience higher rates of child marriages, child labour and domestic violence. Most of these women work in sectors below the threshold of minimum pay and under harmful working conditions, particularly among less affluent groups (between IDR 1,001,000 and IDR 2,000,000). These female workers work in home industries where workplace health and safety standards are exceedingly low, resulting in further causes of health harm. The health problems are exacerbated for the children as well, as many of them work daily, and others up to 21–30 hours a week, under unsafe conditions. This results in physical and mental illnesses that further reduce the general well-being of these communities.

Despite the programs instituted by the Indonesian government aimed at improving the welfare of citizens, access remains unequal. The Jakarta ID card (KTP Jakarta Province) requirement remains a principal barrier that limits residents and migrants from enjoying certain health benefits. For instance, Jakarta's free health initiatives, including state-subsidised medical treatment and care, are only for individuals with a Jakarta residency ID. Thus, those from outside the province who have migrated to Jakarta to pursue better economic opportunities are deprived of these needed healthcare benefits. This unequal system deprives those who most need—women and children from vulnerable groups—the benefits of healthcare programs. Despite a number of initiatives by various groups to offer free medical care for these communities from various entities, there remains an urgent need to improve immigrant fishermen's access to health services in Jakarta. In addition, there is a compelling need for policies to improve the health infrastructure in coastal areas, increase the availability of healthcare services to vulnerable communities and improve public health education, particularly among women and children. These developments may help to close health gaps and improve overall health outcomes.

Way Forward: Collaboration Among Local Governments

The problems arising from the increasing migration of regional fishermen to Jakarta cannot be effectively addressed by the Jakarta local government on its own. A holistic approach is needed, focusing on sustainable development including sustainable fisheries management, economic diversification and inclusive urban development. In addition, collaboration with the local governments of the regions surrounding Jakarta from where the migrants are coming will be required. One example of initiative to relocate fishermen affected by sea level rise (SLR) is the establishment of pilot project called the *Fisherman welfare kampong* in Eretan, Indramayu, by the Ministry of Social Affairs. The project includes establishment of housing and supporting infrastructure as well as communities empowerment program. This program especially focuses on equipping women with skills to pursue alternative livelihood (Tempo, 2025). In addition, according to the interviews, efforts to reduce the migration of fishermen to Jakarta have been undertaken by the local government of Cirebon Regency through training programs aimed at equipping fishermen with skills for alternative livelihood. On the other hand, the local government of Indramayu Regency is better prepared to receive Indramayu's fishermen relocated from Jakarta as there are many ample vacant land to accommodate them. Furthermore, the Indramayu local government is collaborating with the Geospatial Information Agency (BIG) to provide the information for the fishermen on the fish stock location in the Indramayu fishing ground.

In 2017, in an effort to address the problem of illegal settlement of fishermen in Muara Angke, the Jakarta government relocated many who had migrated from Indramayu Regency back to Karangsong, in their home province; this amounted to 16% of the fishermen in Muara Angke. They were relocated to 300 government-built semi-permanent houses; however, and they were required to pay IDR 90 in rent to the local government. Many refused to pay, and based on our observations and interviews with local people who had relocated to this place, many simply returned to Muara Angke because there was insufficient employment in Indramayu.

In seeking to address the problem of migration and illegal settlement in Jakarta, the government needs to focus on approaches that support moving with dignity. A more inclusive social justice approach could assist the government in responding to the impacts of climate change (Nurhidayah & McIlgorm, 2019). The decision to relocate at the individual or micro level is triggered by a range of social, economic and environmental factors, and it is this deeper level, the root causes of this migration, that the government must address. If the cause of migration is employment opportunities, local governments should help provide alternative livelihoods or empowerment to local people and to those who need to be relocated. Incorporating and accommodating the migrant fisher population into their chosen destination is also a solution, but is likely to be costly. Collaboration between local governments in both migrants' places of origin and their intended destination, for receiving more sustainable livelihood, is necessary to provide a more long-term solution.

Conclusion

It is plausible that climate change will trigger even more migration in the future, with the influx of fishermen and their families from surrounding areas such as Cirebon, Tegal and Indramayu, further increasing urbanisation in Jakarta. The rapid and unplanned nature of this urbanisation has outpaced the ability of local authorities to provide adequate housing, infrastructure and services, leading to the emergence of slum areas that lack basic infrastructure and services. As a result, many local communities live in substandard housing conditions with limited or no access to essential facilities such as clean water, sanitation and healthcare.

There is no single solution that can address all the problems associated with climate-change-related migration and urbanisation. A holistic approach that encompasses political recognition of the problem, and collaboration between institutions and local authorities, is essential to address this complex issue. Where the government needs to relocate these migrant communities, it is essential that they be enabled to move with dignity, and empowerment—especially small-scale fishers, women and children—poverty reduction and sustainable development are keys to addressing this issue.

References

Ahmed, I., Chowdhury, M. A., Zzaman, R. U., Ul Islam, S. L., Nahar, S., & Roy, S. K. (2024). Assessing vulnerability of fishermen communities in coastal Bangladesh: A 'climate vulnerability index'- based study in Assasuni Upazila, Satkhira, Bangladesh. *Natural Hazards Research, 4*(4), 562–572. https://doi.org/10.1016/j.nhres.2023.12.018

Akmalia, N. (2016). *Urban fishing the permanent migration of Indramayu fishermen to kampung nelayan Cilincing, North Jakarta*. University of Indonesia.

Barid, V. B. (2025). Interview with Head of Fisheries and Marine Affairs, North Jakarta KPKP Sub-dept, on February 26, 2024 in Jakarta. BRIN.

de Bruin, S., Hoch, J., de Bruijn, J., Hermans, K., Maharjan, A., Kummu, M., & van Vliet, J. (2024). Scenario projections of South Asian migration patterns amidst environmental and socioeconomic change. *Global Environmental Change, 88*(August), 102920. https://doi.org/10.1016/j.gloenvcha.2024.102920

Fathoni, A. (2024). *Interview with the Head of Karangsong Village, Indramayu Regency and the Chairperson of the Indramayu Farmers and Fishermen Union*. BRIN.

Fathony, A. (2024). Interview with The Muara Angke Port Management Unit, on March 4, 2025 in Jakarta. BRIN.

Hanafi, I. (2024). Minister Susi: 10 GT ships do not need a permit. Antara. https://www.antaranews.com/berita/469187/menteri-susi%2D%2Dkapal-10-gt-tak-perlu-izin?utm_source=topnews&utm_medium=home&utm_campaign=news

Jakariya, M., Rahman, A., Sayem, S. M., Saad, S., Alam, M. S., Sarker, S. R., Ali, M. S., & Akter, D. (2020). Development of livelihood vulnerability index for the coastal fishermen communities of Bangladesh using spatial information technique. *Groundwater for Sustainable Development, 11*, 100475. https://doi.org/10.1016/j.gsd.2020.100475

Kuswanto, A. (2024). *Monitoring of the Marine and Estuary Environment of Jakarta Bay in 2024*.

Le Bras, Q., Gascuel, D., Quemper, F., & Levrel, H. (2024). Transition and adaptation: An analysis of how professional fishermen change their practices. *Marine Policy, 164*, 106154. https://doi.org/10.1016/j.marpol.2024.106154

Mbaye, A., Schmidt, J., & Cormier-Salem, M.-C. (2023). Social construction of climate change and adaptation strategies among Senegalese artisanal fishers: Between empirical knowledge, magico-religious practices and sciences. *Social Sciences & Humanities Open, 7*(1), 100360. https://doi.org/10.1016/j.ssaho.2022.100360

Mohamed Shaffril, H. A., Samah, A. A., Samsuddin, S. F., & Ali, Z. (2019). Mirror-mirror on the wall, what climate change adaptation strategies are practiced by the Asian's fishermen of all? *Journal of Cleaner Production, 232*, 104–117. https://doi.org/10.1016/j.jclepro.2019.05.262

N'Souvi, K., Adjakpenou, A., Sun, C., & Ayisi, C. L. (2024). Climate change perceptions, impacts on the catches, and adaptation practices of the small-scale fishermen in Togo's coastal area. *Environmental Development, 49*, 100957. https://doi.org/10.1016/j.envdev.2023.100957

Nurhidayah, L., & Mcllgorm, A. (2019). Coastal adaptation laws and the social justice of policies to address sea level rise: An Indonesian insight. *Ocean an Coastal Management, v*(171), 1–18.

Purnomo, A. H., Kurniawan, T., Farandy, A. R., Apriliani, T., Nurlaili, I., & M., & Sajise, A. J. (2024). Revisiting the climate change adaptation strategy of Jakarta's coastal communities. *Ocean & Coastal Management, 253*, 107112. https://doi.org/10.1016/j.ocecoaman.2024.107112

Rahman, M. S., Huang, W.-C., Toiba, H., Putritamara, J. A., Nugroho, T. W., & Saeri, M. (2023). Climate change adaptation and fishers' subjective well-being in Indonesia: Is there a link? *Regional Studies in Marine Science, 63*, 103030. https://doi.org/10.1016/j.rsma.2023.103030

Roshania, R. P., Cunningham, S. A., Das, A., Bag, T., Giri, R., Mala, G. S., Young, M. F., Srikantiah, S., Mahapatra, T., & Ramakrishnan, U. (2025). To stay or grow? Migration patterns and child growth in rural Bihar, India. *Health and Place, 91*(April 2024), 103395. https://doi.org/10.1016/j.healthplace.2024.103395

Samah, A. A., Shaffril, H. A. M., & Fadzil, M. F. (2019). Comparing adaptation ability towards climate change impacts between the youth and the older fishermen. *Science of the Total Environment, 681*, 524–532. https://doi.org/10.1016/j.scitotenv.2019.05.089

Secretary of Pluit Village. (2024). Interview.

Shaffril, H. A. M., Abu Samah, A., & D'Silva, J. L. (2017a). Adapting towards climate change impacts: Strategies for small-scale fishermen in Malaysia. *Marine Policy, 81*(December 2016), 196–201. https://doi.org/10.1016/j.marpol.2017.03.032

Shaffril, H. A. M., Abu Samah, A., & D'Silva, J. L. (2017b). Climate change: Social adaptation strategies for fishermen. *Marine Policy, 81*(March 2016), 256–261. https://doi.org/10.1016/j.marpol.2017.03.031

Szymkowiak, M., Steinkruger, A., & Furman, K. (2023). Integrating climate adaptation into Gulf of Alaska fishing community planning. *Marine Policy, 156*, 105802. https://doi.org/10.1016/j.marpol.2023.105802

Tempo. (2022). *Ke Kantor Anies Baswedan Bawa Jeriken, Warga Muara Angke Tuntut Air Bersih.* Https://Www.Tempo.Co/Arsip/Ke-Kantor-Anies-Baswedan-Bawa-Jeriken-Warga-Muara-Angke-Tuntut-Air-Bersih-423482

Tempo. (2025). *Mensos Kunjungi Lokasi Banjir Rob di Kampung Eretan Kulon.* https://www.tempo.co/info-tempo/mensos-kunjungi-lokasi-banjir-rob-di-kampung-eretan-kulon-1364445

Chapter 18
The Role of Local Government in Addressing Migration and Climate Change

Yudi Wahyu

Introduction

Climate change challenges are very complex and require collaboration among multiple stakeholders. The cities of Jakarta and Semarang are sinking, with each city submerging nearly 20 cm per year. Both cities are expected to be submerged by sea water by 2050, according to Copernicus Sentinel-6 satellite data (Atha Pradana et al., 2023).

This chapter explores the role of local government in Indonesia in responding to climate change and migration issues, and the policies that have been undertaken to consider their impact on women and children. Based on discussions with coastal communities and local government officials (in four field sites, namely North Jakarta, Semarang, Pekalongan, and Demak) in 2024, climate change is identified by flooding that happens in coastal areas. Flooding in coastal areas is primarily caused by sea level rise, often called high tide floods (*banjir rob*), and land subsidence (*amblesan tanah*). Migration, in this chapter, is identified as relocation, where people who have had their house sunk by sea water need to move to a safer place, and also encompasses immobile populations who have been left behind in unsafe communities.

The Indonesian national government does not have a comprehensive plan and adequate policies regarding this environmental migration situation, which could lead to potential conflicts between the displaced and host communities in the future. The displaced people can be more vulnerable because they have become disconnected from their previous social, cultural, and resource networks (Lisan & Putri, 2020, pp. 5 & 10).

Y. Wahyu (✉)
Disrupting Violence Beacon, Griffith University, Brisbane, QLD, Australia
e-mail: y.fajarmargonowahyu@griffith.edu.au

L. Nurhidayah et al. (eds.), *Climate Change, Labour and Migration in Indonesia*, Springer Climate, https://doi.org/10.1007/978-3-032-15334-0_18

The local governments discussed in this chapter are the Provincial Government of Jakarta (for the city administrative area of North Jakarta) and the Central Java Provincial Government (which refers to four local governments, which are Semarang City, Demak Regency, Pekalongan City, and Pekalongan Regency). The sources of information are from surveys, interviews, roundtable discussions, and stakeholder forums with key informants from government and community members in those four areas conducted in January, April, and September 2024.

Literature Review

This chapter uses a systematic literature review of the ProQuest database, with the search method outlined in Annex 1. The search produced 1799 journal articles, from which 67 were chosen based on their relevance to Indonesia, especially studies done in Jakarta, Central Java, Pekalongan, Demak, and Semarang. Jakarta Province was the most commonly studied location, reflecting its key role in national climate discussions.

Several studies highlight Jakarta's top-down approach to flood management. Meshkani (2024, p. 566) shows that in the case of flood management on the riverbank, Jakarta's local government used a top-down decision-making approach to dealing with disadvantaged communities. Warner and Wiegel (2021, p. 10) document the evolution of the Giant Sea Wall (renamed into "outer sea dike") in order to keep sea flooding out.

Hellman (2015, p. 477) traces its origins to national-level planning by the Ministry of Public Works, which in 2009 produced the policy documents *The Jakarta Master Plan for Preventing Floods* and *The Guide for Mitigating Floods in Jakarta*. The policy direction is that people living in the affected area must be educated or moved.

Van Voorst and Hellman (2015, pp. 800 & 804) critique the use of climate change narratives to justify evictions, arguing that such approaches compound the vulnerability of marginalised urban populations as physically vulnerable to injury and loss of property, socio-economically vulnerable as their livelihood is tied to the living area, and facing the risk of eviction.

The literature also reveals a significant gap in participatory governance. Firman et al. (2011, p. 375) and Mulyani Sunarharum et al. (2014, p. 239) highlight the limited engagement of communities in flood risk planning, despite the proliferation of regulatory frameworks. Wannewitz and Garschagen (2021, pp. 3293, 3297) observe that dominant research paradigms in Jakarta focus on physical and engineering solutions, with insufficient attention to hybrid adaptation strategies, which incorporate both natural and environmental flood drivers (e.g., focusing on climatic, hydrological, and physical factors) and socio-economic flood drivers (e.g., anthropogenic factors that accelerate land subsidence, river clogging, and land conversions).

Regulatory instruments such as the Regional Spatial Plan (*RTRW*) and the Strategic Environmental Assessment (*KLHS*) have been employed to evaluate provincial government responses to climate change and migration in Indonesia. These frameworks were applied in Nurlambang's (2012) study of Jakarta's approach to environmental migration, which found that the Jakarta Spatial Planning Document 2010–2030 afforded limited attention to climate-related concerns. The mitigation and adaptation programmes outlined in the plan offered minimal consideration of the potential for climate-induced migration (Nurlambang, 2012, p. 82).

At the national level, Indonesia has introduced the Low Carbon Development Planning (LCDP) initiative, launched by Bappenas in October 2017 and incorporated into the National Medium-Term Development Plan (RPJMN) 2020–2024. LCDP serves as a strategic framework for integrating climate policy across sectors and regions. At the provincial and regency levels, it guides the development of local climate strategies. In Central Java Province, the LCDP framework has been implemented in Sukoharjo and Cilacap regencies, as well as in Semarang City (Djamal et al., 2022). These efforts are supported by a range of regulatory instruments, including the Regional Long-Term Development Plan (RPJPD), Regional Medium-Term Development Plan (RPJMD), Regent/Mayor Decrees, and Regional Regulations.

Policy Scoping Findings

The analysis of government regulations presented in this chapter is based on a combination of findings from field interviews and a review of relevant policies. The regulatory review focuses on three interrelated dimensions: flood management as an aspect of climate change, relocation as a response to migration, and the social impacts on women and children as a gender-related concern.

Climate-change-induced migration has emerged as a critical area of inquiry within environmental and social policy discourse. Theoretically, two primary drivers underpin this phenomenon: the degradation of environmental conditions due to climate change, and the economic disruption experienced by communities as a result of declining livelihood opportunities (Nurlambang, 2012).

Migration decisions are shaped by varying levels of vulnerability, which encompass exposure to climate-related hazards, sensitivity to environmental stressors, and the capacity to adapt. These factors are not uniformly distributed across populations, and particular attention must be given to the differentiated adaptive capacities of women, whose heightened vulnerability often stems from structural inequalities and limited access to resources (Chindarkar, 2012). Consequently, government policy must be sensitive to these disparities when designing climate adaptation and relocation strategies.

Migration responses to climate stressors can be broadly categorised into three types: in situ adaptation (where communities remain in place and adapt locally), temporary migration, and permanent migration (Chindarkar, 2012). These categories reflect the spectrum of mobility decisions available to affected populations and highlight the need for flexible, context-specific policy interventions.

Jakarta Province

Regional Low Carbon Development Plan (RLCDP)

In response to climate change, the Jakarta Provincial Government introduced Governor Regulation No. 90 of 2021, updating the earlier Regulation No. 131 of 2012 (RLCDP, 2021). This regulation outlines the Regional Low Carbon Development Plan (RLCDP), which aims to reduce greenhouse gas emissions (mitigation activities) and enhance community resilience (adaptation activities).

The RLCDP assesses vulnerability through three indicators—exposure, sensitivity, and adaptive capacity—supported by 17 sub-indicators. It also identifies four climate risks: high rainfall, drought, temperature change, and sea level rise, categorised across a seven-level risk scale from very low to very high.

Adaptation strategies are structured across seven sectors: public health, water resources, coastal and small island management, energy, food security, housing, and climate-resilient infrastructure. Programme locations are selected based on risk levels at the urban village (*kelurahan*) level, with interventions targeting problematic indicators.

However, adaptations targeting women and children remain limited. Programmes that focus on women and children are primarily confined to the health sector, concentrating on basic services such as immunisation and maternal care.[1] Furthermore, the regulation does not have provisions for climate-induced relocation. Housing programmes focus on improving access to affordable housing for low-income residents. Additionally, livelihood support for coastal communities is included within food security initiatives[2] rather than being part of a broader adaptation strategy.

---

[1] These include: (i) healthcare for homeless children, (ii) providing basic and follow-up immunisations for children aged 12–22 months and primary school-aged children, (iii) healthcare services for children under five who have pneumonia, and (iv) standardised health services for babies, school-aged children, the elderly, and pregnant women.

[2] The food security programme involves the distribution of fish and plant seeds, hydroponic vegetable farming, providing facilities and infrastructure for aquaculture fishers, planting productive and medicinal plants, and cultivating edible fish species.

*Regional Spatial Plan (*RTRW*)*

In the Jakarta Provincial Spatial Plan 2024–2044 (Spatial Plan Jakarta, 2024), the strategic approach to tackling tidal flooding involves constructing coastal embankments alongside sea walls as part of the National Capital Integrated Coastal Development (NCICD) initiative along Jakarta's northern coastline, covering areas within North Jakarta City and the Thousand Islands Regency. The plan also includes improvements to the drainage network infrastructure for rivers, canals, and connecting channels.

Concerning local livelihoods, the economic development strategy for the coastal zone and the Thousand Islands Regency focuses on participatory tourism planning; the development of tourism products and markets aligned with the fisheries commodity chain; and the promotion of a "blue economy," defined as economic activities derived from coastal and marine resources.

The policy direction related to migration is centred on housing, specifically through the participatory upgrading of kampung areas. Participation is structured across seven stages: planning, designing, implementation, utilisation, supervision, evaluation, and decision-making. Two strategic approaches are employed: in situ, which focuses on improving existing settlements without relocating residents, and ex situ, which involves relocation as part of broader area enhancement. The programmes developed aim to improve the quality of urban kampungs[3] (*Kampung Kota*) through the application of both approaches.

Programmes specifically targeting women and children are not explicitly identified. Instead, such considerations are embedded within the broader principle of inclusivity. For example, the development of an inclusive, globally oriented fisheries and marine industry hub along the northern coastal area of Jakarta, as well as the establishment of disaster evacuation networks accessible by inclusive mass public transport—ensuring usability for all groups, including older persons and individuals with disabilities.

[3] Urban kampung (*Kampung Kota*), as defined in the regulation, refer to settlements that have emerged without adequate infrastructure planning—particularly in relation to clean water, sanitation, and building standards. These areas are typically characterised by high population density and insecure land tenure.

Central Java Province

Regulation Related to Climate Change

The Government of Central Java Province has articulated its climate resilience strategy through the Regional Climate Change Adaptation Action Plan (*RA API Jateng*, 2023). The province comprises 35 regencies and cities, of which 17 are coastal areas—13 located along the northern coast and 4 along the southern coast.

Five key sectors have been identified as most vulnerable to climate change impacts: (i) water resources, with increasing risks of drought; (ii) agriculture, particularly a decline in rice production; (iii) marine and coastal areas, where wave heights have entered hazardous zones and the Coastal Vulnerability Index (CVI) is notably high—14 regions fall under CVI Class 4 and 2 under Class 5; (iv) public health, with rising cases of dengue fever and pneumonia; and (v) disaster management, addressing floods, landslides, droughts, whirlwinds, fires, and vector-borne diseases.

Adaptation measures for coastal sectors include flood mitigation infrastructure, such as sea walls and other coastal protection structures, including hard engineering solutions like gabions. In terms of livelihoods, the plan outlines training and business mentoring for fishers. Regarding housing, the plan also includes support for disaster-affected residents, facilitating decisions on whether to remain or relocate.

Regarding programme targets for women and children, the climate change adaptation action plan does not explicitly address these groups. Their inclusion appears to be embedded within programme titles, necessitating further investigation to uncover specific provisions. For instance, the Climate Village Programme (*ProKlim*) emphasises gender participation as a critical component for the sustainability of local climate action (ProKlim Jateng, 2019). Additionally, the Regional Medium-Term Development Plan (RPJMD) for 2025–2029 explicitly outlines tactical programmes for training in food processing and financial management for coastal women and female farmers (RPJMD Jateng, 2025).

Other initiatives, such as the provision of floating houses (*Rumah Apung*) (RPJMD Jateng, 2025) and the *Buy Land Get a House* (*Tuku Lemah Oleh Omah*) scheme, implemented during Governor Ganjar Pranowo's administration (2020–2023) (Baihaqi Annizar, 2024; Central Java Provincial Government, 2023), are not explicitly referenced in the adaptation plan for coastal zones.

Based on policy documents and regulations issued by three local governments in Central Java (see Table 18.1), responses to tidal flooding have primarily focused on the development of flood control infrastructure. These measures include the construction of protective structures along coastlines and offshore areas to safeguard vulnerable coastal zones. Notably, such infrastructure projects are closely aligned with national government programmes, indicating a multi-level governance approach to climate adaptation.

Livelihood diversification for coastal communities has also been introduced at the programme level. However, not all initiatives have been prioritised in official

Table 18.1 Highlights of three Central Java local governments' Climate Change Adaptation Action Plan

Local government	Tidal flooding (environmental context)	Livelihoods (economic dimension)	Housing programmes (migration)	Programmes for women and children (gender perspective)
Semarang City	Development of flood control infrastructure (at the urban village level)	Livelihood diversification (at the urban village level)	The development of climate-adaptive housing, such as floating houses and stilt houses (at the urban village level)	Women's participation in climate adaptation training and activities (at the city level)
Pekalongan City	Provision of coastal protection structures and disaster-resilient infrastructure	Diversification of employment opportunities for coastal communities	Reconstruction of climate-resilient coastal settlements	Not explicitly mentioned
Demak Regency	Construction of breakwaters, sea walls, and other coastal defence structures	Training, awareness, and provision of alternative income sources for fisher households* (*located in broader priority list, not in priority list)	Development of adaptive settlements Rehabilitation and construction of housing for disaster victims	Efforts to ensure the fulfilment of rights and protection for children, women, youth, persons with disabilities, and older persons are implemented through caregiving and support services, resilience-building initiatives, and protection from violence—including child marriage and human trafficking

(Sources: RA APID Demak, 2024; RAD API Pekalongan, 2024; RAD API Semarang, 2019)

planning documents—for instance, in Demak Regency, certain proposals did not advance to the priority programme list. Housing rehabilitation for residents affected by tidal flooding has emerged as a critical component of adaptation strategies. This includes the provision of climate-resilient housing, such as floating or stilt houses for those choosing to remain in situ, as well as relocation support for those opting to move to safer areas.

Efforts to enhance adaptive capacity among women and children remain varied. While some local governments explicitly mention gender-responsive interventions, others do not. Furthermore, models of support addressing violence against women and children in coastal areas—including child protection issues—require further development and integration into broader adaptation frameworks.

Regional Spatial Plan (RTRW)

The spatial planning regulations of Central Java Province and its four local governments (see Table 18.2) reveal a consistent approach to coastal flood control, primarily through the construction of sea walls and coastal embankments. This uniformity suggests a strong alignment with national policy, particularly in terms of infrastructure direction and funding sources, indicating that tidal flood management is closely tied to central government initiatives.

Table 18.2 Key features of spatial planning in Central Java Province and its four local governments

Local government	Flood control system (environmental context)	Strategic zones in coastal area (economic dimension)	Housing programmes (migration)	Programmes for women and children (gender perspective)
Central Java Province	Construction of sea walls and coastal embankments	Priority industrial zones: Demak Regency, Semarang CityMaritime industrial zones: - Shipbuilding industry zones: Pekalongan City, Semarang City - Fisheries production centres: Pekalongan City, Semarang City, Demak Regency	Housing and settlement programmes in disaster-prone areas	None explicitly stated
Semarang City	Development of coastal embankments and water retention ponds for tidal flood and flood control	As part of the city's strategic socio-cultural zone, the Kampung Bahari Tambak Lorok area includes: (a) restructuring of fishing villages; (b) development of housing infrastructure and facilities; (c) environmental management	Housing renewal in slum areas through land consolidation and vertical housing development across Semarang City	None explicitly stated
Pekalongan City	Development and maintenance of flood control embankments; wave-breaking structures (revetments); sea current flow guides (groins); and coastal conservation initiatives	Development of a *minapolitan* zone (or fisheries-based urban area) in *Kelurahan* Panjang Wetan, North Pekalongan Subdistrict	Restructuring and revitalisation of informal settlements; housing renewal in slum areas through land consolidation and vertical housing development (*Rusun*)	None explicitly stated

(continued)

Table 18.2 (continued)

Local government	Flood control system (environmental context)	Strategic zones in coastal area (economic dimension)	Housing programmes (migration)	Programmes for women and children (gender perspective)
Pekalongan Regency	Construction of embankments, long storage systems, and flood control infrastructure	Development of formal and informal urban economic sectors in the northern coastal zone	In flood-prone areas, housing development is regulated Land for relocation is prepared within the disaster evacuation network. Relocation land is designated by the Regent	None explicitly stated
Demak Regency	Construction of embankments to prevent coastal abrasion	Strategic zones designated in Sayung Subdistrict include: - Surodadi Beach tourism zone (marine tourism development) - integrated industrial zone in Sayung - industry, trade, and services along primary arterial road corridor	Strategy for managing protected areas includes: - gradual relocation of settlements from flood-prone, tidal flood, abrasion, and protected zones - community education and outreach programmes in flood-prone areas	None explicitly stated

(Sources: RTRW Jateng, 2024; RTRW Demak, 2020; RTRW Kota Pekalongan, 2020; RTRW Kab Pekalongan, 2020; RTRW Semarang, 2021)

Notably, the Spatial Plan of Central Java Province (RTRW Jateng, 2024) designates six municipalities—Pekalongan Regency, Pekalongan City, Semarang City, Batang Regency, Pemalang Regency, and Demak Regency—as Strategic Provincial Zones (*KSP*) vulnerable to tidal flooding and land subsidence. These zones are categorised under strategic areas based on environmental function and carrying capacity. Within the indicative programme framework, three priority areas (Pekalongan Regency, Pekalongan City, and Demak Regency) are targeted for five key activities: (i) delineation of tidal flood risk zones, (ii) public awareness campaigns on disaster-resilient cities, (iii) construction of sea walls and water retention facilities, (iv) development of climate-adaptive housing, and (v) regulation of spatial utilisation.

In support of economic growth in coastal regions, spatial plans at both provincial and local levels prioritise the development of fisheries, shipbuilding, trade, services, and marine tourism. These activities are concentrated in designated strategic zones, reflecting a coordinated effort to leverage coastal resources for sustainable development.

Housing policies for coastal populations are integrated into disaster-prone area management (at the provincial and regency levels) and slum area revitalisation (in urban settings). Among the reviewed Spatial Plan documents, only Pekalongan Regency explicitly includes relocation planning in its spatial document, assigning the Regent authority to designate relocation sites when disaster-affected land becomes uninhabitable (RTRW Kab Pekalongan, 2020).

Regarding community participation, none of the spatial planning documents explicitly mentions programmes targeting women and children, nor do they address inclusivity in a substantive way. Participation is generally framed within three procedural stages: (i) formulation of spatial plans, (ii) spatial utilisation, and (iii) control of spatial utilisation. Although Semarang City outlines 16 specific community participation activities (RTRW Semarang, 2021, Article 160), none explicitly refer to the involvement of women and children, highlighting a gap in gender-responsive planning.

Discussion

Based on policy mapping across six local governments (two provinces and four regencies/cities) above, it is shown that only Demak Regency explicitly targets women and children in its Regional Climate Change Adaptation Action Plan. The absence of explicitly named programmes targeting women and children in most regional planning documents may help explain the limited awareness among these groups regarding government policy on climate adaptation initiatives. For instance, a survey of 315 women and 85 children conducted as part of our KONEKSI study in January 2024—covering five regions (North Jakarta, Pekalongan City, Pekalongan Regency, Semarang City, and Demak Regency)—revealed that only 33% of women respondents were aware of government or non-government programmes involving women and children. In relation to flood management, just 36% of women and 19% of children reported that their views had been taken into consideration. This is illustrated in Fig. 18.1.

Findings from roundtable discussions held with community members across four regions—Jakarta, Pekalongan, Semarang, and Demak—highlight that the foremost priority in responding to climate change is the provision of flood control infrastructure, particularly to address tidal flooding. Simultaneously, communities are experiencing shifts in livelihoods and housing conditions, with many residents needing to raise the elevation of their homes annually, often at considerable personal expense.

For example, during a community roundtable in Demak, participants described a transition in livelihoods from rice farming to aquaculture. However, due to increasing coastal abrasion, fishponds have become ineffective. Those without specialised skills or capital to start a business are compelled to seek alternative employment, often as factory labourers or construction workers.

Women face specific challenges in this transition. Access to training is a key concern, particularly when programmes must be child friendly—allowing women to attend while caring for their children—and flexible enough to support their entry

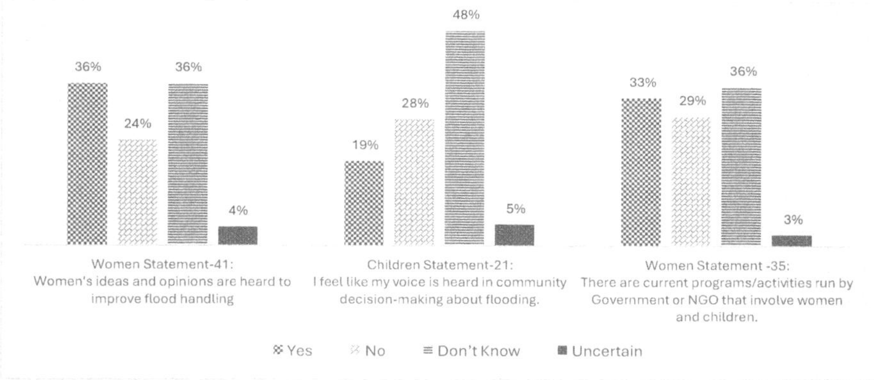

Fig. 18.1 Women's and children's participation

into the workforce. This dual burden should be explicitly addressed in planning documents. Moreover, ongoing support is essential. For instance, after women have acquired production skills—such as transforming plastic waste into decorative items—they continue to require assistance in marketing and selling their products.

The challenges faced by women in Semarang, as revealed through community roundtable discussions, differ somewhat from those in other regions. Many women are employed in garment factories, where long working hours—from early morning until late evening—are common due to frequent overtime. This creates a situation in which children are left unsupervised at home, increasing their exposure to negative social influences. One observed consequence is a growing reluctance among adolescents to continue their education, with some dropping out of school altogether.

For younger children, particularly toddlers, the situation often leads to stunting. Young mothers frequently leave their children in the care of grandparents or informal caregivers who may lack adequate knowledge of nutritious food choices for early childhood development. Female participants in the roundtable highlighted that one of the underlying causes is the absence of labour regulations that are responsive to the needs of women and children—an issue that has persisted for many years.

During community roundtable discussions held in Jakarta and Pekalongan, participants highlighted the economic hardship faced by many families, which often compels children to engage in labour to support their parents. In Jakarta, for instance, children were reported to work as shell peelers, while in Pekalongan, they were tasked with delivering a sheet of batik fabric to various locations and operating overlock sewing machines—a form of work accessible to both boys and girls aged 15 years.

Concerns regarding child stunting were also raised by participants in Jakarta's community roundtable. One intervention involved providing sewing work to mothers of stunted toddlers. This arrangement enabled mothers to bring their children to the workplace, allowing them to monitor feeding practices while simultaneously earning an income. Fathers, in contrast, were typically employed as fishermen or labourers at fish auctions, roles which did not permit active childcare involvement.

Based on findings from the community roundtable discussions, it was evident that residents require considerable time to deliberate before deciding to relocate. Several factors were identified as motivating relocation, including the aspiration for improved living conditions, government facilitation in moving to a new location, and the deteriorating condition of existing homes, particularly those affected by coastal abrasion.

Conversely, resistance to relocation was also expressed by participants. Reasons included being long-term or native residents—some having lived in the area for nearly 35 years—the new location being distant from sources of income or employment, difficulties in securing new jobs, lack of financial resources to support the move, and perceptions of the relocation as a form of forced eviction.

Among the four community roundtable locations, only participants in Jakarta reported the implementation of floating and stilt houses within their residential area, supported by the Ministry of Defence. Additional infrastructure provided included water treatment equipment, enabling the processing of river water for bathing and drinking purposes. This intervention significantly reduced household expenditure on clean water, which previously reached up to IDR 800,000 per month (approximately AUD 80). Further support was anticipated from the Jakarta Clean Water Company (PAM Jaya), which planned to install facilities for clean water for bathing and washing.

Conclusion

This chapter highlights the complex interplay between climate-induced migration and local government responses, particularly in coastal areas vulnerable to tidal flooding. While infrastructure development—such as sea walls and coastal embankments—has been prioritised at both local and national levels, the social dimensions of climate adaptation remain under-addressed.

Findings from policy scoping and community roundtable discussions reveal that most local adaptation programmes do not explicitly target women and children. This gap contributes to the unequal distribution of childcare responsibilities, intensifying the vulnerability of women in affected communities. Moreover, livelihood diversification initiatives for coastal populations often overlook the need for child-friendly training environments, limiting their accessibility for mothers.

Housing support programmes designed to be climate-resilient must also incorporate safety measures for children, particularly in flood-prone settlements where incidents of drowning have occurred. The absence of such considerations underscores the need for more inclusive and context-sensitive adaptation planning.

It is therefore imperative that local governments explicitly integrate and prioritise programmes that address the specific needs of women and children. Doing so will not only enhance the effectiveness of climate adaptation strategies but also promote social equity and resilience within vulnerable communities.

*****

Annex 1 Literature Review Process Using the ProQuest Database

a. This process forms part of the literature review for the KONEKSI study, led by Dr Amy Young and supported by team members Dr Ana Borges, Dr Febi Dwirahmadi, and Yudi Wahyu.
b. The author has made modifications tailored to the specific requirements of this chapter.
c. Update on February 27, 2025.
d. The keyword used in Advanced Search, with the options "Anywhere except full text—NOFT" and limited to "peer reviewed," can be seen below:

((local*) OR (govern*) OR (Jakarta) OR (pekalongan) OR (semarang) OR (demak) OR (policy AND response*) OR (climate AND change) OR (global AND warming) OR (sea-level AND rise) OR (sea AND level AND rise) OR (saltwater AND intrusion) OR (increas* AND storm AND intens*) OR (increas* AND rain*) OR (climate AND justice) OR (climate AND vulnerab*) OR (increas* AND flood*) OR (climate AND inequality) OR (coast* AND flood*) OR (warm*) OR (temp*))

AND

(((migrat*) OR (relocat*) OR (evict*) OR (displace*) OR (refugee*)) OR ((forced AND labour) OR (forced AND labor) OR (slave*)) OR ((work* AND safety) OR (occupation* AND safety) OR (occupation* AND health) OR (employment AND condition*) OR (social AND vulnerab*) OR (social AND isolat*) OR (poverty)))

AND

Indonesia*

e. Inclusion and exclusion criteria (apply filter on "Subject" and "Language"), resulting in 1799 journal articles.
f. Manually selecting journal articles with a focus on *Indonesia with study based on Jakarta, Central Java, Pekalongan, Demak and Semarang*, and resulting in 67 journal articles.
g. The 67 journal articles are placed in *Zotero* as reference management tools.

Inclusion	Exclusion
Reference to climate change and/or natural disasters	Not a journal article or published conference proceedings*
Reference to migration, relocation, displaced person or employment (labour)	No reference to at least one of these terms
Reference to politics, local government	
Published in English or Bahasa Indonesia	Published in any other language
Related to government policy response, either national or local level	No government-policy-related issue such as methods or estimation of sea level rise
Reference to Indonesia	

References

Atha Pradana, Yew, Y., Demetriou, C., Prayitno, H., Delgado, R., González, P., & Ratnasari, D. U. (2023). It's sinking: Coastal cities of Jakarta and Semarang, Indonesia. *Prehospital and Disaster Medicine*, *38*(S1), s94. ProQuest One Academic. https://doi.org/10.1017/S1049023X23002649

Baihaqi, A. (2024, October 8). The Central Java Provincial Government's Initiative to Safeguard Coastal Residents Against Tidal Inundation. *Tirto.Id*. https://tirto.id/ikhtiar-pemprov-jateng-lindungi-warga-pesisir-dari-banjir-rob-g4xL

Central Java Provincial Government. (2023, May 11). *In 2023, Ganjar Completed Assistance for 615 Homes through the "Tuku Lemah Oleh Omah" Scheme*. https://jatengprov.go.id/publik/2023-ganjar-tuntaskan-bantuan-615-rumah-lewat-tuku-lemah-oleh-omah/

Chindarkar, N. (2012). Gender and climate change-induced migration: Proposing a framework for analysis. *Environmental Research Letters, 7*(2), 025601. https://doi.org/10.1088/1748-9326/7/2/025601

Climate Change Adaptation Action Plan for Central Java Province. (2023, August 29). https://www.slideshare.net/slideshow/rencana-aksi-adaptasi-perubahan-iklim-bappeda-provinsijateng/273659888

Demak Regency Regent Regulation Number 71 of 2024 Concerning Regional Climate Change Adaptation Action Plan, Pub. L. No. 71/2024 (2024).

Djamal, R., Handayani, A., & Risandewi, T. (2022). The readiness of local regulations in supporting low carbon development in Central Java Indonesia. *IOP Conference Series: Earth and Environmental Science, 1016*(1), 012044. https://doi.org/10.1088/1755-1315/1016/1/012044

Firman, T., Surbakti, I. M., Idroes, I. C., & Simarmata, H. A. (2011). Potential climate-change related vulnerabilities in Jakarta: Challenges and current status. *Habitat International, 35*(2), 372–378. https://doi.org/10.1016/j.habitatint.2010.11.011

Governor Regulation of Central Java Number 51 of 2019 Concerning the Climate Village Programme in Central Java, Pub. L. No. 51/2019 (2019).

Hellman, J. (2015). Living with floods and coping with vulnerability. *Disaster Prevention and Management: An International Journal, 24*(4), 468–483. https://doi.org/10.1108/DPM-04-2014-0061

Jakarta Provincial Regulation Number 90 Year 2021 on the Regional Low Carbon Development Plan (RLCDP), Pub. L. No. 90/ 2021 (2021).

Lisan, A. R. K., & Putri, R. F. (2020). Strengthening national adaptation in dealing with Mega Sea level rise migration in Indonesia. *IOP Conference Series: Earth and Environmental Science, 451*(1), 012011. https://doi.org/10.1088/1755-1315/451/1/012011

Meshkani, T. (2024). Climate change and flooding: Governmental responses to displacement and relocation in Jakarta's informal neighborhoods. *International Journal of Disaster Resilience in the Built Environment, 15*(4), 551–570. https://doi.org/10.1108/IJDRBE-02-2023-0037

Mulyani Sunarharum, T., Sloan, M., & Susilawati, C. (2014). Re-framing planning decision-making: Increasing flood resilience in Jakarta. *International Journal of Disaster Resilience in the Built Environment, 5*(3), 230–242. https://doi.org/10.1108/IJDRBE-02-2014-0015

Nurlambang, T. (2012). *Public Policy Matters on Climate Change and Migration in Indonesia* (Climate Change, Migration and Human Security in Southeast Asia, pp. 74–83). S. Rajaratnam School of International Studies; JSTOR. http://www.jstor.org.libraryproxy.griffith.edu.au/stable/resrep05890.9

Pekalongan Mayor Regulation Number 34a of 2024 Concerning Regional Action Plan for Climate Change Adaptation, Pub. L. No. 34A/2024 (2024).

Regional Regulation of Central Java Province No. 7 of 2025 Concerning the Regional Medium-Term Development Plan for the Years 2025–2029, Pub. L. No. 7/2025 (2025).

Regional Regulation of Central Java Province Number 8 of 2024 Concerning the Spatial Plan of Central Java Province for 2024–2044, Pub. L. No. 8/2024 (2024).

Regional Regulation of Demak Regency Number 1 of 2020 Concerning Amendments to Regional Regulation of Demak Regency Number 6 of 2011 Concerning the Spatial Plan of Demak Regency for 2011–2031, Pub. L. No. 1/2020 (2020).

Regional Regulation of Pekalongan City Number 9 of 2020 Concerning Amendments to Regional Regulation of Pekalongan City Number 30 of 2011 Concerning the Spatial Plan of Pekalongan City for 2009–2029, Pub. L. No. 9/2020 (2020).

Regional Regulation of Pekalongan Regency Number 3 of 2020 Concerning the Spatial Plan of Pekalongan Regency for 2020–2040, Pub. L. No. 3/2020 (2020).

Regional Regulation of Semarang City Number 5 of 2021 Concerning Amendments to Regional Regulation Number 14 of 2011 Concerning the Spatial Plan of Semarang City for 2011–2031, Pub. L. No. 5/2021 (2021).

Regional Regulation of the Jakarta Province No. 7 of 2024 on the Regional Spatial Plan for 2020–2044 (2024).

Semarang City Regional Action Plan for Climate Change Adaptation. (2019). Pemerintah Kota Semarang. https://ikupi.org/rencana-aksi-daerah-adaptasi-perubahan-iklim-rad-api-kota-semarang-2019/

van Voorst, R., & Hellman, J. (2015). One risk replaces another: Floods, evictions and policies on Jakarta's Riverbanks. *Asian Journal of Social Science*, *43*(6), 786–810. JSTOR.

Wannewitz, M., & Garschagen, M. (2021). Review article: Mapping the adaptation solution space – lessons from Jakarta. *Natural Hazards and Earth System Sciences, 21*(11), 3285–3322. https://doi.org/10.5194/nhess-21-3285-2021

Warner, J. F., & Wiegel, H. (2021). Displacement induced by climate change adaptation: The case of 'Climate Buffer' infrastructure. *Sustainability, 13*(16), 9160. https://doi.org/10.3390/su13169160

Part IV
Conclusion

Chapter 19
Gender and Climate Change: Transdisciplinary Reflections

Amy Young, Ana Borges Jelinic, Jennifer Boddy, Patrick O'Leary, Donna McAuliffe, Elena Marchetti, Susan Harris Rimmer, and Sara E. Davies

> For us, the process of peeling the layers of problems that arise due to climate change is like peeling an onion because many new issues emerge, accompanied by our tears.

As the project that initiated this volume came to an end, the research team, which had grown throughout the project, joined together to produce a series of reflections. Reflecting together takes a level of vulnerability, and it demonstrated the trust that had grown between the team. This chapter embeds some common themes amongst the reflections in key learnings from the volume. The impacts of climate change on the communities of coastal Java can seem overwhelming, and disempowering.

A. Young (✉)
Childlight East Asia and Pacific Hub, School of Social Science, University of New South Wales, Sydney, Australia
e-mail: amy.young@unsw.edu.au

A. B. Jelinic · E. Marchetti
Disrupting Violence Beacon, Griffith University, Brisbane, Australia

J. Boddy
Dean of Sustainability, Griffith University, Brisbane, QLD, Australia

P. O'Leary
Disrupting Violence Beacon, Griffith University, Brisbane, QLD, Australia

ARC Centre of Excellence for the Elimination of Violence Against Women, Brisbane, QLD, Australia

D. McAullife
School of Health Sciences and Social Work, Griffith University, Brisbane, QLD, Australia

S. H. Rimmer
School of Law, Griffith University, Brisbane, QLD, Australia

S. E. Davies
ARC Centre of Excellence for the Elimination of Violence Against Women, Brisbane, QLD, Australia

L. Nurhidayah et al. (eds.), *Climate Change, Labour and Migration in Indonesia*, Springer Climate, https://doi.org/10.1007/978-3-032-15334-0_19

However, this chapter provides some practical guidance and calls for action, based on research, that can facilitate change. The authors also reflect on how the challenges and opportunities presented in the process of undertaking transdisciplinary research on the intersecting issues of gender, climate change, labour and migration are reflective of the challenges and opportunities in developing effective and sustainable solutions.

Collaborative Activism

The research that guided this volume is underpinned by a transdisciplinary research framework, which enabled collaboration and consensus across the diverse disciplines of the authors. Similarly, the steps to address the impacts of social change driven by climate change will need collaboration across disciplines, ministries, governments and communities. The recommendations espoused by the authors in this volume call for collaboration across government agencies to make sure the stories gathered are acted upon. Collaboration is sometimes seen as a buzzword, or a recommendation that favours 'talking' over 'action'. This framing negates that action without co-ordination, and an understanding of the intersectional challenges that arise from climate change, will not be effective. Authors throughout the volume have called for collaboration, not only across academic disciplines or ministries, but also with community.

> This research has been very eye-opening for me, not only to understand what happens at the ground (i.e., individual-household-community level) but also to realize more about the very strategic role of policy and the urgent needs of operationalized collaborative initiatives in dealing with various climate change impact. Indeed, as academics, we do research and publications and make policy recommendations. But if it is not concretely followed by continued facilitation, advocacy, and collaborative action that reaches out to vulnerable people (especially women and children), all the output and deliverables will essentially go only nowhere.

This reflection captures the responsibilities of the authors to continue to use the research to advocate for their participants. Collaborative anthropology and activist anthropology are examples of disciplines that over the past 20 years have placed advocacy as a core component of research. This builds on, and goes beyond, the participatory research methods and community consultation undertaken in this study, and aims to elevate the voices of community in discussions that impact their lives.

Transdisciplinary approaches have become increasingly valued in academia, and provide a framework for multifaceted discussions outside of academia. Working across disciplines, while rewarding, can also be challenging, as discussed in length in Fam and O'Rouke's (2021) volume on lessons that can be learnt from failed interdisciplinary and transdisciplinary projects. This volume certainly does not report on a failed project, however there were important points of learning from challenges the research team faced that can be used to guide collaborations across policy areas. The first was working across different institutional policies. Each of the three partner organisations, two universities and a government research institution, held differing governance structures, financial policies and ethical frameworks. While research principles had been agreed upon by the research team to ensure equitable and collaborative decision-making, internal institutional pressures and priorities sometimes made equitable decision-making untenable. For example, as the Australian institution was named as lead on the contract, the Australian institution's ethical approval was placed above local Indonesian research ethics when collecting data. Undertaking ethical research, especially with children, was of great importance to the research team, and there is congruence between Australian and Indonesian ethical codes. However, there were also points where cultural differences made the abstract ethical values developed in Australia difficult to implement in local Indonesian villages. The idea of privacy, or of locals not knowing who is taking part in community roundtables, is almost impossible when the discussions are held in a community hall with people coming and going outside. Further, taking official group photos at the end of roundtables was something that the community valued, however all photos ran the risk of ensuring that participants could be identified as someone who has taken part in the research. Having an ethical code is integral, yet so is ensuring that it can be applied in such a way that allows the community to make decisions on the way that they wish to interact with the research team. This example highlights the challenges that can arise when institutional policies complicate transdisciplinary research. The research team were able to navigate institutional policies to ensure that the project ran as planned, however it did require ongoing dialogue, creativity and activism to advocate for the project within institutional constraints.

Beyond practical constraints, transdisciplinary ways of working required the team to examine the ways that social work, public health, urban planning, law and engineering could work together with local climate change champions to design and undertake the research. This required members of the research team from different disciplines to adopt the same principles towards each other as those used to guide engagement with participants. Adapted from Moore et al.'s (2016) guide to engagement, four principles guided the collaboration: respect, inclusion, awareness of power dynamics and shared understanding of purpose. Awareness of power dynamics across titles, genders, nationalities and disciplines was crucial to open communication. This saw team members encouraged to communicate in the language they felt most comfortable with, and to communicate through a channel that was most suitable for them. The shared understanding of purpose was crucial to bringing the research team together, and in delivering the work outlined in this volume. However, there were also areas where our purposes didn't translate across disciplinary and linguistic boundaries. Most notably the idea of forced labour, translated into Bahasa Indonesia as compelled labour, became contested throughout the project. Roundtables in the project were held in Bahasa Indonesia, with English language speakers from the Australian team present as observers, with translators. Roundtables were guided by case-studies developed from the interviews and surveys with women and children. They reflected the stories told in the data to initiate discussion. Not all participants saw the examples of children working as problematic. In one instance, a participant turned to the Australian researchers and stated that while they may see children working as an issue, it is necessary for Indonesian communities. The grey areas between compelled labour and children contributing to family life through assisting in family businesses are the areas that require ongoing dialogue and examination of power dynamics between children and communities. In collectivist societies, such as Indonesia, children are provided with protections from the community, however there are also community responsibilities that may see individual wellbeing placed at risk. These are the nuanced conversations that transdisciplinary approaches can guide, and through which responsive policy can be developed.

The roundtables with community and government provided a framework for diverse opinions to come together on the topics covered in this volume. Community members reflected that it was the first time that they have had the opportunity to focus on women and children in climate change discussions, and it was a moment of skill transition between researchers, as can be seen in the following reflection.

> I gained new knowledge about the 'Roundtable' methodology, which has proven effective in encouraging in-depth and inclusive discussions. This methodology allows all participants, from academics, government and local communities, to share their perspectives and knowledge. Through this process, I realised that cross-sector collaboration is important in finding solutions to climate change problems. Open discussions and mutual respect for each perspective are the keys to finding a more holistic and sustainable approach. This experience not only enriched my research skills but also deepened my understanding of the importance of collective collaboration in facing the challenges of climate change impacts.

Of the eight roundtables held, each had its own distinctive feel. For example, the community roundtable in Demak drew attention to the immediacy of the climate crisis. The event was held in a community hall which can be cut off during certain tidal periods. There was anxiety amongst the research team at roads to the hall being cut off, and participants waded through water to get home following data collection in the village. The discussion itself highlighted the physical risks to children from flood waters, resulting in children drowning after falling or hitting their heads, and the domestic violence concerns in the village. For the research team, this illustrates their role in elevating the priorities of a community who need an avenue to stakeholders who can address their immediate and longer term concerns for survival.

Local Champions Included on a Global Scale

> I hope this research can be a glimmer of hope in helping the government produce policies that are in accordance with needs and can help affected communities.

All aspects of the research, from developing the proposal, initial meetings, data collection and dissemination and advocacy included collaboration with local women's groups.

> One of the most memorable moments was meeting local champions, inspiring individuals who are relentlessly fighting to face the challenges of climate change in their communities.

> Their dedication and resilience have opened my eyes to the fact that real action at the local level is the key to preserving the environment and creating a more sustainable future.

All local champions in the research were women, who were passionate about their communities and addressing climate change through creating economic opportunities. Gender plays a role in climate activism worldwide, and with increased visibility of climate activists have come increased experiences of gender-based violence targeting climate activists (UN, 2025). Climate change itself can be presented as an issue of gender-based violence, given the disproportionate impacts on women (Wonders, 2018). MacGregor (2014) notes that the universalisation of climate-change-focused studies has led to a lack of focus on structural factors, such as gender and the impacts of historic factors and hierarchical power structures. Wonders (2018) advocates for the integration of slow violence into types of gender-based violence, such as intimate partner violence, to allow for the understanding of slow changes in society as producers of gendered risks and insecurity. Slow violence, as outlined by Nixon (2011), examines violence over spatial and temporal spans. In focusing on both communities who have been left behind, and those migrated, this volume has given insight into the impacts of women and children of climate change and migration, including structural exclusion of health and education systems. Nevertheless, this volume speaks to the hope of women-led initiatives in communities and the differences that these make. As one researcher described it:

> I am moved by the continuous small actions conducted by the women in the research locations. Despite the fact that they face decent quality of life, and influenced by patriarchy and extended family concept, the energy to believe that they could contribute to their neighbourhood indeed has transformed into a collective better strategy.

What Can Be Learnt?

Another memorable thing is being able to interact with people who struggle to live in their homes even though sea water enters every day. It can't be imagined, but it is enough to make me feel what they feel.

The authors in this volume call for clarity in collaboration when developing policy responses to climate change. Clarity is gained from embedding local realities in policy development. For example, our research highlights the need to include sea-level rise and land subsistence into disaster management practices. The recognition that slow-onset disasters have on the lives of the community needs to be recognised in legislation to guide appropriate responses. Slow-onset disaster responses can be integrated into specific climate change legislation, and less specific yet related legislations, such as labour regulations. Social protection and social adaptation strategies need to include a recognition of the intersectional impacts of climate change. Flexibility is needed to ensure that responses can be adapted to local contexts, and that community can shape responses to their own needs. The challenge of climate change is both local and global, current and future. Given this, local-champion-led responses, that are inclusive of children, can make a difference on a global scale.

References

Fam, D., & O'Rourke, M. (2021). *Interdisciplinary and transdisciplinary failures: Lessons learned from cautionary tales*. Routledge Taylor and Francis.

MacGregor, S. (2014). Only resist: Feminist ecological citizenship and the post-politics of climate change. *Hypatia, 29*(3), 617–633.

Moore, T., McArthur, M., Heerde, J., Roche, S., & O'Leary, P. (2016). *Our safety counts: Children and young people's perceptions of safety and institutional responses to their safety concerns (Report)*. Institute of Child Protection Studies, Australian Catholic University. https://doi.org/10.24268/fhs.8349

Nixon, R. (2011). *Slow violence and the environmentalism of the poor*. Harvard University Press.

United Nations (U). (2025). *Colliding Crises: How the climate crises fuels gender-based violence*. Retrieved 6/1/2026 from: https://spotlightinitiative.org/publications/colliding-crises-how-climate-crisis-fuels-gender-basedviolence

Wonders, N. (2018). Climate change, the production of gendered insecurity and slow intimate partner violence. In K. Fitz-Gibbon, J. McCulloch, & S. Walklate (Eds.), *Intimate partner violence, risk and security: Securing women's lives in a global world*. Routledge, Taylor and Francis.

Chapter 20
Climate Change, Labour, and Migration: Developing Innovative and Holistic Responses to Environmental Disturbances

Laely Nurhidayah, Amy Young, and Wiwandari Handayani

Introduction

Sea level rise and worsening land subsidence are global threats to most coastal communities. Valuable and rare ecosystems can be affected by long-term sea level rise (Roy et al., 2023). The implications of these environmental disturbances extend far into issues of labour and migration. These impacts are especially severe in developing countries such as Indonesia, where existing inequalities exacerbate vulnerabilities. Women, children, and those experiencing poverty bear a disproportionate burden from coastal inundation.

This is a complex problem that requires a comprehensive and integrated approach—one that draws not only from the natural sciences but also from the social sciences. From a social science perspective, key areas of concern include health, education, economic empowerment, social protection, relocation, disaster risk reduction (DRR), infrastructure, nature-based solutions, and emerging issues related to protection, accommodation, and retreat strategies.

A key takeaway from this research is that there is no one-size-fits-all solution to climate adaptation. It requires the trialling of innovative solutions, supported by continuous evaluation and adjustment. Studying the impacts of sea level rise and land subsidence on local communities highlights a fundamental concern: the right

L. Nurhidayah (✉)
Research Center for Law National Research and Innovation Agency, Jakarta, Indonesia
e-mail: lael003@brin.go.id

A. Young
Childlight East Asia and Pacific Hub, School of Social Science, University of New South Wales, Kensington, NSW, Australia

W. Handayani
University of Diponegoro, Semarang, Indonesia

L. Nurhidayah et al. (eds.), *Climate Change, Labour and Migration in Indonesia*, Springer Climate, https://doi.org/10.1007/978-3-032-15334-0_20

to a safe and healthy environment—one that enables people to live productively and in harmony with nature.

The local communities consulted in the research study that underpinned this volume emphasized the urgent need for government protection against the impacts of environmental changes. In addition, they expressed a strong desire for improved economic opportunities in the face of a changing climate. Their voices reflect the need for adaptation strategies that not only ensure physical safety but also promote long-term livelihood security and social well-being.

Adaptive Law and Policy Framework

An adaptive legal and policy framework is essential to strengthen the protection of people affected by climate change, particularly in vulnerable sectors and regions. As the impacts of climate change—such as sea level rise and land subsidence—intensify, existing laws and policies must evolve to respond to increasingly complex challenges that intersect with social, economic, and environmental systems.

In the labour sector, climate change is already displacing workers, especially in agriculture, fisheries, and informal coastal economies. To respond to this disruption, adaptive social protection mechanisms must be developed and expanded. These mechanisms should include unemployment benefits, re-skilling and up-skilling programs, climate-resilient livelihood support, and emergency assistance for those who lose their jobs due to environmental degradation or displacement. Ensuring that these protections are accessible, inclusive, and gender-sensitive is critical to avoid deepening existing inequalities. This issue is further explored in detail by Mareta et al. in Chap. 13, where there is discussion of the intersection of climate change, labour vulnerabilities, and the need for responsive social protection systems.

In addition, retreat policies and planned relocation must be explicitly addressed within national and local legislation. These strategies should not only focus on the physical movement of communities from high-risk areas but also ensure that relocation processes are just, participatory, and uphold the rights and dignity of affected populations. Legal frameworks must mandate proper consultation with communities, provide adequate compensation, and guarantee access to essential services, land, housing, and livelihoods in the new locations. Nurhidayah and Borges reflect on this very issue in Chap. 3, and call for enhanced co-ordination between relevant institutions.

Critically, sea level rise and land subsidence should be formally recognized and categorized as disasters under national disaster management laws. This reclassification would enable the mobilization of emergency funds, the inclusion of these hazards in risk assessments and early warning systems, and the integration of climate adaptation into disaster risk reduction (DRR) strategies. It would also enhance coordination between institutions responsible for environmental management, disaster response, and social welfare. Ultimately, a responsive and forward-looking legal framework must be holistic—bridging environmental science, human rights, social protection, and climate justice—to ensure that no one is left behind in the face of climate-related displacement and livelihood disruption.

Integrating Science, Social Policy, and Community-Based Solutions for Resilient Coastal Adaptation

The case studies of Jakarta, Semarang, Pekalongan, and Demak show that technological development alone is not sufficient to address the impacts of sea level rise. Scientific approaches must be integrated with an understanding of social characteristics, community vulnerabilities, livelihoods, local needs, and traditional knowledge. Infrastructure projects implemented without proper impact assessments, community consultations, and consideration of local contexts have, in some cases, blocked fishermen's access to the sea. This disruption not only threatens livelihoods but also violates communities' right to food, especially when no compensation is provided, as Suntoro and colleagues reported in Chap. 7.

Community-based solutions for building resilience were also highlighted throughout the volume, including in Suntoro et al.'s, and Hidayati et al.'s contributions. Initiatives such as mangrove restoration and planting, urban farming, and waste banks represent innovative, participatory responses developed by local communities to address the impacts of climate change and restore degraded environments. These grassroots efforts not only enhance ecological resilience but also contribute to local livelihoods, food security, and environmental awareness.

Mangrove restoration, for example, helps protect coastal areas from erosion, storm surges, and sea level rise, while simultaneously supporting biodiversity and fisheries. Urban farming empowers communities to produce their own food sustainably, reducing dependence on external systems and strengthening food resilience. Waste banks promote environmental cleanliness and circular economy practices by encouraging communities to collect, sort, and recycle waste, often providing additional income for low-income households.

These community-driven solutions demonstrate the importance of local knowledge, collective action, and social innovation in responding to climate change. However, to scale up their impact and ensure long-term sustainability, they require consistent support from governments, non-governmental organizations, the private sector, and other stakeholders. This includes providing technical assistance, funding, capacity-building, policy support, and platforms for knowledge-sharing.

By investing in and empowering community-based adaptation, stakeholders can foster more inclusive, effective, and context-specific strategies for climate resilience—anchored in the lived realities and strengths of the people most affected.

Gender-Responsive Climate Policy

Women and children are among the most severely affected groups by climate change, particularly in the context of sea level rise. In many coastal communities, women work in informal sectors to supplement household income, often supporting their husbands who work as fishermen. As fishing becomes increasingly unpredictable due to the changing climate, women play a critical role in sustaining family livelihoods. Their efforts are essential not only for economic survival but also for ensuring household food security and preventing malnutrition and stunting in children.

Children are also significantly impacted, especially through disruptions to their education. More frequent flooding and damaged infrastructure often force school closures or make it difficult for children to attend regularly. These educational disruptions can have long-term consequences on their development and future opportunities. Dewi et al. provide insight from Pekalongan on the impacts on educational outcomes.

Given these vulnerabilities, it is essential that women and children receive greater priority in climate adaptation and disaster risk reduction strategies. Their specific needs, experiences, and contributions must be recognized and incorporated into planning and policy processes. Mechanisms such as *Musrenbang Perempuan*—a women-focused participatory planning forum—play a crucial role in amplifying women's voices and ensuring inclusive decision-making at the community level, as Suntoro et al. advocate for in Chap. 14.

Furthermore, disaster risk reduction policies and programs must include targeted support for women and children. This includes early warning systems that are accessible to all, safe and inclusive evacuation shelters, gender-responsive emergency response plans, and post-disaster recovery efforts that address the unique needs of women and children.

Ultimately, building climate resilience must be inclusive, intersectional, and gender-sensitive, ensuring that those who are most vulnerable are also empowered to lead and shape the solutions.

Human Rights and Social Justice

Sea level rise and land subsidence pose serious threats to human rights, human security, and social justice. Despite these profound impacts, policy responses often fail to adequately address the social justice dimensions of climate change. Vulnerable populations—particularly those living in informal or marginalized settlements—are frequently excluded from formal protection mechanisms and adaptation programs.

As discussed in the Chap. 3 by Nurhidayah et al., residents of illegal or unregistered settlements are typically not entitled to state-provided housing or relocation support. However, in recognition of their basic human rights, some local governments have taken adaptive measures to accommodate these communities by providing alternative housing solutions, such as stilt houses or floating homes. These efforts, while not always formalized in national policy, reflect an emerging acknowledgement of the need to uphold human dignity and the right to adequate housing, even for those outside legal frameworks (OHCHR, 2025).

Recognizing and protecting the rights of local communities—regardless of their legal status—are essential for ensuring an equitable response to climate change. Policy frameworks must go beyond technical solutions and incorporate principles of social justice, equity, and inclusion. This includes legal recognition of vulnerable groups, participatory planning processes, and tailored support that reflects the lived realities of those most affected by environmental degradation.

Further, as highlighted in Chap. 16 by Nur Hayati et al., the relocation process in Demak raises serious human rights concerns that must be addressed by the government. Issues such as lack of proper consultation, insufficient compensation, and disruption of livelihoods underscore the need for rights-based approaches to relocation and retreat policies (UNGA, 2025).

Ultimately, climate adaptation must not only be about physical resilience but also about protecting the rights and livelihoods of all people, particularly those who have been historically marginalized or excluded.

Reflection and Policy Recommendations

This research offers a brief glimpse into the complex challenges surrounding coastal adaptation in the face of climate change. While it highlights key issues such as sea level rise, land subsidence, labour, human mobility, and social vulnerability, much more work is needed to fully understand the scope and scale of these problems. In particular, further research is essential to identify which communities require urgent attention for planned relocation, and which areas should be prioritized for protection and investment in resilience.

A transdisciplinary approach—one that integrates insights from environmental science, social science, law, economics, and local knowledge—is crucial to developing effective, equitable, and sustainable solutions. Such an approach can help design strategies that not only safeguard the environment but also protect community livelihoods, uphold human rights, and promote social justice. Ultimately, finding a "win-win" solution requires collaboration across sectors and disciplines, and most importantly, meaningful engagement with the communities who are directly affected.

Through this research, it is hoped that governments will place greater focus on social science perspectives to complement technical and engineering approaches. By doing so, adaptation measures—such as the construction of giant sea walls—can be better evaluated in terms of their social, economic, and human rights implications. A stronger grounding in social science can also help ensure that adaptation strategies are informed by local realities, community needs, and inclusive decision-making processes.

Finally, this research offers valuable lessons and contributes to both academic literature and policymaking by highlighting the challenges, best practices, and innovations emerging from the Indonesian context. These insights can serve as a data-driven foundation for informing state practices at the international level in responding to sea level rise. By documenting local experiences, community-based solutions, and existing policy gaps, this study provides a meaningful contribution towards shaping more inclusive, effective, and context-sensitive climate adaptation strategies.

References

OHCHR. (2025). *The Human Rights to Adequate Housing special Rapporteur on the right to adequate housing*. OHCR, https://www.ohchr.org/en/special-procedures/sr-housing/human-right-adequate-housing

Roy, P., Pal, S. C., Chakrabortty, R., Chowdhuri, I., Saha, A., & Shit, M. (2023). Effects of climate change and sea-level rise on coastal habitat: Vulnerability assessment, adaptation strategies and policy recommendations. *Journal Environmental Management., 15*(330), 117187. https://doi.org/10.1016/j.jenvman.2022.117187. Epub 2023 Jan 6. PMID: 36610196.

United Nations General Assembly. (2025). *Human Rights Council Sixtieth session 8 September–3 October 2025, Climate justice, sustainability and the right to development, Thematic study by the Expert Mechanism on the Right to Development*. https://docs.un.org/en/A/HRC/60/95

Index

L. Nurhidayah et al. (eds.), *Climate Change, Labour and Migration in Indonesia*, Springer Climate, https://doi.org/10.1007/978-3-032-15334-0